RELIABILITY AND AVAILABILITY OF CLOUD COMPUTING

IEEE Press
445 Hoes Lane
Piscataway, NJ 08854

Technical Reviewers

Xuemei Zhang
Principal Member of Technical Staff
Network Design and Performance Analysis
AT&T Labs

Rocky Heckman, CISSP
Architect Advisor
Microsoft

RELIABILITY AND AVAILABILITY OF CLOUD COMPUTING

Eric Bauer
Randee Adams

IEEE PRESS

A JOHN WILEY & SONS, INC., PUBLICATION

cover image: © iStockphoto
cover design: Michael Rutkowski

Published by John Wiley & Sons, Inc., Hoboken, New Jersey.
Published simultaneously in Canada.

Library of Congress Cataloging-in-Publication Data:
Bauer, Eric.
Reliability and availability of cloud computing / Eric Bauer, Randee Adams.
p. cm.
ISBN 978-1-118-17701-3 (hardback)
1. Cloud computing. 2. Computer software–Reliabillity. 3. Computer software–Quality control. 4. Computer security. I. Adams, Randee. II. Title.
QA76.585.B394 2012
004.6782–dc23

2011052839

Printed in the United States of America.

10 9 8 7 6 5 4 3 2 1

To our families and friends
for their continued encouragement and support.

CONTENTS

FIGURES

TABLES

EQUATIONS

INTRODUCTION

Cloud computing is a new paradigm for delivering information services to end users, offering distinct advantages over traditional IS/IT deployment models, including being more economical and offering a shorter time to market. Cloud computing is defined by a handful of essential characteristics: on-demand self service, broad network access, resource pooling, rapid elasticity, and measured service. Cloud providers offer a variety of service models, including infrastructure as a service, platform as a service, and software as a service; and cloud deployment options include private cloud, community cloud, public cloud and hybrid clouds. End users naturally expect services offered via cloud computing to deliver at least the same service reliability and service availability as traditional service implementation models. This book analyzes the risks to cloud-based application deployments achieving the same service reliability and availability as traditional deployments, as well as opportunities to improve service reliability and availability via cloud deployment. We consider the service reliability and service availability risks from the fundamental definition of cloud computing—the essential characteristics—rather than focusing on any particular virtualization hypervisor software or cloud service offering. Thus, the insights of this higher level analysis and the recommendations should apply to all cloud service offerings and application deployments. This book also offers recommendations on architecture, testing, and engineering diligence to assure that cloud deployed applications meet users' expectations for service reliability and service availability.

Virtualization technology enables enterprises to move their existing applications from traditional deployment scenarios in which applications are installed directly on native hardware to more evolved scenarios that include hardware independence and server consolidation. Use of virtualization technology is a common characteristic of cloud computing that enables cloud service providers to better manage usage of their resource pools by multiple cloud consumers. This book also considers the reliability and availability risks along this evolutionary path to guide enterprises planning the evolution of their application to virtualization and on to full cloud computing enablement over several releases.

AUDIENCE

The book is intended for IS/IT system and solution architects, developers, and engineers, as well as technical sales, product management, and quality management professionals.

ORGANIZATION

The book is organized into three parts: *Part I, "Basics," Part II, "Analysis,"* and *Part III—, "Recommendations."* Part I, "Basics," defines key terms and concepts of cloud computing, virtualization, service reliability, and service availability. Part I contains three chapters:

- *Chapter 1, "Cloud Computing."* This book uses the cloud terminology and taxonomy defined by the U.S. National Institute of Standards and Technology. This chapter defines cloud computing and reviews the essential and common characteristics of cloud computing. Standard service and deployment models of cloud computing are reviewed, as well as roles of key cloud-related actors. Key benefits and risks of cloud computing are summarized.
- *Chapter 2, "Virtualization."* Virtualization is a common characteristic of cloud computing. This chapter reviews virtualization technology, offers architectural models for virtualization that will be analyzed, and compares and contrasts "virtualized" applications to "native" applications.
- *Chapter 3, "Service Reliability and Service Availability."* This chapter defines service reliability and availability concepts, reviews how those metrics are measured in traditional deployments, and how they apply to virtualized and cloud based deployments. As the telecommunications industry has very precise standards for quantification of service availability and service reliability measurements, concepts and terminology from the telecom industry will be presented in this chapter and used in Part II, "Analysis," and Part III, "Recommendations."

Part II, "Analysis," methodically analyzes the service reliability and availability risks inherent in application deployments on cloud computing and virtualization technology based on the essential and common characteristics given in Part I.

- *Chapter 4, "Analyzing Cloud Reliability and Availability."* Considers the service reliability and service availability risks that are inherent to the essential and common characteristics, service model, and deployment model of cloud computing. This includes implications of service transition activities, elasticity, and service orchestration. Identified risks are analyzed in detail in subsequent chapters in Part II.
- *Chapter 5, "Reliability Analysis of Virtualization."* Analyzes full virtualization, OS virtualization, paravirtualization, and server virtualization and coresidency using standard reliability analysis methodologies. This chapter also analyzes the software reliability risks of virtualization and cloud computing.
- *Chapter 6, "Hardware Reliability, Virtualization, and Service Availability."* This chapter considers how hardware reliability risks and responsibilities shift as applications migrate to virtualized and cloud-based hardware platforms, and how hardware attributed service downtime is determined.
- *Chapter 7, "Capacity and Elasticity."* The essential cloud characteristic of rapid elasticity enables cloud consumers to dispense with the business risk of

locking-in resources weeks or months ahead of demand. Rapid elasticity does, however, introduce new risks to service quality, reliability, and availability that must be carefully managed.

- *Chapter 8, "Service Orchestration Analysis."* Service orchestration automates various aspects of IT service management, especially activities associated with capacity management. This chapter reviews policy-based management in the context of cloud computing and considers the associated risks to service reliability and service availability.
- *Chapter 9, "Geographic Distribution, Georedundancy, and Disaster Recovery."* Geographic distribution of application instances is a common characteristic of cloud computing and a best practice for disaster recovery. This chapter considers the service availability implications of georedundancy on applications deployed in clouds.

Part III, "Recommendations," considers techniques to maximize service reliability and service availability of applications deployed on clouds, as well as the design for reliability diligence to assure that virtualized applications and cloud based solutions meet or exceed the service reliability and availability of traditional deployments.

- *Chapter 10, "Applications, Solutions and Accountability."* This chapter considers how virtualized applications fit into service solutions, and explains how application service downtime budgets change as applications move to the cloud. This chapter also proposes four measurement points for service availability, and discusses how accountability for impairments in each of those measurement points is attributed.
- *Chapter 11, "Recommendations for Architecting a Reliable System."* This chapter covers architectures and techniques to maximize service availability and service reliability via virtualization and cloud deployment. A simple case study is given to illustrate key architectural points.
- *Chapter 12, "Design for Reliability of Virtualized Applications."* This chapter reviews how design for reliability diligence for virtualized applications differs from reliability diligence for traditional applications.
- *Chapter 13, "Design for Reliability of Cloud Solutions."* This chapter reviews how design for reliability diligence for cloud deployments differs from reliability diligence for traditional solutions.
- *Chapter 14, "Summary."* This gives an executive summary of the analysis, insights, and recommendations on assuring that reliability and availability of cloud-based solutions meet or exceed the performance of traditional deployment.

ACKNOWLEDGMENTS

The authors were greatly assisted by many deeply knowledgeable and insightful engineers at Alcatel-Lucent, especially: Mark Clougherty, Herbert Ristock, Shawa Tam, Rich Sohn, Bernard Bretherton, John Haller, Dan Johnson, Srujal Shah, Alan McBride,

Lyle Kipp, and Ted East. Joe Tieu, Bill Baker, and Thomas Voith carefully reviewed the early manuscript and provided keen review feedback. Abhaya Asthana, Kasper Reinink, Roger Maitland, and Mark Cameron provided valuable input. Gary McElvany raised the initial architectural questions that ultimately led to this work. This work would not have been possible without the strong management support of Tina Hinch, Werner Heissenhuber, Annie Lequesne, Vickie Owens-Rinn, and Dor Skuler.

Cloud computing is an exciting, evolving technology with many avenues to explore. Readers with comments or corrections on topics covered in this book, or topics for a future edition of this book, are invited to send email to the authors (Eric.Bauer@Alcatel-Lucent.com, Randee.Adams@Alcatel-Lucent.com, or pressbooks@ieee.org).

Eric Bauer
Randee Adams

I

BASICS

1

CLOUD COMPUTING

The U.S. National Institute of Standards and Technology (NIST) defines cloud computing as follows:

> Cloud computing is a model for enabling ubiquitous, convenient, on-demand network access to a shared pool of configurable computing resources (e.g., networks, servers, storage, applications, and services) that can be rapidly provisioned and released with minimal management effort or service provider interaction [NIST-800-145].

This definition frames cloud computing as a "utility" (or a "pay as you go") consumption model for computing services, similar to the utility model deployed for electricity, water, and telecommunication service. Once a user is connected to the computing (or telecommunications, electricity, or water utility) cloud, they can consume as much service as they would like whenever they would like (within reasonable limits), and are billed for the resources consumed. Because the resources delivering the service can be shared (and hence amortized) across a broad pool of users, resource utilization and operational efficiency can be higher than they would be for dedicated resources for each individual user, and thus the price of the service to the consumer may well be lower from a cloud/utility provider compared with the alternative of deploying and

Reliability and Availability of Cloud Computing, First Edition. Eric Bauer and Randee Adams.

operating private resources to provide the same service. Overall, these characteristics facilitate outsourcing production and delivery of these crucial "utility" services. For example, how many individuals or enterprises prefer to generate all of their own electricity rather than purchasing it from a commercial electric power supplier?

This chapter reviews the essential characteristics of cloud computing, as well as several common characteristics of cloud computing, considers how cloud data centers differ from traditional data centers, and discusses the cloud service and cloud deployment models. The terminologies for the various roles in cloud computing that will be used throughout the book are defined. The chapter concludes by reviewing the benefits of cloud computing.

1.1 ESSENTIAL CLOUD CHARACTERISTICS

Per [NIST-800-145], there are five essential functional characteristics of cloud computing:

1. on-demand self service;
2. broad network access;
3. resource pooling;
4. rapid elasticity; and
5. measured service.

Each of these is considered individually.

1.1.1 On-Demand Self-Service

Per [NIST-800-145], the essential cloud characteristic of "on-demand self-service" means "a consumer can unilaterally provision computing capabilities, such as server time and network storage, as needed automatically without requiring human interaction with each service's provider." Modern telecommunications networks offer on-demand self service: one has direct dialing access to any other telephone whenever one wants. This behavior of modern telecommunications networks contrasts to decades ago when callers had to call the human operator to request the operator to place a long distance or international call on the user's behalf. In a traditional data center, users might have to order server resources to host applications weeks or months in advance. In the cloud computing context, on-demand self service means that resources are "instantly" available to service user requests, such as via a service/resource provisioning website or via API calls.

1.1.2 Broad Network Access

Per [NIST-800-145] "broad network access" means "capabilities are available over the network and accessed through standard mechanisms that promote use by heterogeneous thin or thick client platforms (e.g., mobile phones, laptops, and PDAs)." Users expect to access cloud-based services anywhere there is adequate IP networking, rather than requiring the user to be in a particular physical location. With modern wireless networks, users expect good quality wireless service anywhere they go. In the context of cloud computing, this means users want to access the cloud-based service via whatever wireline or wireless network device they wish to use over whatever IP access network is most convenient.

1.1.3 Resource Pooling

Per [NIST-800-145], the essential characteristic of "resource pooling" is defined as: "the provider's computing resources are pooled to serve multiple consumers using a multi-tenant model, with different physical and virtual resources dynamically assigned and reassigned according to consumer demand." Service providers deploy a pool of servers, storage devices, and other data center resources that are shared across many users to reduce costs to the service provider, as well as to the cloud consumers that pay for cloud services. Ideally, the cloud service provider will intelligently select which resources from the pool to assign to each cloud consumer's workload to optimize the quality of service experienced by each user. For example, resources located on servers physically close to the end user (and which thus introduce less transport latency) may be selected, and alternate resources can be automatically engaged to mitigate the impact of a resource failure event. This is essentially the utility model applied to computing. For example, electricity consumers don't expect that a specific electrical generator has been dedicated to them personally (or perhaps to their town); they just want to know that their electricity supplier has pooled the generator resources so that the utility will reliably deliver electricity despite inevitable failures, variations in load, and glitches.

Computing resources are generally used on a very bursty basis (e.g., when a key is pressed or a button is clicked). Timeshared operating systems were developed decades ago to enable a pool of users or applications with bursty demands to efficiently share a powerful computing resource. Today's personal computer operating systems routinely support many simultaneous applications on a PC or laptop, such as simultaneously viewing multiple browser windows, doing e-mail, and instant messaging, and having virus and malware scanners running in the background, as well as all the infrastructure software that controls the keyboard, mouse, display, networking, real-time clock, and so on. Just as intelligent resource sharing on your PC enables more useful work to be done cost effectively than would be possible if each application had a dedicated computing resource, intelligent resource sharing in a computing cloud environment enables more applications to be served on less total computing hardware than would be required with dedicated computing resources. This resource sharing lowers costs for the data center hosting the computing resources for each application, and this enables lower prices to be charged to cloud consumers than would be possible for dedicated computing resources.

1.1.4 Rapid Elasticity

[NIST-800-145] describes "rapid elasticity" as "capabilities can be rapidly and elastically provisioned, in some cases automatically, to quickly scale out, and rapidly released to quickly scale in. To the consumer, the capabilities available for provisioning often appear to be unlimited and can be purchased in any quantity at any time."

Forecasting future demand is always hard, and there is always the risk that unforeseen events will change plans and thereby increase or decrease the demand for service. For example, electricity demand spikes on hot summer afternoons when customers crank up their air conditioners, and business applications have peak usage during business hours, while entertainment applications peak in evenings and on weekends. In addition, most application services have time of day, day of week, and seasonal variations in traffic volumes. Elastically increasing service capacity during busy periods and releasing capacity during off-peak periods enables cloud consumers to minimize costs while meeting service quality expectations. For example, retailers might experience heavy workloads during the holiday shopping season and light workloads the rest of the year; elasticity enables them to pay only for the computing resources they need in each season, thereby enabling computing expenses to track more closely with revenue. Likewise, an unexpectedly popular service or particularly effective marketing campaign can cause demand for a service to spike beyond planned service capacity. End users expect available resources to "magically" expand to accommodate the offered service load with acceptable service quality. For cloud computing, this means all users are served with acceptable service quality rather than receiving "busy" or "try again later" messages, or experiencing unacceptable service latency or quality.

Just as electricity utilities can usually source additional electric power from neighboring electricity suppliers when their users' demand outstrips the utility's generating capacity, arrangements can be made to overflow applications from one cloud that is operating at capacity to other clouds that have available capacity. This notion of gracefully overflowing application load from one cloud to other clouds is called "cloud bursting."

1.1.5 Measured Service

[NIST-800-145] describes the essential cloud computing characteristic of "measured service" as "cloud systems automatically control and optimize resource use by leveraging a metering capability at some level of abstraction appropriate to the type of service (e.g., storage, processing, bandwidth, and active user accounts). Resource usage can be monitored, controlled, and reported, providing transparency for both the provider and the consumer of the utilized service." Cloud consumers want the option of usage-based (or pay-as-you-go) pricing in which their price is based on the resources actually consumed, rather than being locked into a fixed pricing arrangement. Measuring resource consumption and appropriately charging cloud consumers for their actual resource consumption encourages them not to squander resources and release unneeded resources so they can be used by other cloud consumers.

1.2 COMMON CLOUD CHARACTERISTICS

NIST originally included eight common characteristics of cloud computing in their definition [NIST-B], but as these characteristics were not essential, they were omitted from the formal definition of cloud computing. Nevertheless, six of these eight common characteristics do impact service reliability and service availability, and thus will be considered later in this book.

- *Virtualization.* By untethering application software from specific dedicated hardware, virtualization technology (discussed in Chapter 2, "Virtualization") gives cloud service providers control to manage workloads across massive pools of compute servers.
- *Geographic Distribution.* Having multiple geographically distributed data center sites enables cloud providers flexibility to assign a workload to resources close to the end user. For example, for real-time gaming, users are more likely to have an excellent quality of experience via low service latency if they are served by resources geographically close to them than if they are served by resources on another continent. In addition, geographic distribution in the form of georedundancy is essential for disaster recovery and business continuity planning. Operationally, this means engineering for sufficient capacity and network access across several geographically distributed sites so that a single disaster will not adversely impact more than that single site, and the impacted workload can be promptly redeployed to nonaffected sites.
- *Resilient Computing.* Hardware devices, like hard disk drives, wear out and fail for well-understood physical reasons. As the pool of hardware resources increases, the probability that some hardware device will fail in any week, day, or hour increases as well. Likewise, as the number of online servers increases, so does the risk that software running on one of those online server instances will fail. Thus, cloud computing applications and infrastructure must be designed to routinely detect, diagnose, and recover service following inevitable failures without causing unacceptable impairments to user service.
- *Advanced Security.* Computing clouds are big targets for cybercriminals and others intent on disrupting service, and the homogeneity and massive scale of clouds make them particularly appealing. Advanced security techniques, tools, and policies are essential to assure that malevolent individuals or organizations don't penetrate the cloud and compromise application service or data.
- *Massive Scale.* To maximize operational efficiencies that drive down costs, successful cloud deployments will be of massive scale.
- *Homogeneity.* To maximize operational efficiencies, successful cloud deployments will limit the range of different hardware, infrastructure, software platforms, policies and procedures they support.

1.3 BUT WHAT, EXACTLY, IS CLOUD COMPUTING?

Fundamentally, cloud computing is a new business model for operating data centers. Thus, one can consider cloud computing in two steps:

1. What is a data center?
2. How is a cloud data center different from a traditional data center?

1.3.1 What Is a Data Center?

A data center is a physical space that is environmentally controlled with clean electrical power and network connectivity that is optimized for hosting servers. The temperature and humidity of the data center environment are controlled to enable proper operation of the equipment, and the facility is physically secured to prevent deliberate or accidental damage to the physical equipment. This facility will have one or more connections to the public Internet, often via redundant and physically separated cables into redundant routers. Behind the routers will be security appliances, like firewalls or deep packet inspection elements, to enforce a security perimeter protecting servers in the data center. Behind the security appliances are often load balancers which distribute traffic across front end servers like web servers. Often there are one or two tiers of servers behind the application front end like second tier servers implementing application or business logic and a third tier of database servers. Establishing and operating a traditional data center facility—including IP routers and infrastructure, security appliances, load balancers, servers' storage and supporting systems—requires a large capital outlay and substantial operating expenses, all to support application software that often has widely varying load so that much of the resource capacity is often underutilized.

The Uptime Institute [Uptime and TIA942] defines four tiers of data centers that characterize the risk of service impact (i.e., downtime) due to both service management activities and unplanned failures:

- *Tier I.* Basic
- *Tier II.* Redundant components
- *Tier III.* Concurrently maintainable
- *Tier IV.* Fault tolerant

Tier I "basic" data centers must be completely shut down to execute planned and preventive maintenance, and are fully exposed to unplanned failures. [UptimeTiers] offers "Tier 1 sites typically experience 2 separate 12-hour, site-wide shutdowns per year for maintenance or repair work. In addition, across multiple sites and over a number of years, Tier I sites experience 1.2 equipment or distribution failures on an average year." This translates to a data center availability rating of 99.67% with nominally 28.8 hours of downtime per year.

Tier II "redundant component" data centers include some redundancy and so are less exposed to service downtime. [UptimeTiers] offers "the redundant components of Tier II topology provide some maintenance opportunity leading to just 1 site-wide shutdown each year and reduce the number of equipment failures that affect the IT operations environment." This translates to a data center availability rating of 99.75% with nominally 22 hours of downtime per year.

Tier III "concurrently maintainable" data centers are designed with sufficient redundancy that all service transition activities can be completed without disrupting service. [UptimeTiers] offers "experience in actual data centers shows that operating better maintained systems reduces unplanned failures to a 4-hour event every 2.5 years. . . ." This translates to a data center availability rating of 99.98%, with nominally 1.6 hours of downtime per year.

Tier IV "fault tolerant" data centers are designed to withstand any single failure and permit service transition type activities, such as software upgrade to complete with no service impact. [UptimeTiers] offers "Tier IV provides robust, Fault Tolerant site infrastructure, so that facility events affecting the computer room are empirically reduced to (1) 4-hour event in a 5 year operating period. . . ." This translates to a data center availability rating of 99.99% with nominally 0.8 hours of downtime per year.

1.3.2 How Does Cloud Computing Differ from Traditional Data Centers?

Not only are data centers expensive to build and maintain, but deploying an application into a data center may mean purchasing and installing the computing resources to host that application. Purchasing computing resources implies a need to do careful capacity planning to decide exactly how much computing resource to invest in; purchase too little, and users will experience poor service; purchase too much and excess resources will be unused and stranded. Just as electrical power utilities pool electric power-generating capacity to offer electric power as a service, cloud computing pools computing resources, offers those resources to cloud consumers on-demand, and bills cloud consumers for resources actually used. Virtualization technology makes operation and management of pooled computing resources much easier. Just as electric power utilities gracefully increase and decrease the flow of electrical power to customers to meet their individual demand, clouds elastically grow and shrink the computing resources available for individual cloud consumer's workloads to match changes in demand. Geographic distribution of cloud data centers can enable computing services to be offered physically closer to each user, thereby assuring low transmission latency, as well as supporting disaster recovery to other data centers. Because multiple applications and data sets share the same physical resources, advanced security is essential to protect each cloud consumer. Massive scale and homogeneity enable cloud service providers to maximize efficiency and thus offer lower costs to cloud consumers than traditional or hosted data center options. Resilient computing architectures become important because hardware failures are inevitable, and massive data centers with lots of hardware means lots of failures; resilient computing architectures assure that those hardware failures cause minimal service disruption. Thus, the difference between a traditional data center and a cloud computing data center is primarily the business model along with the policies and software that support that business model.

1.4 SERVICE MODELS

NIST defines three service models for cloud computing: infrastructure as a service, platform as a service, and software as a service. These cloud computing service models logically sit above the IP networking infrastructure, which connects end users to the

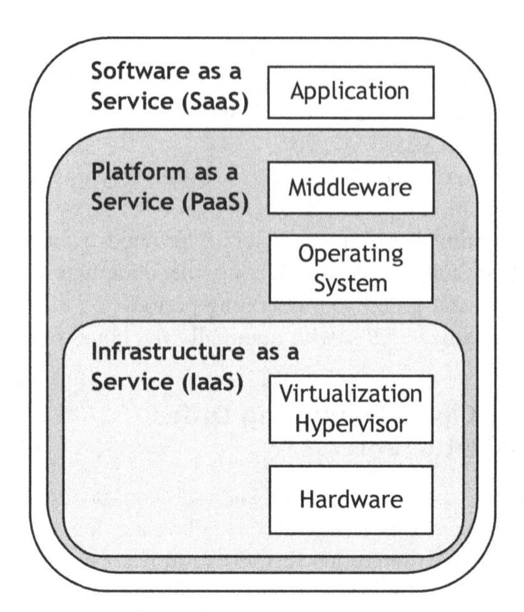

Figure 1.1. Service Models.

applications hosted on cloud services. Figure 1.1 visualizes the relationship between these service models.

The cloud computing service models are formally defined as follows.

- *Infrastructure as a Service (*IaaS*).* "[T]he capability provided to the consumer is to provision processing, storage, networks, and other fundamental computing resources where the consumer is able to deploy and run arbitrary software, which can include operating systems and applications. The consumer does not manage or control the underlying cloud infrastructure but has control over operating systems, storage, deployed applications, and possibly limited control of select networking components (e.g., host firewalls)" [NIST-800-145]. IaaS services include: compute, storage, content delivery networks to improve performance and/or cost of serving web clients, and backup and recovery service.
- *Platform as a Service (*PaaS*).* "[T]he capability provided to the consumer is to deploy onto the cloud infrastructure consumer-created or acquired applications created using programming languages and tools supported by the provider. The consumer does not manage or control the underlying cloud infrastructure including network, servers, operating systems, or storage, but has control over the deployed applications and possibly application hosting environment configurations" [NIST-800-145]. PaaS services include: operating system, virtual desktop, web services delivery and development platforms, and database services.
- *Software as a Service (*SaaS*).* "[T]he capability provided to the consumer is to use the provider's applications running on a cloud infrastructure. The consumer does not manage or control the underlying cloud infrastructure including

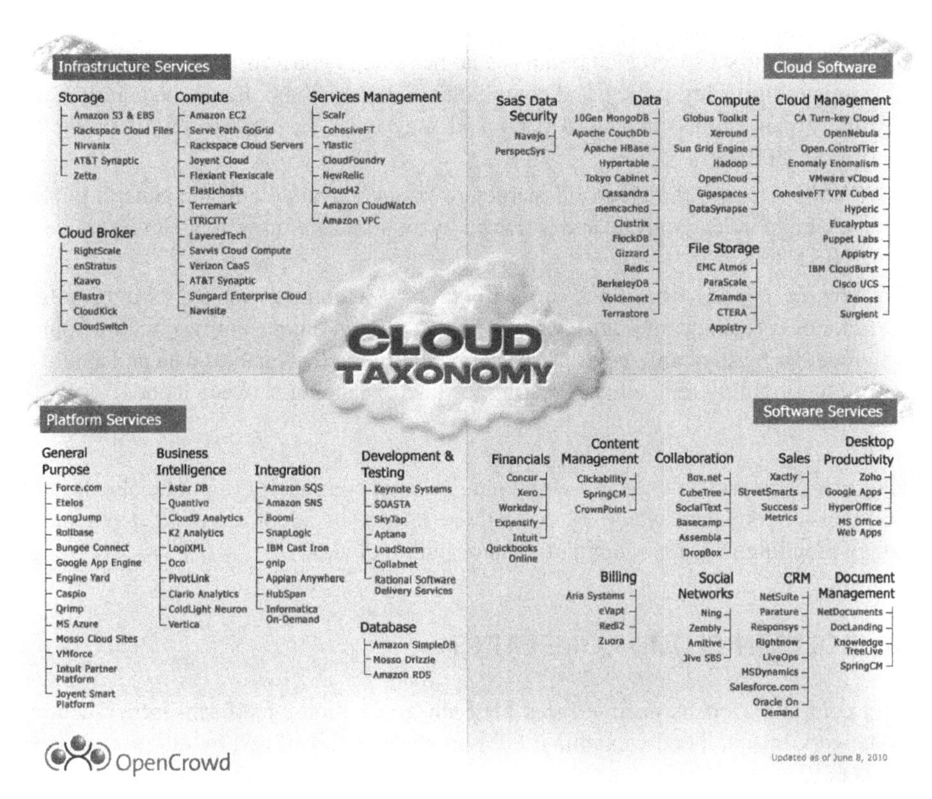

Figure 1.2. OpenCrowd's Cloud Taxonomy.

Source: Copyright 2010, Image courtesy of OpenCrowd, opencrowd.com.

network, servers, operating systems, storage, or even individual application capabilities, with the possible exception of limited user-specific application configuration settings" [NIST-800-145]. SaaS applications include: e-mail and office productivity; customer relationship management (CRM), enterprise resource planning (ERP); social networking; collaboration; and document and content management.

Figure 1.2 gives concrete examples of IaaS, PaaS, and SaaS offerings.

1.5 CLOUD DEPLOYMENT MODELS

NIST recognizes four cloud deployment models:

- *Private Cloud.* "the cloud infrastructure is operated solely for an organization. It may be managed by the organization or a third party and may exist on premise or off premise." [NIST-800-145]

- *Community Cloud.* "the cloud infrastructure is shared by several organizations and supports a specific community that has shared concerns (e.g., mission, security requirements, policy, and compliance considerations). It may be managed by the organizations or a third party and may exist on premise or off premise" [NIST-800-145].
- *Public Cloud.* "the cloud infrastructure is made available to the general public or a large industry group and is owned by an organization selling cloud services" [NIST-800-145].
- *Hybrid Cloud.* "the cloud infrastructure is a composition of two or more clouds (private, community, or public) that remain unique entities but are bound together by standardized or proprietary technology that enables data and application portability (e.g., cloud bursting for load-balancing between clouds)" [NIST-800-145].

Cloud service providers typically offer either private, community or public clouds, and cloud consumers select which of those three to use, or adopt a hybrid deployment strategy blending private, community and/or public clouds.

1.6 ROLES IN CLOUD COMPUTING

Cloud computing opens up interfaces between applications, platform, infrastructure, and network layers, thereby enabling different layers to be offered by different service providers. While NIST [NIST-C] and some other organizations propose new roles of *cloud service consumers*, *cloud service distributors*, *cloud service developers and vendors*, and *cloud service providers*, the authors will use the more traditional roles of suppliers, service providers, cloud consumers, and end users, as illustrated in Figure 1.3.

Specific roles in Figure 1.3 are defined below.

- *Suppliers* develop the equipment, software, and integration services that implement the cloud-based and client application software, the platform software, and the hardware-based systems that support the networking, compute, and storage that underpin cloud computing.
- *Service providers* own, operate, and maintain the solutions, systems, equipment, and networking needed to deliver service to end users. The specific service provider roles are defined as:
    - *IP network service providers* carry IP communications between end user's equipment and IaaS provider's equipment, as well as between IaaS data centers. Network service providers operate network equipment and facilities to provide Internet access and/or wide area networking service. Note that while there will often be only a single infrastructure, platform, and software service provider for a particular cloud-based application, there may be several different network service providers involved in IP networking between the IaaS

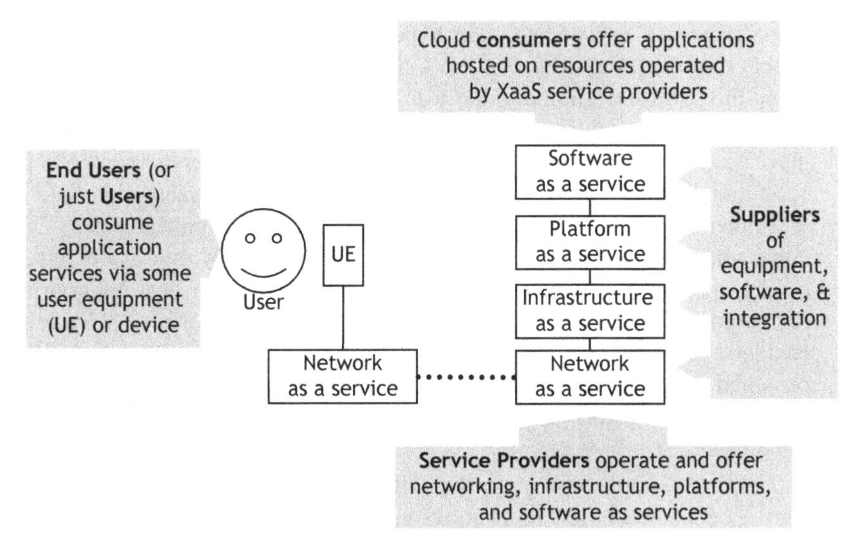

Figure 1.3. Roles in Cloud Computing.

service provider's equipment and end users' equipment. Internet service providers and Internet access providers are examples of network service providers. While IP networking service is not explicitly recognized in NIST's service model, these service providers have a crucial role in delivering end-to-end services to cloud users and can thus impact the quality of experience for end users.

- *IaaS providers* "have control of hardware, hypervisor and operating system, to provide services to consumers. For IaaS, the provider maintains the storage, database, message queue or other middleware, or the hosting environment for virtual machines. The [PaaS/SaaS/cloud] consumer uses that service as if it was a disk drive, database, message queue, or machine, but they cannot access the infrastructure that hosts it" [NIST-C]. Most IaaS providers focus on providing complete computing platforms for consumers' VMs, including operating system, memory, storage, and processing power. Cloud consumers often pay for only what they use, which fits nicely into most companys' computing budget.
- *PaaS providers* "take control of hardware, hypervisor, OS and middleware, to provide services. For PaaS, the provider manages the cloud infrastructure for the platform, typically a framework for a particular type of application. The consumer's application cannot access the infrastructure underneath the platform" [NIST-C]. PaaS providers give developers complete development environments in which to code, host, and deliver applications. The development environment typically includes the underlying infrastructure, development tools, APIs, and other related services.

  - *SaaS providers* "rely on hardware, hypervisor, OS, middleware, and application layers to provide services. For SaaS, the provider installs, manages and maintains the software. The provider does not necessarily own the physical infrastructure in which the software is running. Regardless, the consumer does not have access to the infrastructure; they can access only the application"[NIST-C]. Common SaaS offerings include desktop productivity, collaboration, sales and customer relationship management, and documentation management.
- *Cloud consumers*, (or simply "consumers") are generally enterprises offering specific application services to end users by arranging to have appropriately configured software execute on XaaS resources hosted by one or more service providers. Cloud consumers pay service providers for cloud XaaS resources consumed. End users are typically aware only of the enterprise's application; the services offered by the various XaaS service providers are completely invisible to end users.
- *End users* (or simply *users*) use the software applications hosted on the cloud. Users access cloud-based applications via IP networking from some user equipment, such as a smartphone, laptop, tablet, or PC.

There are likely to be several different suppliers and service providers supporting a single cloud consumer's application to a community of end users. The cloud consumer may have some supplier role in developing and integrating the software and solution. It is possible that the end users are in the same organization as the one that offers the cloud-based service to end users.

1.7 BENEFITS OF CLOUD COMPUTING

The key benefit of cloud computing for many enterprises is that it turns IT from a capital intensive concern to a pay-as-you-go activity where operating expenses track usage—and ideally computing expenses track revenue. Beyond this strategic capital expense to operating expense shift, there are other benefits of cloud computing from [Kundra] and others:

- *Increased Flexibility.* Rapid elasticity of cloud computing enables resources engaged for an application to promptly grow and later shrink to track the actual workload so cloud consumers are better able to satisfy customer demand without taking financial risks associated with accurately predicting future demand.
- *Rapid Implementation.* Cloud consumers no longer need to procure, install, and bring into service new compute capacity before offering new applications or serving increased workloads. Instead, they can easily buy the necessary computing capacity "off the shelf" from cloud service providers, thereby simplifying and shortening the service deployment cycle.
- *Increased Effectiveness.* Cloud computing enables cloud consumers to focus their scarce resources on building services to solve enterprise problems rather

than investing in deploying and maintaining computing infrastructure, thereby increasing their organizational effectiveness.

- *Energy Efficiency.* Cloud service providers have the scale and infrastructure necessary to enable effective sharing of compute, storage, networking, and data center resources across a community of cloud consumers. This not only reduces the total number of servers required compared with dedicated IT resources, but also reduces the associated power, cooling, and floor space consumed. In essence, intelligent sharing of cloud computing infrastructure enables higher resource utilization of a smaller overall pool of resources compared with dedicated IT resources for each individual cloud consumer.

1.8 RISKS OF CLOUD COMPUTING

As cloud computing essentially outsources responsibility for critical IS/IT infrastructure to a service provider, the cloud consumer gives up some control and is confronted with a variety of new risks. These risks range from reduced operational control and visibility (e.g., timing and control of some software upgrades) to changes in accountability (e.g., provider service level agreements) and myriad other concerns. This book considers only the risks that service reliability and service availability of virtualized and cloud-based solutions will fail to achieve performance levels the same as or better than those that traditional deployment scenarios have achieved.

2

VIRTUALIZATION

Virtualization is the logical abstraction of physical assets, such as the hardware platform, operating system (OS), storage devices, data stores, or network interfaces. Virtualization was initially developed to improve resource utilization of mainframe computers, and has evolved to become a common characteristic of cloud computing. This chapter begins with a brief background of virtualization, then describes the characteristics of virtualization and the lifecycle of a virtual machine (VM), and concludes by reviewing popular use cases of virtualization technology.

2.1 BACKGROUND

The notion of virtualization has been around for decades. Dr. Christopher Strachey from Oxford University used the term virtualization in his book *Time Sharing in Large Fast Computers* in the 1960s. Computer time sharing meant that multiple engineers could share the computers and work on their software in parallel; this concept became known as multiprogramming. In 1962, one of the first supercomputers, the Atlas Computer, was commissioned. One of the key features of the Atlas Computer was the supervisor,

Reliability and Availability of Cloud Computing, First Edition. Eric Bauer and Randee Adams.

responsible for allocating system resources in support of multiprogramming. The Atlas Computer also introduced the notion of virtual memory that is the separation of the physical memory store from the programs accessing it. That supervisor is considered an early OS. IBM quickly followed suit with the M44/44X project that coined the term VM. Virtual memory and VM technologies enabled programs to run in parallel without knowledge of the existence of the other executing programs. Virtualization was used to partition large mainframe computers into multiple VMs, providing the ability for multiple applications and processes to run in parallel, and thus better utilize hardware resources. With the advent of less expensive computers and distributed computing, this ability to maximize the utilization of hardware became less necessary.

The proliferation of computers in the 1990s created another opportunity for virtualization to improve resource utilization. VMware and others constructed virtualization products to enable myriad applications running on many lightly utilized computers to be consolidated onto a smaller number of servers. This server consolidation dramatically reduced hardware-related operating expenses, including data center floor space, cooling, and maintenance. By decoupling applications from the underlying hardware resources that support them to enable efficient resource sharing, virtualization technology enables the cloud computing business model that is proliferating today.

2.2 WHAT IS VIRTUALIZATION?

A simple analogy of virtualization is the picture-in-picture feature of some televisions and set top boxes because it displays a small virtual television image on top of another television image, thereby allowing both programs to play simultaneously. Computer virtualization is like this in that several applications that would normally execute on dedicated computer hardware (analogous to individual television channels) are actually run on a single hardware platform that supports virtualization, thereby enabling multiple applications to execute simultaneously.

Virtualization can be implemented at various portions of the system architecture:

- *Network virtualization* entails virtual IP management and segmentation.
- *Memory virtualization* entails the aggregation of memory resources into a pool of single memory and managing the memory on behalf of the multiple applications using it.
- *Storage virtualization* provides a layer of abstraction for the physical storage of data at the device level (referred to as block virtualization) or at the file level (referred to as file virtualization). Block virtualization includes technologies such as storage area network (SAN) and network attached storage (NAS) that can efficiently manage storage in a central location for multiple applications across the network rather than requiring the applications to manage their own storage on a physically attached device.
- *Processor virtualization* enables a processor to be shared across multiple application instances.

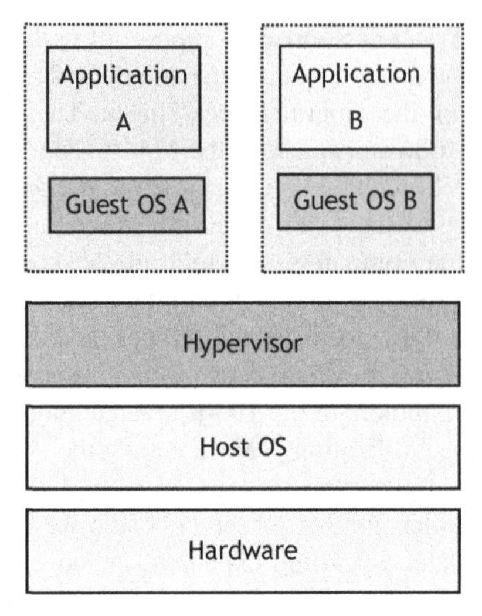

Figure 2.1. Virtualizing Resources.

Virtualization decouples an application from the underlying physical hardware, including CPU, networking, memory, and nonvolatile data storage or disk. Application software experiences virtualization as a VM, which is defined by [OVF] as "an encapsulation of the virtual hardware, virtual disks, and the metadata associated with it." Figure 2.1 gives a simple depiction of a typical virtualized server. One of the key components of virtualization is the hypervisor (also called the VM monitor (VMM); these terms will be used interchangeably in this chapter), which supports the running of multiple OSs concurrently on a single host computer. The hypervisor is responsible for managing the applications' OSs (called the guest OSs) and their use of the system resources (e.g., CPU, memory, and storage). Virtual machines (VMs) are isolated instances of the application software and Guest OS that run like a separate computer. It is the hypervisor's responsibility to support this isolation and manage multiple VM's running on the same host computer.

A virtual appliance is a software image delivered as a complete software stack installed on one or more VMs, managed as a unit. A virtual appliance is usually delivered as Open Virtualization Format (OVF) files. The purpose of virtual appliances is to facilitate the deployment of applications. They often come with web interfaces to simplify virtual appliance configuration and installation.

2.2.1 Types of Hypervisors

There are two types of hypervisors (pictured in Figure 2.2):

- *Type 1.* The hypervisor runs directly on the hardware (aka, bare metal) to control the hardware and monitor the guest OSs, which are on a level above the hypervisor. Type 1 represents the original implementation of the hypervisor.

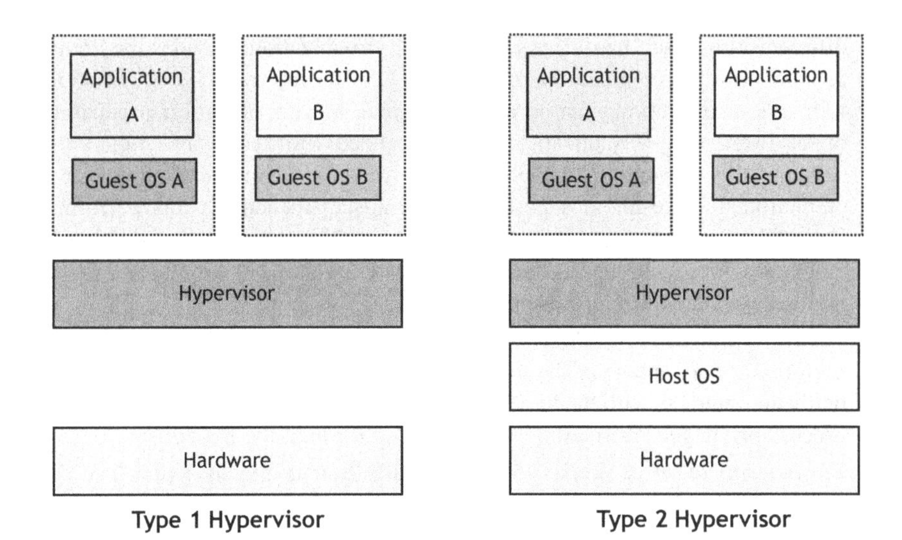

Figure 2.2. Type 1 and Type 2 Hypervisors.

- *Type 2.* The hypervisor runs on top of an existing OS (referred to as the host OS) to monitor the guest OSs, which are running at a third level above the hardware (above the host OS and hypervisor).

2.2.2 Virtualization and Emulation

In the industry, the terms virtualization and emulation are sometimes used interchangeably, but they actually refer to two separate technologies. Emulation entails making one system behave like another to enable software that was written to run on a particular system to be able to run on a completely different system with the same interfaces and produce the same results. Emulation does increase the flexibility for software to move to different hardware platforms, but it does usually have a significant performance cost. Virtualization provides a decoupling of an entity from its physical assets. VMs represent isolated environments that are independent of the hardware they are running on. Some virtualization technologies use emulation while others do not.

2.3 SERVER VIRTUALIZATION

There are three types of server virtualization:

- *Full virtualization* allows instances of software written for different OSs (referred to as guest OSs) to run concurrently on a host computer. Neither the application software nor the guest OS needs to be changed. Each VM is isolated from the others and managed by a hypervisor or VMM, which provides emulated hardware to the VMs so that application and OS software can seamlessly run on different virtualized hardware servers. Full virtualization provides the ability to support multiple applications on multiple OSs on the same server. In addition failovers or migrations can be performed onto servers on different generations of hardware. Full virtualization can be realized with hardware emulation that supports this separation of the hardware from the applications; however, this emulation does result in a performance impact. To address this performance impact, hardware-assisted virtualization is available to manage the isolation. This emulation does incur a performance overhead that may be partially addressed by hardware-assisted virtualization.

- *Hardware-assisted virtualization* is similar to full virtualization but has the added performance advantage of the processors being virtualization aware. The system hardware interacts with the hypervisors and also allows the guest OSs to directly process privileged instructions without going through the hypervisor.
- *Paravirtualization* is similar to full virtualization in that it supports VMs on multiple OSs; however, the guest OSs must be adapted to interface with the hypervisor. Paravirtualization provides a closer tie between the guest OS and the hypervisor. The benefit is better performance since emulation is not required; however, in order to realize this tighter interface between the guest OS and the hypervisor, changes must be made to the guest OS to make the customized API calls. Some products support paravirtualization with hardware assist to further improve performance.
- *OS virtualization* supports partitioning of the OS software into individual virtual environments (sometimes referred to as containers), but they are limited to running on the same host OS. OS virtualization provides the best performance since native OS calls can be made by the guest OS. The simplicity is derived from the requirement that the guest OS be the same OS as the host; however, that is also its disadvantage. OS virtualization cannot support multiple OSs on the same server; however, it can support hundreds of instances of the containers on a single server.

2.3.1 Full Virtualization

Full virtualization (depicted in Figure 2.3) uses a VM monitor (or hypervisor) to manage the allocation of hardware resources for the VMs. No changes are required of the guest OS. The hypervisor emulates the privileged operation and returns control to the guest OS. The VMs contain the application software, as well as its OS (referred to as the Guest OS). With full virtualization, each VM acts as a separate computer, isolated from other VMs co-residing on that hardware. Since the hypervisor runs on bare metal, the various Guest OSs can be different; this is unlike OS virtualization, which requires the virtual environments to be based off an OS consistent with the host OS.

2.3.1.1 Hardware-Assisted Virtualization. Hardware-assisted virtualization provides optimizations using virtualization aware processors. Virtualization-aware processors are those that know of the presence of the server virtualization stack and can

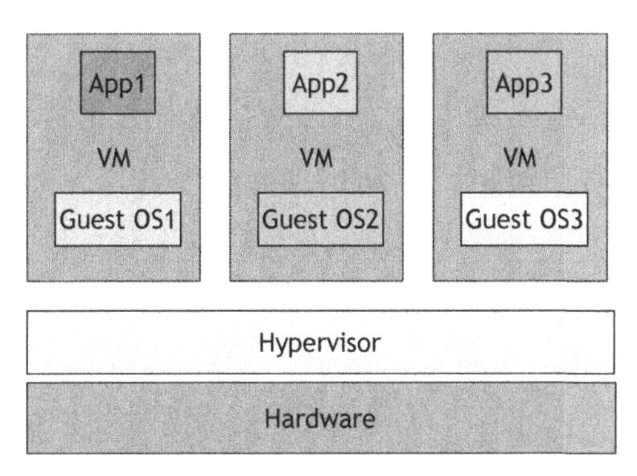

Figure 2.3. Full Virtualization.

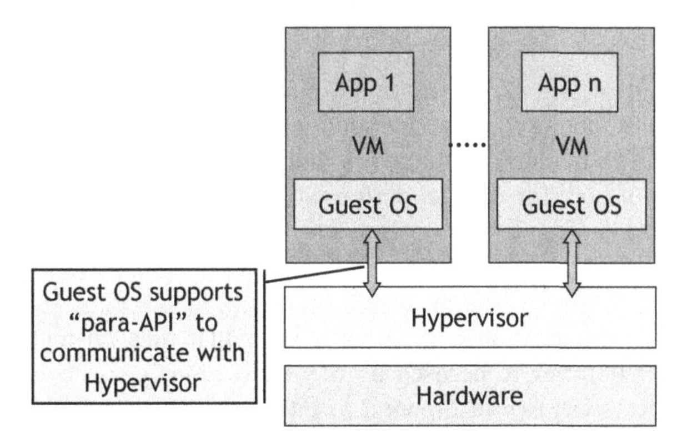

Figure 2.4. Paravirtualization.

therefore do things, such as interact directly with the hypervisors or dedicate hardware space to VMs. The hypervisor still provides isolation and control of the VMs and allocation of the system resources, but the guest OSs can process privileged instructions without going through the hypervisor. Intel and AMD are two of the main providers who support hardware-assisted virtualization for their processors.

2.3.2 Paravirtualization

Paravirtualization (illustrated in Figure 2.4) has a slightly different approach from full virtualization that is meant to improve performance and efficiency. The hypervisor actually multiplexes (or coordinates) all application access to the underlying host computer resources. A hardware environment is not simulated; however, the guest OS is executed in an isolated domain, as if running on a separate system. Guest OS software needs to be specifically modified to run in this environment with kernel mode drivers and application programming interfaces to directly access the parts of the hardware

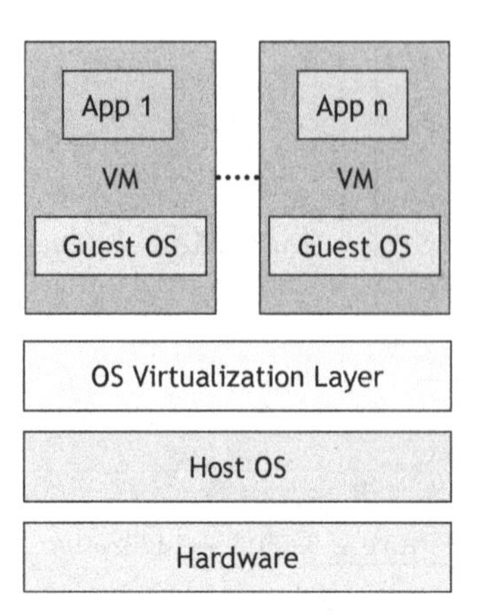

Figure 2.5. Operating System Virtualization.

such as storage and memory. There are some products that support a combination of paravirtualization (particularly for network and storage drivers) and hardware assist that take the best of both for optimal performance.

2.3.3 OS Virtualization

Operating system virtualization consists of a layer that runs on top of the host OS providing a set of libraries to be used by the applications to isolate their use of the hardware resources as shown in Figure 2.5. Each application or application instance can have its own file system, process table, network configuration, and system libraries. Each isolated instance is referred to as a virtual environment or a container. Since the virtual environment or container concept is similar to that of a VM, for consistency, the term "virtual machine" will be used in subsequent comparisons. The kernel provides resource management features to limit the impact of one container's activities on the other containers. OS virtualization does not support OSs other than the host OS. Note that Figure 2.5 indicates Guest OSs for the application; however, in the case of OS virtualization, the Guest OSs must be the same OS as the host operation. The use of the term Guest OS is to provide consistency with the other server virtualization types. There is very little overhead associated with OS virtualization, since it uses the native OS calls and does not need emulation.

2.3.4 Discussion

The three primary types of server virtualization all provide a partitioning of applications into their own VMs (or virtual environments) and use a hypervisor to perform as the host OS that manages the hardware resources on behalf of the applications. In all three types, there is no need to make any changes to the application software itself; the application software will behave as if it had exclusive access to all of the underlying hardware resources. The virtualization types differ in: complexity, ability to support one or more OSs, performance, and level of access to hardware resources. Those differences are summarized in Table 2.1. Some examples of products that support the virtualization type are also included in the table for reference.

2.4 VM LIFECYCLE

While different virtualization technologies and different implementations support somewhat different VM lifecycles, the Distributed Management Task Force [DSP1057] recommends the following standard VM states:

- *Defined.* The virtual system is defined (or configured) but is not enabled to perform tasks, and thus does not consume any resources. The application software is *not* running in this state.
- *Active.* The virtual system is instantiated, and generally resources are enabled to perform tasks. The application software is running or runnable in this state.
- *Paused.* The virtual system and its virtual resources are disabled from performing tasks; however, the virtual system and its virtual resources are still instantiated; resources remain allocated. The application software is *not* running in this state and is considered temporarily inactive (or quiescent).
- *Suspended.* The virtual system and its virtual resources are disabled from performing tasks and the state of the virtual system and its virtual resources are saved to nonvolatile data storage. Resources may be de-allocated. The state is considered enabled but offline.

The transitions between these states are illustrated in Figure 2.6 and defined as follows:

- *Define* (indicated as "create" in Figure 2.6) entails the definition of a new VM.
- *Activate* represents a transition from the *defined* state to the *active* state, entailing the allocation of resources and the enabling of the system. Systems can transition from *paused* or *suspended* to active with this transition.
- *Deactivate* is the deallocation of resources and disabling of the virtual system from *activate*, *paused*, or *suspended* to defined.
- *Pause* entails the disabling of the virtual system moving from *active* to *paused.*
- *Suspend* entails the disabling of the virtual system and the moving of the state of the virtual system and its resources to nonvolatile data storage transitioning from *active* or *paused* to *suspended.*
- *Shut down* entails the notification of the system that it needs to shut down. The software then terminates its tasks and itself and then performs the same steps as deactivate.
- *Reboot* entails a soft boot transitioning from *active*, *paused*, or *suspended* to *active*. The system remains instantiated and resources remain allocated.

TABLE 2.1. Comparison of Server Virtualization Technologies

Virtualization Type	Heterogeneous OS Support	OS Modification needed	Performance Impact	Hardware Assist Supported	Advantages	Disadvantages	Example Products
Full virtualization	Yes	No	Overhead from emulation-improved with hw-assisted virtualization	yes	Full VM portability	Performance impact with emulation especially on I/O; addressed with hw-assisted virtualization	VMware, KVM AMD-V
Paravirtualization	Yes	Yes	Faster than full virtualization; can be enhanced with hw assist	Yes	Better performance than full virtualization	Requires guest OS modifications	Xen, KVM, VMware, AMD-V, and Hyper-V
OS virtualization	No	No	Faster than paravirtualization	No	Best performance, scalability	Only supports a single OS	Linux-VServer, Sun LDOM, and OpenVZ

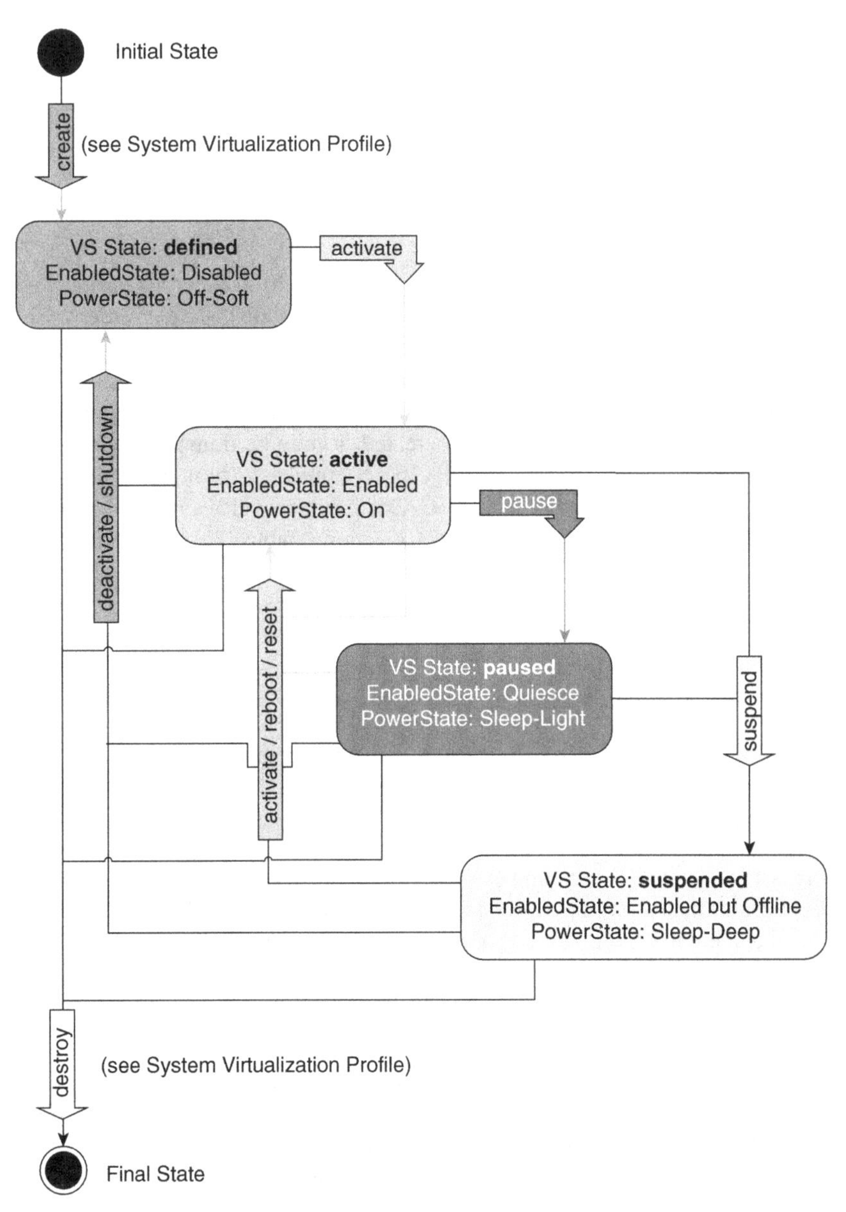

Figure 2.6. Virtualized Machine Lifecycle State Transitions.
Source: Distributed Management Task Force.

- *Reset* entails a hard boot from *active*, *paused*, or *suspended* to *active*. State information saved during *suspend* may be lost with the hard boot.

Per [DSP1057], the virtual system transitions are defined in Table 2.2.

2.4.1 VM Snapshot

A snapshot is a mechanism to preserve a copy of the VM at a certain instant in time and can include its memory contents, settings, and virtual disk state in order to restore the VM and its resources to the point at which the snapshot was taken. Since the snapshot needs to capture state information, a pause option may be available to facilitate the snapshot, but other options, such as copy-on-write, may be used to avoid the pause. Once the snapshot has been successfully created, it may be immediately activated or stored for later activation. Snapshots may be continuously built (at intervals) with incremental changes from the last snapshot. A snapshot provides a means of recovering from a failure in one version of the VM to a more stable (i.e., prefailure) version. Snapshot recovery is useful when an update to a VM causes issues, such as system instability. Since it represents an older version of the VM, it does not offer seamless service recovery for the user in the event of a failure, since it will not have the most recent state and session information. Snapshots can be created, applied, and destroyed when no longer needed. Snapshots are often used for backup and data recovery routines.

2.4.2 Cloning VMs

Cloning is a mechanism for making a duplicate copy of a VM (referred to as the parent). This is useful when multiple copies are needed of the same VM, such as setting up equivalent test environments for a group of testers or students. The two types of clones are:

1. Once the VM has been copied from its parent, it is completely independent of the parent VM. Any changes to the parent do not impact the clone. Some products refer to this as a *full clone*. Full clones perform better than linked clones because they are independent, but they take longer to set up.
2. Once the VM has been copied from its parent, it shares virtual disks with the parent, and thus to function properly, the cloned VM must maintain access to its parent. Some products refer to this as a *linked clone*.

Cloning is the most efficient way to make a copy of a VM (since it requires less time than a full installation of a VM and its guest OS) that is activated to take over for a failed VM, to increase the number of VMs to increase capacity, or to be used in scenarios as described in Section 2.4.3 for service transition or disaster recovery.

TABLE 2.2. Virtual Machine Lifecycle Transitions

Observation of Virtual System Transition	Requirement	"From" Virtual System State	"To" Virtual System State	RequestedState Property and Parameter Value	RequestPowerStateChange ( ): Property Value
Observation of state transitions not supported	n/a	n/a	n/a	**12 (Not Applicable)**	n/a
"**define**" (Optional)	Optional	No CIM_ComputerSystem instance	"Defined"	Not applicable.	
"**activate**" (Optional)	Optional	"Defined" "Paused" "Suspended"	"Active"	**2 (Enabled)**	2 (On)
"**deactivate**" (Optional)	Optional	"Active" "Paused" "Suspended"	"Defined"	**3 (Disabled)**	8 (Off–Soft)
"**pause**" (Optional)	Optional	"Active"	"Paused"	**9 (Quiesce)**	3 (Sleep–Light)
"**suspend**" (Optional)	Optional	"Active" "Paused"	"Suspended"	**6 (Offline)**	4 (Sleep–Deep)
"**shut down**" (Optional)	Optional	"Active" "Paused" "Suspended"	"Defined"	**4 (Shut Down)**	8 (Off–Soft)
"**reboot**" (Optional)	Optional	"Active" "Paused" "Suspended"	"Active"	**10 (Reboot)**	5 (Power Cycle [Off–Soft])
"**reset**" (Optional)	Optional	"Active" "Paused" "Suspended"	"Active"	**11 (Reset)**	9 (Power Cycle [Off–Hard])
Information about recent or pending state transitions not available	Optional	n/a	n/a	**5 (No Change)**	n/a

2.4.3 High Availability Mechanisms

High availability mechanisms ensure that an application is continuously available to its users. This generally entails redundant components and a heartbeat mechanism that quickly detects failure of an active component and automatically recovers service to a redundant component. In order to provide the level of fault detection, isolation, and recovery required for high availability systems, many virtualization software vendors include mechanisms for high availability, data synchronization, and the use of clusters. A cluster is a group of tightly coupled computers that work as a whole to support continuous service availability even in the event of failures via automatic failover and load balancing among members of the cluster. Clusters are used for higher availability or scaling purposes. High availability mechanisms associated with virtualization are responsible for monitoring and controlling VMs. If a failure is detected by the high availability mechanism, then the VM will be restarted on the same or on a different computer within its cluster depending on the nature of the failure; state information is not generally preserved. Some enhancements to the high availability mechanisms do provide data synchronization, including state information to ensure no loss of service or data during the recovery. One such mechanism maintains a shadow copy of the application in lockstep so that when a failure is detected by the high availability mechanism, the shadow copy of the application can take over with no loss of data or disruption for the user.

2.5 RELIABILITY AND AVAILABILITY RISKS OF VIRTUALIZATION

Chapter 5, "Reliability Analysis of Virtualization," offers a traditional reliability analysis of virtualization technology and its impact on high availability architectures. Chapter 6, "Hardware Reliability, Virtualization and Service Availability," discusses the impact of hardware failures on virtualized systems. Chapter 12, "Design for Reliability of Virtualized Applications," explains how traditional system design for reliability can be tailored for virtualized applications.

3

SERVICE RELIABILITY AND SERVICE AVAILABILITY

Failures are inevitable in complex systems. Both native and virtualized systems are subject to the same fundamental error and failure scenarios: hardware fails, latent residual software defects are activated, electrical power is disrupted, and so on. Failures and other impairments can impact the service delivered to users in three primary ways:

- Service response times can degrade, producing service latency impairments.
- Isolated service requests can fail to respond correctly within an acceptable time, producing service reliability impairments.
- Repeated service requests can fail, producing service availability impairments.

Not only do virtualization technology and cloud computing introduce additional risks that can impair service reliability and service availability, but measurement and accountability of impairments change subtly with cloud computing. This chapter explains the concepts and details behind traditional metrics and accountabilities. Part II, "Analysis," will consider how these measurements change with cloud computing, and Part III, "Recommendations," will consider how accountabilities and key quality indicators may shift in cloud computing.

Reliability and Availability of Cloud Computing, First Edition. Eric Bauer and Randee Adams.

This chapter begins by reviewing errors and failures, and then considers service availability, service reliability, and service latency. A brief background on redundancy and high availability is given. The chapter concludes with a discussion of the reliability considerations of streaming media services.

3.1 ERRORS AND FAILURES

Residual software or hardware defects will occasionally be activated to produce errors, and some errors will escalate and catastrophically impact system operation, thus causing critical failures. If a system doesn't recover from the initial failure promptly, then a cascade of secondary failures may be triggered. These concepts are illustrated with an ordinary pneumatic tire on an automobile or truck in Figure 3.1 (from [Bauer10]). A nail on the road presents a hazard or fault that can be activated by driving over it, thereby puncturing the tire to create a hole that leaks air (an error). Over time this air leak will cause a repairable tire failure, commonly called a "flat tire." If the driver doesn't stop driving on a failed tire quickly enough, then the tire will become irreparably damaged. If the driver continues driving on a flat tire even after the tire is damaged, then the wheel rim will eventually be damaged.

These failures may be minor like transmission of a single IP packet being corrupted or arriving out of sequence, or they could be major like a software failure that crashes a critical process and requires automatic or manual actions to recover service. Just as there are a myriad of potential failure scenarios, there is a range of service impacts that can accrue from those failures. The primary characteristic of the service impact of a failure is the duration of service disruption. Very brief or one-shot transient events can often be mitigated by simple mechanisms, like automatic retry/retransmission mechanisms and the impact of these events may result in slightly longer service latency for affected transactions. Longer service disruptions are likely to cause service to degrade so much that the event is more visible to users. For example: if too many packets in a streaming video playback are lost, then the user will see pixilation or other video anomalies; if too many packets are lost from an audio program or call, then the user will hear degraded audio or periods of silence. If degraded service persists for more than a few seconds, then most users will deem the service or session to have failed, and will abandon it, or, in many cases, users will retry these requests, thereby increasing the load on the system. Figure 3.2 visualizes the failure escalation from transient condition to service unavailability for a canonical application. Failures with service impact of tens or hundreds of milliseconds are often viewed as "transient conditions" which

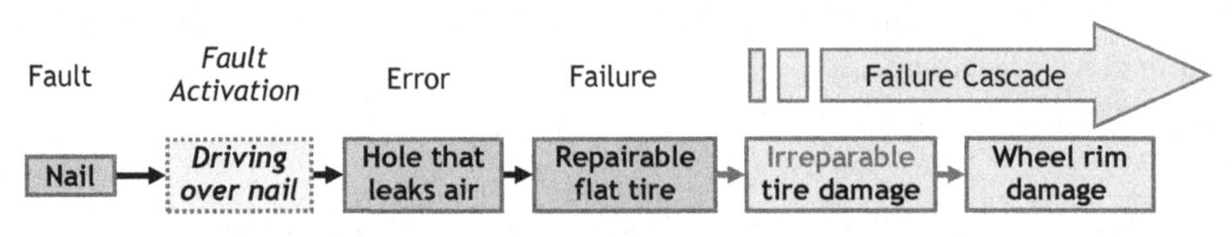

Figure 3.1. Fault Activation and Failures.

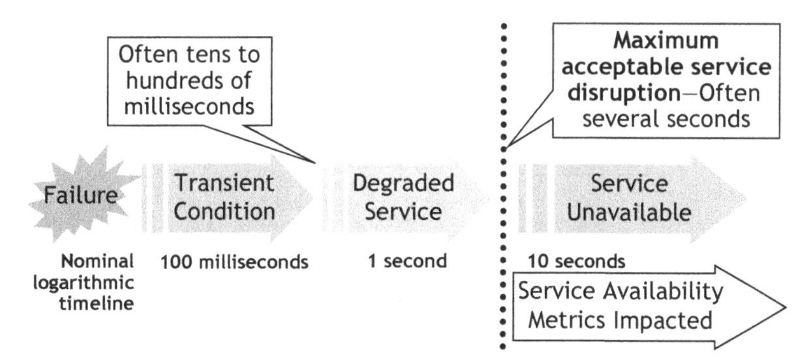

Figure 3.2. Minimum Chargeable Service Disruption.

may be detectable by video and audio users but may not be detectable to web or data users. As discussed in Section 3.4, "Service Reliability," depending on exactly how service reliability metrics are defined and computed, transient conditions may trigger impairments in service reliability metrics. Failures of several hundreds of milliseconds to several seconds are likely to be noticeable to most end users, and thus will be perceived as periods of degraded user service and should be a cause of concern to enterprises and service providers. Service disruptions of longer than a few seconds are likely to be considered service outages and thus accrue downtime and impact service availability metrics.

Note that users accessing an application via a wireless device may attribute service impairments to their wireless access, especially if they are moving or accessing the network during a particularly busy period or from a busy location. Thus, wireless users may implicitly attribute some portion of application service failures to the wireless access network rather than the application itself.

3.2 EIGHT-INGREDIENT FRAMEWORK

The eight-ingredient framework, or 8i, developed by Bell Labs [Rauscher06] is a useful model for methodically considering all potential system vulnerabilities. The 8i framework ingredients are: software, hardware, power, environment, payload, network, human, and policy. Systems are built from *hardware* that hosts application and platform *software*. The hardware depends directly on electrical *power* and a suitable operating *environment* (e.g., acceptable temperature and humidity). Systems interact with users and other systems via IP *networks* carrying application *payloads* structured according to standard or proprietary protocols. The systems are supported by *human* maintenance engineers or operators who follow documented or undocumented procedures and *policies*. Figure 3.3 graphically illustrates a typical system in the 8i context. Each of these ingredients plays a crucial role; each ingredient has vulnerabilities and is subject to faults, errors, and failures.

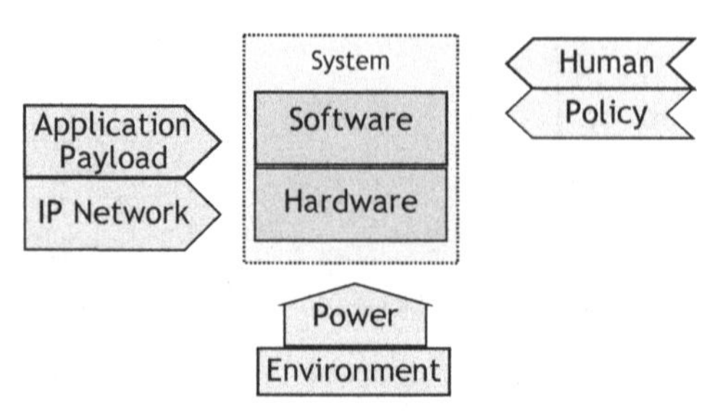

Figure 3.3. Eight-Ingredient ("8i") Framework.

Now let's look at the eight ingredients in detail:

- *Hardware.* Systems are built of physical objects made from electronic and physical components, cables, connectors, fasteners, and so on. Hardware is prone to well-known physical failure mechanisms and equipment suppliers are expected to deliver hardware with acceptably low hardware failure rates throughout the equipment's designed service life. Delivering hardware with acceptably low failure rates is achieved by following well-known hardware design for reliability and manufacturing quality practices. Hardware should be designed to give software visibility, and ideally automatic notification, of hardware failures so system software can promptly report the failure and activate automatic recovery mechanisms. Operationally, the system must rapidly detect hardware failures and isolate them to the appropriate Field Replaceable Unit (FRU) so that maintenance engineers can promptly replace the failed FRU to minimize simplex exposure time and restore the system to full operational status. Hardware failures are considered in Chapter 6, "Hardware Reliability, Virtualization and Service Availability."
- *Software.* Software enables the system's hardware to deliver valuable services to users. Software is prone to programming defects, as well as specification, architecture, design, and integration flaws that cause the system to behave incorrectly in certain situations. Software failures are considered in Chapter 5, "Reliability Analysis of Virtualization."

The remaining six ingredients are external to the system itself, but are crucial in actual system operation:

- *Power.* Appropriate AC or DC power and proper electrical grounding is required for electronic systems to function. This category includes overvoltage and voltage spikes caused by lightening, power crosses, and short circuits for systems that are externally powered. Battery-powered systems are vulnerable to somewhat different hazards such as battery exhaustion and wear out.
- *Environment.* Hardware systems are sensitive to ambient environmental conditions, including: temperature, relative humidity, elevation (because air density is

critical for cooling), dust, and corrosive gases (e.g., the air pollution that causes acid rain). The system's physical design should assure that the system operates properly when maintained within its specified environmental parameters. The system must be properly located and installed so that there is sufficient physical clearance for maintenance engineers to work on and around the equipment to manipulate cabling and hardware units. The environment must be physically secured to prevent deliberate or accidental damage to the system, including physical security attacks and theft of hardware.

- *Network.* Systems fit into a networked solution context in which IP packets are passed to and from other systems. The "network" ingredient transports the application's "payload" ingredient. Crucial facilities (like optical transport fibers and copper cables) and elements (like routers and Ethernet switches, themselves subject to 8i vulnerabilities) can fail, thus disrupting prompt and reliable delivery of application payloads.
- *Payload.* Systems interact with users and other systems via messages or streams of application data passed via IP network facilities and infrastructure. As many of the elements that a particular system communicates with are likely to be different types of systems, often from other equipment suppliers, it is essential that network elements be tolerant of messages or data streams that might be somewhat different than expected. The information passed may be different because other elements interpret protocol specifications differently, or because they have enhanced the protocol in a novel way.
- *Human.* Human beings use, operate, and maintain systems. Humans who perform routine and emergency maintenance on systems present a notable risk because wrong actions (or inaction) can disable or damage the system. Wrong actions by humans can occur for many reasons including:
  - documented procedure is wrong, absent, or unavailable;
  - man-machine interface was poorly designed, thereby making proper execution of procedures more confusing, awkward, or error-prone;
  - human was not properly trained; and
  - human makes a mistake because they are under stress, rushed, confused or tired.
- *Policy.* To successfully operate a complex system, it is essential to have business policies and processes that organize workflows and govern operations and behavior. Operational policies are required for all of the elements and interoperation with other systems and end users, as well as for employees and customers. These policies often include adopting industry standards, regulatory compliance strategies, maintenance and repair strategies, service level agreements, and so on. Enterprises define specific policies, such as standards compliance and "profiles," that define discretionary values for protocols that permit multiple options that must be supported by system suppliers. Several policies impact system failure rates, including:
  - policies for deploying critical software patches and updates;
  - policies for skills and training of maintenance engineers and other staff;

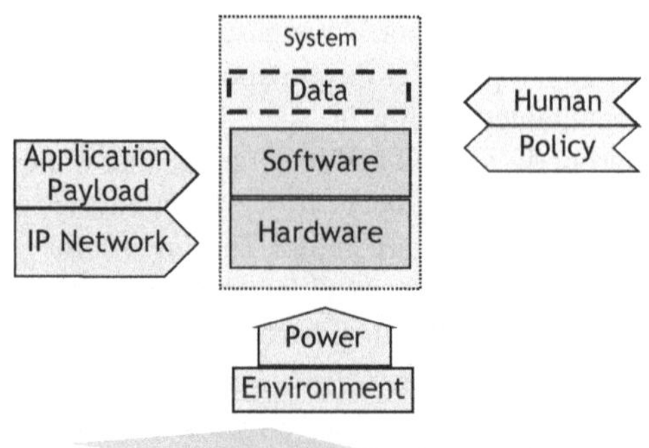

Figure 3.4. Eight-Ingredient Plus Data Plus Disaster (8i + 2d) Model.

- security policies that assure networks and systems are hardened against cyber security attacks.

Several policies impact outage recovery times, including:

- emergency outage recovery policies;
- outage escalation policies; and
- technical support agreements with hardware and software suppliers.

Readers will note that the formal 8i model conspicuously omitted the data ingredient that represents all the configuration, application, and user information that is required to assure proper service delivery. While the 8i inventors felt data could simply be lumped with software, the authors of this book consider application, configuration, and user data to be an independent and co-equal ingredient with software and hardware. Another pragmatic, if somewhat inelegant, extension of the 8i model is to explicitly consider force majeure or disaster events. The importance of considering force majeure or disaster events separately from ordinary ingredient failures is that disaster events can impact multiple ingredients (e.g., an earthquake might impact commercial power, the structural integrity of the data center environment and external IP networking infrastructure) and simultaneously overwhelm ordinary redundancy mechanisms. As a result, disaster events are typically mitigated via distinct business continuity and disaster recovery plans. This book will refer to the 8i model overlaid with a data ingredient and disaster event risk as "8i + 2d." This "8i + 2d" model is depicted in Figure 3.4.

3.3 SERVICE AVAILABILITY

When a failure event persists for more than a few seconds, it is likely to impact not only isolated user service requests, but also the user-initiated retries of those failed requests. Brief service impact events may cause individual transactions or sessions to fail, thus prompting the user to retry the transaction or session (e.g., redialing a dropped call); if the first (and perhaps second) retried attempt fails because service is still impacted, then the event will often be considered a service outage, and thus impact service availability metrics. When service is impacted so long that retried user operations fail—thereby causing users to abandon their efforts to access the service—the service is generally deemed unavailable.

3.3.1 Service Availability Metric

Service availability can be quantified via the simple Equation 3.1 as service uptime divided by the sum of service uptime and service downtime.

$$\text{Availability} = \frac{\text{Uptime}}{\text{Uptime} + \text{Downtime}}.$$

Equation 3.1. Basic Availability Formula

Note that the values of Uptime and Downtime used to calculate availability can be either predicted via mathematical modeling (e.g., an architecture based Markov availability model) or via actual field measurements (e.g., from outage trouble tickets or via service probes). Few enterprises explicitly calculate uptime since it would require a constant or at least periodic monitoring and validation of system health; most enterprises carefully track service downtime. Equation 3.2 calculates availability based on service downtime, as well as the total time the target system(s) was expected to be in service (i.e., the minutes during the measurement period that systems were expected to be online so planned downtime is excluded).

$$\text{Availability} = \frac{\text{TotalInServiceTime} - \text{Downtime}}{\text{TotalInServiceTime}}.$$

Equation 3.2. Practical System Availability Formula

TotalInServiceTime is the sum of minutes per month (or other reporting period) that the systems in the population were expected to be operational; Downtime is the minutes of service unavailability prorated by the percentage of capacity or functionality impacted during the outage.

IT Infrastructure Library (ITIL) offers the simple formula of Equation 3.3 for computing availability:

$$\text{Availability (\%)} = \frac{\text{AgreedServiceTime} - \text{Downtime}}{\text{AgreedServiceTime}} \times 100\%.$$

Equation 3.3. Standard Availability Formula

Note that ITIL [ITILv3SD] explicitly uses “AgreedServiceTime” to highlight that that many systems have scheduled maintenance periods during which service can be offline for maintenance. Thus, AgreedServiceTime explicitly excludes planned downtime from consideration in availability calculations; planned downtime is discussed in Section 3.3.7.

While the simple ITIL definition of availability may be adequate for many enterprise applications, the telecommunications industry has evolved far more sophisticated service availability measurements, which are documented in the TL 9000 Measurements Handbook. This book will use these richer TL 9000 service availability measurement concepts. Service availability in the telecommunications industry is formally defined by [TL9000] as: "the ability of a unit to be in a state ready to perform a required function at a given instant in time or for any period within a given time interval, assuming that the external resources, if required, are provided." The unit against which service availability is traditionally normalized is the individual system[1] or network element.[2]

3.3.2 MTBF and MTTR

Many readers will be familiar with Equation 3.4, which uses mean time between failure (MTBF) and mean time to repair (MTTR) to estimate availability.

$$\text{Availability} = \frac{\text{MTBF}}{\text{MTBF} + \text{MTTR}}.$$

Equation 3.4. Estimation of System Availability from MTBF and MTTR

This simple equation is easily understood by considering Figure 3.5. MTTR is the time to return a system to service and MTBF is the time the system is expected to be up or online before it fails (again). This means that the system will nominally be online and

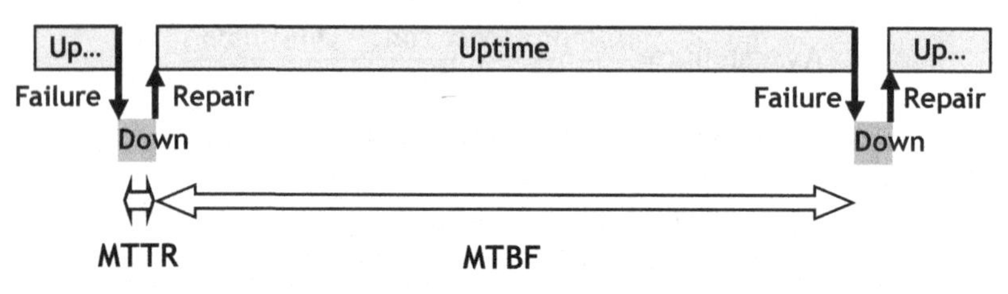

Figure 3.5. MTBF and MTTR.

[1] System is formally defined by [TL9000] as: "A collection of hardware and/or software items located at one or more physical locations where all of the items are required for proper operation. No single item can function by itself."

[2] *Network element* is formally defined by [TL9000] as: "A system device, entity or node including all relevant hardware and/or software components located at one location. The Network Element (NE) must include all components required to perform the primary function of its applicable product category. If multiple FRUs, devices, and/or software components are needed for the NE to provide its product category's primary function, then none of these individual components can be considered an NE by themselves. The total collection of all these components is considered a single NE. Note: While an NE may be comprised of power supplies, CPU, peripheral cards, operating system and application software to perform a primary function, no individual item can be considered an NE in its own right."

up for MTBF hours before failing, and MTTR hours will nominally be required to repair the system and bring it back online. Thus, Equation 3.4 is equivalent to the simple availability formula of Equation 3.1 where *MTBF* is used as an estimate of Uptime and MTTR is used as an estimate of Downtime.

While most readers will have seen predicted MTBF values offered by suppliers and standard MTTR values from industry standards, suppliers, and service providers, they may not realize that large service providers and enterprises with large populations of systems in service will often compute actual MTTR and MTBF based on actual performance of deployed equipment, operational policies, and staff. As with mileage estimates for automobiles "*your mileage may vary*," but standard MTBF and MTTR values—like standard mileage estimates—are a useful baseline when evaluating systems and planning deployments.

The simpler Equation 3.4 is not generally appropriate for systems that include any redundancy for two related reasons:

- *Redundancy Should Enable Service to Be Restored Far Faster Than the Time It Takes to Repair the Failed Element.* Operationally, it should be much faster to switch service to a redundant element (e.g., in seconds) rather than to repair the failed element, which could take hours. Very fast recovery times contribute to a very small mean time to restore service (MTTRS), which can boost service availability.
- *Redundant Elements Are Arranged So That Single Failures Will Not Cause Service Disruption.* For example, various RAID configurations enable individual hard disk failures to be masked from application software, and failure of redundant fans or power supplies should not impact service. On systems with redundancy, only a fraction of the failures that will eventually require maintenance actions (MTBF) will cause service impact, and hence be considered critical, thus improving the mean time between critical failures (MTBCF).

3.3.3 Service and Network Element Impact Outages

Complex and redundant systems can generally experience outages that either directly impact user service, or outages that only cause a loss of redundancy or other impact that does not directly impact user service. Events impacting user service are called *service impact outages* and are defined by [TL9000] as: *a failure where end-user service is directly impacted.* Outages that impact primary functionality (what ITIL calls vital business function or VBF) of a network element, up to and including user service impact, are called *network element impact outages* and are defined by [TL9000] as: a *failure where a certain portion of a network element functionality/capability is lost/down/out of service for a specified period of time.*

The distinction between service impact outages and network element impact outages is especially important for systems with high availability mechanisms to measure the period that a failed component is unavailable and that the system is operating with no available redundancy, and hence the system is at risk of a prolonged service impact outage if another failure occurs before the network element impact outage has

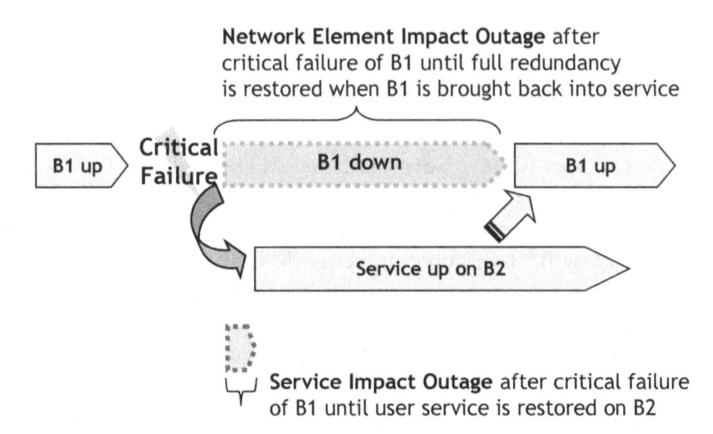

Figure 3.6. Service and Network Element Impact Outages of Redundant Systems.

been resolved, so the system is restored to normal redundancy. Figure 3.6 illustrates this by highlighting both service impact and network element impact outages. Although the network element impact outage when B1 is unavailable might be hours or longer, since user service is rapidly recovered to element B2, the service impact outage period for users is far shorter than the period that B1 is unavailable. Note that until B1 is repaired and brought up, service is nonredundant or simplex exposed, so a second failure cannot be automatically recovered and thus would produce an extended service outage. Thus, well-run enterprises will repair failed components promptly to minimize the simplex exposure time of critical services.

3.3.4 Partial Outages

Larger systems supporting applications with rich functionality for varied user communities and IS/IT maintenance engineers are often far more likely to experience a partial functionality or partial capacity outage than they are to be totally and completely down. For example, one of several software processes can fail and directly impact either the users served by that process or the functionality served by that process. If that failure impacts some or the entire primary functionality offered to some or all of the system offered for an unacceptably long duration, then the event is considered an outage. However, if the event impacts some but not all users or some but not all primary functionality, then the event is a partial outage and the outage should be prorated by the percentage of capacity or functionality lost. To properly consider partial outages, the telecommunications industry and sophisticated enterprises use prorated partial outage downtime formulas, like Equation 3.5.

$$\text{Availability} = \frac{\text{TotalInServiceTime} - \text{TotalProratedDowntime}}{\text{TotalInServiceTime}}.$$

Equation 3.5. Recommended Service Availability Formula

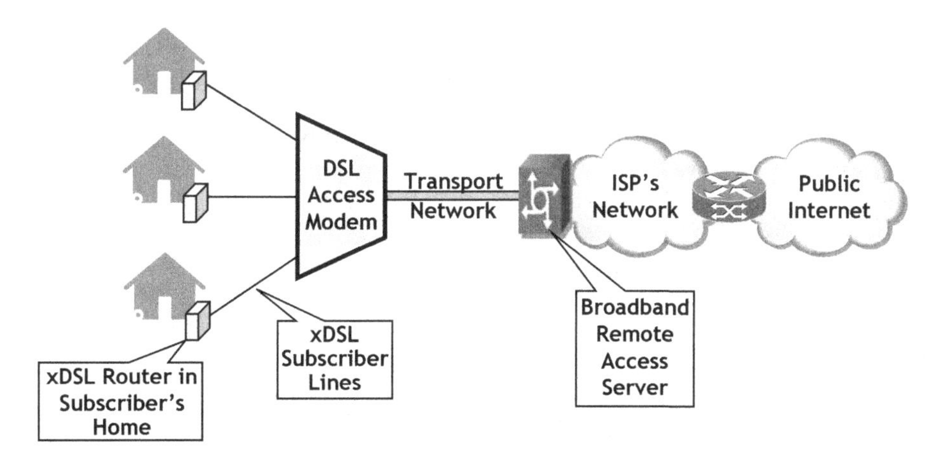

Figure 3.7. Sample DSL Solution.

TotalInServiceTime is the sum of minutes per month (or other reporting period) that the systems in the population were expected to be operational; TotalProratedDowntime is the minutes of service unavailability prorated by the percentage of capacity or functionality impacted during the outage.

As an example of prorating partial capacity loss outages, consider a digital subscriber loop access module (DSLAM) that offers high-speed Internet service over copper wires to subscribers. Figure 3.7 illustrates a sample solution: xDSL routers in subscribers' homes connect via copper wires to the DSLAM, which aggregates the traffic onto high-capacity transport network connections to the Internet service provider's (ISP) broadband remote access server (BRAS), which eventually connects to the public Internet. For engineering and commercial reasons, DSLAMs are generally implemented with many individual line cards, each of which services many dozen subscribers by directly terminating the copper wires that connect to the xDSL routers in subscribers' homes. If a single line card fails completely, then all subscribers served by that line card will be without Internet service until the line card is replaced; subscribers served by other line cards in the DSLAM will not be impacted. If the DSLAM was provisioned to serve 1000 subscribers in the 31-day month of January, then the TotalInServiceTime would be 1000 subscribers times 31 days in January times 24 hours per day times 60 minutes per hour for a total of 44,640,000 subscriber-minutes of in-service time. If the DSLAM experienced a single line card failure in the month which impacted 100 subscribers for 4 hours, then the TotalProratedDowntime is 100 subscribers times 4 hours of downtime times 60 minutes per hour for a total of 24,000 impacted subscriber-minutes for the month. Monthly availability for this sample DSLAM can then be computed via Equation 3.6.

$$\text{Availability} = \frac{44{,}640{,}000 - 24{,}000}{44{,}640{,}000} = \frac{44{,}616{,}000}{44{,}640{,}000} = 99.9462\%.$$

Equation 3.6. Sample Partial Outage Calculation

Readers can easily see how this calculation can be efficiently scaled up to cover many thousands of DSLAMs in a typical service provider's network:

- TotalInServiceTime aggregates the nominal subscriber-minutes (or subscriber-seconds) of service that were expected during the measurement period (i.e., excluding planned maintenance periods);
- TotalProratedDowntime aggregates the subscriber-minutes (or subscriber-seconds) of service impact in the measurement period.

3.3.5 Availability Ratings

Service availability ratings are commonly quantified as the number of nine's or by availability system type; "five 9's"or "high availability" is a common availability expectation for critical systems. Table 3.1 below gives the maximum service downtime for common availability ratings, per [GR2841].

Mature organizations will recognize that different enterprise information services and applications require different availability ratings. While these ratings will vary somewhat based on organizational needs and customer expectations, the standard

TABLE 3.1. Service Availability and Downtime Ratings, per [GR2841]

Number of 9's	Service Availability (%)	System Type	Annualized Down Minutes	Quarterly Down Minutes	Monthly Down Minutes	Practical Meaning
1	90	Unmanaged	52,596.00	13,149.00	4,383.00	Down 5 weeks per year
2	99	Managed	5,259.60	1,314.90	438.30	Down 4 days per year
3	99.9	Well managed	525.96	131.49	43.83	Down 9 hours per year
4	99.99	Fault tolerant	52.60	13.15	4.38	Down 1 hour per year
5	99.999	High availability	5.26	1.31	0.44	Down 5 minutes per year
6	99.9999	Very high availability	0.53	0.13	0.04	Down 30 seconds per year
7	99.99999	Ultra availability	0.05	0.01	–	Down 3 seconds per year

criticality definitions from the U.S. Federal Aviation Administration's National Airspace System's reliability handbook are probably fairly typical:

- **CRITICAL** (service availability rating of 99.999%) *"Loss of this capability would raise to an unacceptable level, the risk associated with providing safe and efficient [system] operations*" [FAA-HDBK-006A].
- **ESSENTIAL** (service availability rating of 99.9%) *"Loss of this capability would significantly raise the risk associated with providing safe and efficient [system] operations"* [FAA-HDBK-006A].
- **ROUTINE** (service availability rating of 99%) *"Loss of this capability would have a minor impact on the risk associated with providing safe and efficient [system] operations*" [FAA-HDBK-006A].

There is also a "safety critical" category with service availability rating of seven 9s for life-threatening risks and services where *"loss would present an unacceptable safety hazard during the transition to reduced capacity operations*" [FAA-HDBK-006A]. Reliability and availability of "safety critical" services are beyond the scope of this book.

3.3.6 Outage Attributability

It is often convenient to consider attributability of the events that impact service availability and service reliability. The telecommunications industry factors outage attributability into three broad and generally applicable categories: product-attributable events, customer- or service-provider attributable events, and external-attributable events. These three orthogonal categories are defined as follows:

- **Product-attributed outages** are defined in [TL9000] as *"an outage primarily triggered by*
  a) system design, hardware, software, components or other parts of the system,
  b) scheduled outage necessitated by the design of the system,
  c) support activities performed or prescribed by [a supplier] *including documentation, training, engineering, ordering, installation, maintenance, technical assistance, software or hardware change actions, etc.,*
  d) procedural error caused by the [supplier],
  e) the system failing to provide the necessary information to conduct a conclusive root cause determination, or
  f) one or more of the above."
- **Service provider-attributable** (or customer-attributable) **outages** are defined by [TL9000] as *"an outage that is primarily attributable to the customer's* [service provider's] *equipment or support activities triggered by*
  a) customer's [service provider] *procedural errors,*
  b) office environment, for example power, grounding, temperature, humidity, or security problems, or
  c) one or more of the above.

Outages are also considered customer [service provider] *attributable if the customer* [service provider] *refuses or neglects to provide access to the necessary information for the* [supplier] *to conduct root cause determination.*"

- **External-attributable outages** are defined in [TL9000] as "outages caused by natural disasters such as tornadoes or floods, and outages caused by third parties not associated with the [service provider] or the [supplier] such as commercial power failures, third-party contractors not working on behalf of the [supplier] or [service provider]."

Outages are also often attributed to root cause categories such as hardware, software and procedural or human. While hardware- and software-attributable outages are fairly straightforward, many readers may not be familiar with the technical definition of procedural error. [TL9000] offers the following definition of procedural error:

"*An error that is the direct result of human intervention or error. Contributing factors can include but are not limited to*

a) deviations from accepted practices or documentation,
b) inadequate training,
c) unclear, incorrect, or out-of-date documentation,
d) inadequate or unclear displays, messages, or signals,
e) inadequate or unclear hardware labeling,
f) miscommunication,
g) non-standard configurations,
h) insufficient supervision or control, or
i) user characteristics such as mental attention, physical health, physical fatigue, mental health, and substance abuse.

Examples of a Procedural Error include but are not limited to

a) removing the wrong fuse or circuit pack,
b) not taking proper precautions to protect equipment, such as shorting out power, not wearing ESD strap, etc.,
c) unauthorized work,
d) not following Methods of Procedures (MOPs)
e) not following the steps of the documentation,
f) using the wrong documentation,
g) using incorrect or outdated documentation,
h) insufficient documentation,
i) translation errors,
j) user panic response to problems,
k) entering incorrect commands,
l) entering a command without understanding the impact, or
m) inappropriate response to a Network Element alarm."

3.3.7 Planned or Scheduled Downtime

Information- and computer-based systems occasionally require planned or preventive maintenance to: upgrade or update software, firmware, or hardware; grow or alter the system's hardware configuration; physically move the equipment or alter its network configuration; and so on. [TL9000] defines *scheduled outage* as follows:

> *"Results from a scheduled or planned maintenance, installation, or manual initialization. This includes such activities as parameter loads, software/firmware changes, and NE growth/update, cutover (for example, switch replacement or absorption), hardware or software growth, preventive maintenance, routine or scheduled diagnostics, data table change, software patching or updates, software generic upgrade, program backup, and data backup."*

Typically enterprises and service providers will plan scheduled outages to occur during so-called maintenance windows (i.e., when system usage will be light) to minimize any user impact. Maintenance windows are traditionally scheduled in the middle of the night where the equipment is physically located, such as between midnight and 4 a.m. local time. As global businesses now operate in several time zones across a region, continent, or the planet, it has become more challenging to pick low usage periods. While maintenance engineers will often require time for preparation work, followed by time to execute all of the steps of the Method of Procedure, and additional time for postwork activities, any period of user service impact should be minimal. For example, while it will invariably take time to download a security or software patch and run the installation program, there should be no service impact during that time. Planned service impact is possible when the system software is gracefully restarted to activate the updated software. Ideally, traffic will have been drained from the system prior to the graceful restart, such as by redirecting user service to an alternate system or by instructing all users to log off prior to the planned restart. Some systems even support "rolling upgrade" strategies in which components are restarted individually so that traffic (perhaps at lower capacity) can continuously be served as portions of the system are gracefully upgraded and restarted.

Periods of service unavailability due to scheduled outages are generally excluded from service availability metrics provided that the duration of service impact is not longer than expected (e.g., the canonical service disruption time expected for successful execution of the particular Method of Procedure). Should the period of service impact be significantly longer than expected (e.g., due to a failed procedure execution, or a hardware or software failure occurring during execution of the procedure when the system was simplex exposed), then the excess service downtime may be recorded as a service outage and impact service availability metrics. Planned activities are considered in this book under the IT Service Management category of Service Transition as described in Section 4.5.4, "Service Transition."

3.4 SERVICE RELIABILITY

The term reliability is sometimes used in the industry as a superset of service availability and various other topics, such as Microsoft's statement "the reliability [service management function] ensures that service capacity, service availability, service continuity, data integrity and confidentiality are aligned to the business needs in a cost effective manner" [Microsoft]. Rather than adopting a very broad and general definition of reliability, this book will use the narrow definition of reliability given by TL 9000.

3.4.1 Service Reliability Metrics

"*Reliability*" is defined as "the ability of an item to perform a required function under stated conditions for a stated time period" [TL9000]. Service reliability characterizes the ability of a system to provide acceptable service, which means correct or accurate service delivered within an acceptable time. Service reliability is essentially the portion of service requests that are successfully served (i.e., are not defective) within the maximum acceptable service latency. Service reliability can be expressed positively in "number of 9s" style via the formula in Equation 3.7.

$$\text{Service Reliability} = \frac{(\text{Successful Responses})}{\text{Total Requests}} \times 100\%.$$

Equation 3.7. Service Reliability Formula

Since most services are very reliable, it is more convenient to focus on the much smaller number of unreliable service events or service defects. These service defects are conveniently normalized as defective transactions or operations per million attempts. Defects per million (DPM) attempts can be computed via Equation 3.8.

$$\text{DPM} = \frac{(\text{Total Requests} - \text{Successful Responses})}{\text{Total Requests}} \times 1{,}000{,}000$$

$$= \frac{\text{Unsuccessful Requests}}{\text{Total Requests}} \times 1{,}000{,}000.$$

Equation 3.8. DPM Formula

Equation 3.9 converts DPM to service reliability probability, and Equation 3.10 converts service reliability to DPM.

$$\text{Service Reliability} = \frac{(1{,}000{,}000 - \text{DPM})}{1{,}000{,}000} \times 100\%.$$

Equation 3.9. Converting DPM to Service Reliability

$$\text{DPM} = (100\% - \text{Service Reliability}) \times 1{,}000{,}000.$$

Equation 3.10. Converting Service Reliability to DPM

For example, if users attempt to send 123,459,789 instant messages to online subscribers via a particular messaging service during a measurement period, and all but 4321 messages are successfully received by the intended recipient within the maximum acceptable service latency, then the DPM in this measurement period is computed via Equation 3.11.

$$\text{DPM} = \frac{4,321}{123,456,789} \times 1,000,000 = 0.0000350 \times 1,000,000 = 35.$$

Equation 3.11. Sample DPM Calculation

3.4.2 Defective Transactions

"Defective service transaction" is defined by [TL9000] as "a transaction where there was a failure to meet one or more internal and/or defined customer requirements concerning the performance of the service." Most application protocols provide return codes that can be used to classify nonsuccessful requests into application failures (nominally attributable to the application supplier, service provider, or enterprise) and request failures (nominally attributable to the user or user equipment). For example, consider the return codes from the IETF's session initiation protocol (SIP), commonly used for voice over IP and video over IP applications. Failures nominally attributed to software, hardware, or network infrastructure failures include:

- Server failure responses, like `500 Server Internal Error` ("*The server encountered an unexpected condition that prevented it from fulfilling the request*" [RFC3261]) or `503 Service Unavailable` ("*The server is temporarily unable to process the request due to a temporary overloading or maintenance of the server*" [RFC3261]).
- *Replies with service latency of greater than the maximum acceptable service latency requirement*, as discussed in Section 3.5, "Service Latency."
- *Request time outs*—requests with no response that clients time out after the maximum number of retries.

Note that persistent server failure responses, unacceptably long response latency, or request time out will cause a system to be deemed unavailable.

Proper application operation and business policies may cause some requests to fail, but those failures are not defective service transactions. Failed transactions nominally attributed to defective user requests include:

- *Invalid request*, such as `404 Not Found` ("*The server has definitive information that the user does not exist at the domain specified in the Request-URI*" [RFC3261]), or attempting to log on to a service with incorrect or unauthorized credentials.
- *Improperly formatted request*, such as `400 Bad Request` ("*The request could not be understood due to malformed syntax*" [RFC3261]).
- *Business policies*, such as `403 Forbidden` ("*The server understood the request, but is refusing to fulfill it*" [RFC3261]), or trying to withdraw too much money from an automated teller machine.
- *Application architecture, configuration, or deployment*, such as `501 Not Implemented` ("*The server does not support the functionality required to fulfill the request*" [RFC3261]).

Note that some operations might erroneously be failed (e.g., `403 Forbidden`) because of a provisioning or configuration error. In effect, the application is correctly processing faulty configuration data resulting in service being unavailable to some—or even all—users. Application software bugs could also cause an incorrect error code to be returned, so one must always be cautious when interpreting error codes.

Thus, application responses should be carefully reviewed to decide exactly which return codes indicate defective transactions that should impact service metrics, and which are considered correct application operation.

3.5 SERVICE LATENCY

Most network-based services execute some sort of transactions on behalf of client users. For example, web applications return web pages in response to HTTP GET requests (and update pages in response to HTTP PUT requests), telecommunications networks establish calls in response to user requests, gaming servers respond to user inputs, media servers stream content based on user requests, and so on. Transaction latency directly impacts the quality of experience of end users; according to [Linden], 500 millisecond increases in service latency causes a 20% traffic reduction for Google.com, and a 100 millisecond increase in service latency causes a 1% reduction in sales for Amazon.com.

The latency between the time an application receives a request (e.g. an HTTP GET) and the time the application sends the response (e.g., a web page) will inevitably vary for reasons, including:

- *Network Bandwidth.* As all web users know, web pages load slower over lower bandwidth (aka, "speed") network connections; DSL is better than dial-up, and fiber to the home is better than DSL. Likewise, insufficient network bandwidth between resources in the cloud—as well as insufficient access bandwidth to users—causes service latency to increase.
- *Caching.* Responses served from cached memory are typically much faster than requests that require one or more disk reads.
- *Disk Geometry.* Unlike random access memory (RAM), in which it takes the same amount of time to access any memory location, disk storage inherently has nonuniform data access times because of the need to move the disk head to the physical disk location to access stored data. Disk heads move in two independent directions:
    - rotationally as the disk storage platters spin; and
    - track-to-track, as the disk heads seek between concentric data storage tracks.

  The physical layout of file systems and databases are often optimized to minimize incremental latency for rotational and track-to-track latency to access likely data, but inevitably some data operations will require more time than others due to physical layout of data on the disk.
- *Disk Fragmentation.* Disk fragmentation causes data to be stored in noncontiguous disk blocks. As reading noncontiguous disk blocks requires time-consuming

disk seeks between disk reads or writes, additional latency is introduced when operating on fragmented portions of files.

- *Request Queuing.* Queuing is a common engineering technique to improve operational efficiency by permitting requests that arrive at the instant that the system is busy processing another request to be queued for service rather than simply rejecting the requests outright. Assuming that the system is engineered properly, request queuing enables the offered load to be served promptly (although not instantly) without having to deploy system hardware for the busiest traffic burst (e.g., the busiest millisecond). In essence, request queuing enables one to trade (expensive) system hardware capacity for increased service latency.
- *Variations in Request Arrival Rates.* There is inevitably some randomness in the arrival rates of service requests, and this moment to moment variation is superimposed on daily, weekly, and seasonal usage patterns. When offered load is higher, request queues will be deeper and hence queuing delays will be greater.
- *Unanticipated Usage and Traffic Patterns.* Database and software architectures are configured and optimized for certain usage scenarios and traffic mixes, such as cache sizes, configuration parameters, and similar optimizations. As usage and traffic patterns vary significantly from nominal expectations, the configured settings may no longer be optimal, and thus performance will degrade from nominal.
- *Network Congestion or Latency.* Bursts or spikes in network activity can cause the latency for IP packets traversing a network to increase.

Figure 3.8 illustrates the service latency distribution for a sample service. In this example, the median (50th percentile) latency is 130 milliseconds, meaning that half of the responses are faster than 130 milliseconds, and half are slower than 130 milliseconds. The distribution tail is naturally much longer above 130 milliseconds because while there are physical limits to the minimum response latency, delays can accumulate for myriad reasons. For this sample solution, the 95th percentile latency is 230

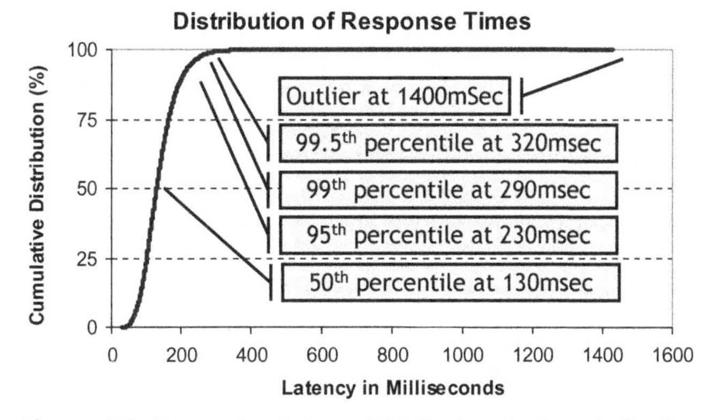

Figure 3.8. Transaction Latency Distribution for Sample Service.

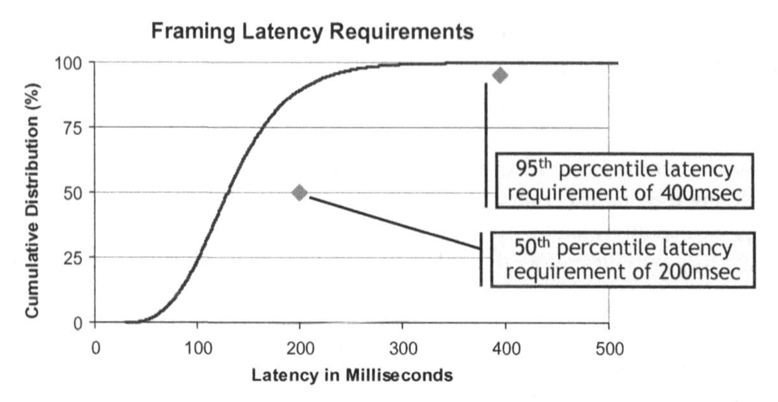

Figure 3.9. Requirements Overlaid on Service Latency Distribution for Sample Solution.

milliseconds (less than twice the 50th percentile latency), and the 99.5th percentile is 320 milliseconds (less than three times the 50th percentile latency). The longest service latency in this modest sample data set was 1400 milliseconds.

The statistical distribution will vary somewhat from application to application based on system architecture and other factors. Several latency data sets for different transactions or operations can be captured, and one can test to determine what mathematical distribution gives an acceptable model of service latency for the target solution. Having assumed a normal (or other) mathematical distribution, one can estimate the latency of any arbitrary point on the distribution with the appropriate mathematical formula based on two reference points, such as the 50th and 95th percentile service latency requirements or measured values.

A common service quality rule of thumb is that the 95th percentile latency should be no more than about twice the 50th percentile latency. Figure 3.9 shows how the data set of Figure 3.8 demonstrates compliance to a maximum 50th percentile service latency requirement of 200 milliseconds along with a 95th percentile latency target of twice that, or 400 milliseconds. Since two points can characterize the parameters of a particular mathematical distribution, two points can be used to specify the service latency performance requirement of a system. Figure 3.9 overlays 200 millisecond 50th percentile and 400 millisecond 95th percentile service latency requirements onto the sample data set, and the reader can instantly see that the actual performance is substantially better than these requirements. As the shape of a system's response latency distribution should remain relatively consistent, one can see how 50th percentile and 95th percentile requirements can easily be evaluated for arbitrary latency data sets, thus often making 2 point requirements an efficient specification technique for service latency.

While the 50th and 95th percentile latency requirements should be specified to be within the range of service latency that is acceptable to users, there is inevitably some latency value that is unacceptably slow and above which the user will consider the request a failure, even if it does eventually complete successfully. This is easily illustrated with web servers. Web browsers include a "cancel" or "stop" button that enables

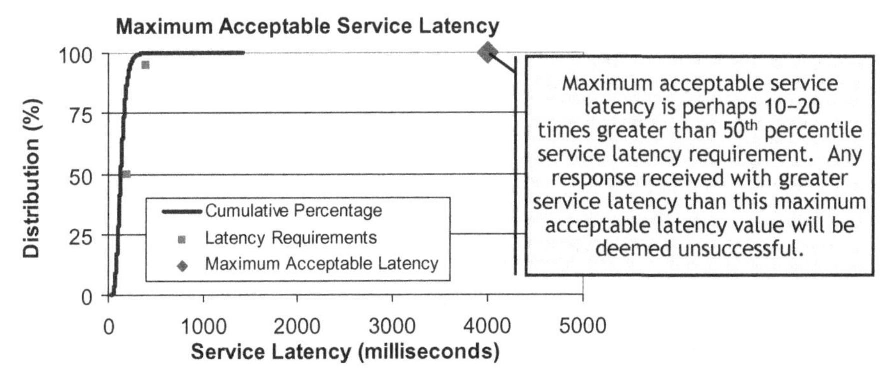

Figure 3.10. Maximum Acceptable Service Latency.

frustrated users to abandon an operation (e.g., web page retrieval or update) and a "reload" button to retry the request. Undoubtedly, all readers have stopped a painfully slow web page load from an apparently nonresponsive server, and hit "reload" in the hope that retrying the request will be more successful. This same behavior applies to other service requests, such as telephone call attempts, where most users will abandon a call attempt that doesn't return audible ring back within 4 seconds of the last digit being dialed or "send/call" being pressed. While the actual maximum time an individual user will wait for a particular web page to load or telephone call to go through or some particular application transaction to complete will vary, the best practice is to set a maximum acceptable service latency requirement. In principle, even a successful response that takes longer than the maximum acceptable service latency time will be deemed unsuccessful by the user, presumably because they have abandoned the request (e.g., hit "cancel" on their browser or ended the call) in frustration. Figure 3.10 overlays a maximum acceptable service latency of 4 seconds (4000 milliseconds) onto the service latency requirements of Figure 3.9. Note that the maximum service latency in this example is 20 times the 50th percentile requirement and 10 times the 95th percentile requirement. The maximum service latency should only rarely be exceeded, and those exceptions are generally captured as impairments to service reliability metrics, which were discussed in Section 3.4, "Service Reliability." Note that the spread between the 50th percentile, 95th percentile, and maximum acceptable service latency is highly application dependent. For example, while it may be acceptable for a web-based application to occasionally take 10 or 20 times longer to respond to an individual request than typical, it is not acceptable for the latency of real-time gaming, voice calling, or video streaming to suddenly increase by an order of magnitude or more. As readers know from personal experience, snappier, more responsive services are more appealing, and thus are more likely to be satisfactory to users. The relationship between service latency and user satisfaction is application dependent, and different applications will have different service latency targets, and different maximum acceptable latency expectations.

Note that different transaction types may have significantly different service latency profiles, and thus should be tracked separately against separate requirements. For example, transactions that query or retrieve data are typically faster than operations that update data and establishing a session or call takes longer than terminating a session or disconnecting a call. Thus, best practice is to specify two maximum service latency points (e.g., 50th percentile and 95th percentile) and absolute maximum acceptable service latency for each major type of transaction (e.g., session/connection establishment, query, update, and session/connection termination).

3.6 REDUNDANCY AND HIGH AVAILABILITY

Critical hardware, software and other failures are inevitable. Figure 3.11 illustrates the service impact of a critical failure of standalone (nonredundant) system B1:

1. Initially system B1 is "up" and service is available to users (i.e., "up")
2. A critical failure occurs (e.g., hardware failure or software crash), and service is unavailable (i.e., "down") to users while maintenance engineers troubleshoot and repair system B1
3. When B1 is repaired and returned to service, service is once more available ("up").

Thus, a critical failure of a standalone system with no internal redundancy has typical outage duration of minutes or hours for a maintenance engineer to troubleshoot the problem and repair the system. Enterprises will often purchase hardware maintenance contracts to guarantee a maximum response time to assure that spare hardware is promptly available to minimize outage duration for hardware-attributed failures.

Unplanned outage durations of hours or several minutes are unacceptable for critical services, so critical systems will be deployed with redundancy and high availability middleware so that critical failures can be mitigated via rapid automatic failure detection and recovery. High availability middleware will automatically detect a critical failure, identify the failed unit and shift traffic to a redundant element so that service is rapidly recovered. Assuming a critical system is built from two redundant units B1 and B2, Figure 3.12 illustrates how redundancy and high availability mechanisms

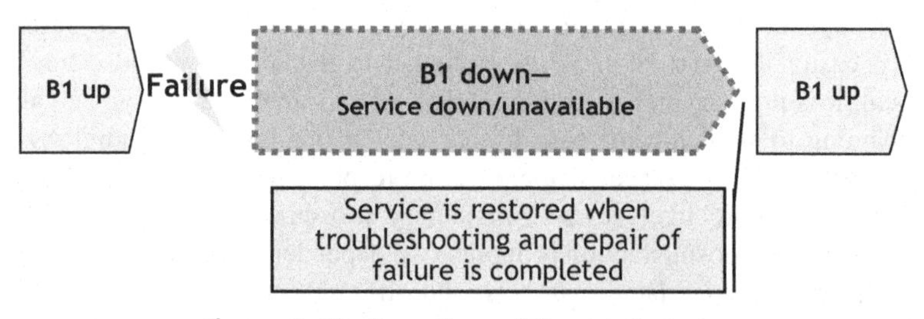

Figure 3.11. Downtime of Simplex Systems.

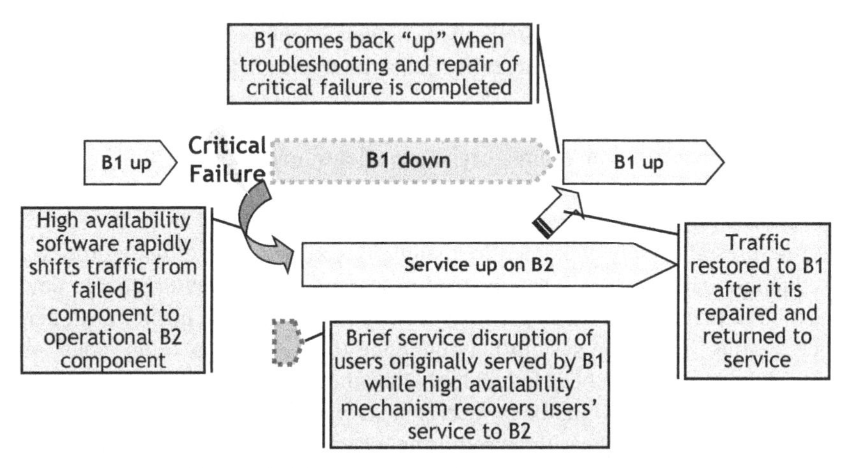

Figure 3.12. Downtime of Redundant Systems.

mitigate service downtime. Unit B1 is serving some or all of the active traffic when it experiences a critical failure. High availability software and mechanisms detect the failure of B1 and redirect traffic previously served by B1 to redundant element B2; after successfully switching over B1's traffic to B2, service is once again available for all users. The critical failure of element B1 must still debugged and repaired, but user service has been restored much, much faster than it would take to troubleshoot and repair the failure of B1. After B1 is repaired, the high availability mechanism (e.g., load balancer) can then shift traffic back onto element B1, or B2 can continue as the active or primary system and repaired B1 can serve as standby or alternate system.

3.6.1 Redundancy

Redundancy in traditional computer-based systems is most often implemented at three levels:

1. *Software Process.* Protected mode operating systems like Linux make software processes excellent recoverable units because they can explicitly be terminated and restarted without impacting any other processes.
2. *Field Replaceable Unit (FRU) Hardware.* As the primary unit of hardware repair, it is common to use the FRU as the primary unit of hardware redundancy as well. For instance, compute blade FRUs are convenient units of redundancy, especially if blade FRUs can be removed, replaced (or simply reseated to diagnose hardware versus software failures), and restarted while system is operational.
3. *Network Element.* Some services are provided across a cluster or pool of individual network elements (e.g., individual rack mount server instances) with a mechanism to balance or direct the traffic load across the pool of individual systems. For example, any operational domain name server (DNS) network

element can respond to a DNS query request. Deploying more DNS servers on a network increases the probability that at least one DNS server will be available to serve user requests.

Multiple instances of the same thread type are often implemented within processes to improve throughput and performance, especially on multicore processors. While these threads are technically redundant, they are not generally useful for robust recovery because failures are rarely contained to a single thread within a software process.

Redundant hardware, such as a pool of processors or an array of storage devices, may also be implemented on a single FRU (or even integrated circuit) for cost, density, or other reasons. While the system may continue to operate in a "limp along" mode after one element in the pool has failed, highly available systems are designed to permit each FRU to be replaced while the system is in service and to then restore service onto the replaced FRU gracefully without requiring a system reboot. Thus, highly available systems should support FRU-level redundancy to maintain service while a FRU is being replaced, reinitialized and gracefully re-accepting service once it has been successfully recovered and re-activated.

As the principle of redundancy can be applied from simple components like fans in a fan tray, to complex systems like data centers in disaster recovery scenarios, to completely different branches of engineering like cables in a suspension bridge, the terminology varies somewhat across industries. Fundamentally there are two common logical redundancy arrangements: load sharing and active-standby.

1. *Load Shared.* In load-shared redundancy arrangements, all operational units are actively serving users. By convention, "N" refers to the number of units required to carry the full engineered service load of the system, and "K" refers to the number of redundant units configured, and hence this configuration is often called "$N + K$ load sharing." The smallest load shared configuration has a single unit capable of carrying the full engineered load ($N = 1$) and a single redundant unit ($K = 1$); this minimal "1 + 1 load sharing" arrangement is typically referred to as "active–active." By keeping the "redundant" unit active, there is a lower probability of undetected or "silent" failure of the redundant unit in active–active configurations compared with active–standby arrangements. For example, commercial airplanes are designed with $N + 1$ engine redundancy so that if one engine fails on takeoff, the airplane can successfully takeoff, maneuver, and land. Another version of "$N + K$" exists in which a "K" unit is put into service only when one of the "N" units fails. At that point, it assumes the traffic previously being handled by the failed "N" unit. Since the "K" unit is not kept active when not in use, the recovery time is slightly longer than the load-shared model; however, it has the advantage of requiring less hardware than the active-standby model in most cases.
2. *Active–Standby.* As the name suggests, one of the units is actively serving users at a time, and the redundant unit is in a standby state not actively serving users. In high availability arrangements, the redundant unit is typically powered on with platform and application software booted to a predefined state. Depending on the application and software platform architecture, the redundant unit may

be ready to take over for a failed active unit in seconds or longer. The terms "hot," "warm," and "cold" are often loosely used to characterize the readiness of the application software on the standby unit. While the precise interpretation of hot, warm, and cold failover varies between applications and industries, common interpretations of these terms are:

- "cold standby" application software (and perhaps operating system) needs to be started on a processor to recover service after failure of active unit;
- "warm standby" application software is running on standby unit, but volatile data are periodically (rather than continuously) synchronized with active so time is required to rebuild latest system state before standby unit can recover service after failure of active; and
- "hot standby" application is running on standby unit and volatile data are kept current so standby unit can recover service rapidly after failure of active.

Since standby units are not actively delivering service, there is a risk that a hardware or software failure has occurred on the standby but has not yet been detected by the monitor software that runs when the unit is in standby. Hot and warm standby systems should periodically execute diagnostic self-test software to verify that hardware and software remains in full working order. The best practice is to routinely switchover service to standby units while the active unit is fully functional (and during a maintenance period in case issues arise) to assure that standby units remain fully operational and ready to recover service from a failure of an active unit. This should expose any previously undetected hardware or software problems when the previously active unit is fully operational to recover service if necessary.

Hybrid redundancy arrangements are sometimes used for applications with titles like primary/secondary or master/backup, in which some functions (e.g., queries) might be distributed across any operational element but other operations (e.g., updates) are only served by the "primary" or "master" instance. If the primary or master fails, then an automatic selection process designates one of the secondary or backup instances to be the new primary or master.

3.6.2 High Availability

Hardware and software failures are inevitable. Highly available systems are designed so that no single failure causes unacceptable service disruption. To accomplish this, systems must be designed to detect, isolate, and recover from failures very rapidly. Traditionally, this means that failure detection, containment and isolation, and recovery must be both automatic and highly reliable, and hardware redundancy must be engineered into the system to rapidly recover from hardware failures.

A basic robustness strategy for a highly available system is illustrated in Figure 3.13. Consider each step in Figure 3.13 separately:

1. *Failure.* Hardware, software, or other failures will inevitably occur.
2. *Automatic Failure Detection.* Modern systems are designed to detect failures via myriad mechanisms ranging from direct hardware mechanisms, like parity

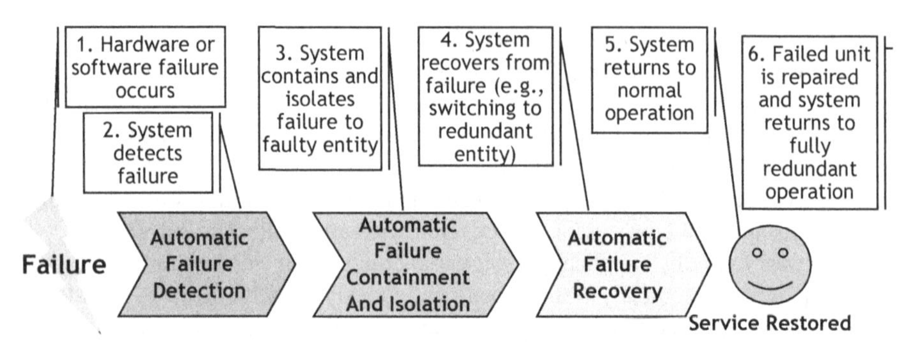

Figure 3.13. Simplified View of High Availability.

checks, to direct software mechanisms like return codes or expiration of time outs, to environmental sensors like temperature or moisture sensors, to sophisticated indirect mechanisms, like integrity audits and throughput monitors. Highly available systems will have several tiers of failure detection so that if one detection tier misses the initial failure event, then another tier will catch it sometime later.

3. *Automatic Failure Containment and Isolation.* The system must contain the failure extent so that the failure does not cascade to affect more users or services than necessary. The system must then correctly diagnose or isolate the failure to the appropriate recoverable module so that proper recovery action can be initiated. Fault isolation should be as fast as possible so that failure recovery action can be promptly activated to shorten service outage, but not so hasty as to incorrectly isolate the failure and activate a wrong recovery action. In addition to prolonging the outage event, activating the wrong recovery mechanism (e.g., switching over or restarting the wrong software module) may unnecessarily affect end users who were not impacted by the failure event itself. The situation when a failure is not isolated to the correct recoverable or repairable module is called a "diagnostic failure."
4. *Automatic Failure Recovery.* After isolating the failure to the proper recoverable module, highly available systems will then automatically activate a recovery action, such as switching service to a redundant module.
5. *Service Restored.* System returns to normal operation when service is restored onto the redundant module.
6. *Full Redundancy Restored.* Replacement of the failed hardware module, repair of failed software module, or other action to correct the primary failure is then completed promptly to restore the system to fully protected (i.e., redundant) status.

In high availability systems, failure detection, isolation, and recovery occur automatically, and the duration of impact to service should be less than the maximum acceptable service disruption latency. Typical high availability systems will automatically detect, isolate, and recover from failures in seconds, but some special-purpose systems like optical transmission equipment will detect, isolate, and recover from failures in milliseconds.

If a failure is not automatically detected by the system, then a so-called "silent failure" situation will exist in which service is not delivered but recovery actions are not activated because neither the system nor the human maintenance engineers are aware of the failure. Silent failures can occur on either active elements (e.g., a server that is nominally available but is not actually accepting new user requests) or redundant/standby elements (e.g., an underinflated or flat spare tire may be lurking in an automobile trunk for months or years before being detected). Depending on system architecture and the specific failure, these silent failures may directly impact users (e.g., a server is down, but the operations team doesn't know it) or they may not immediately impact users but put the system into a vulnerable/simplex state (e.g., spare tire is flat, but the driver doesn't know it). Implementing multiple tiers of failure detection (e.g., guard timers, keepalives, throughput monitors, and so on) and routine switchover/execution of standby/redundant units is the best practice for mitigating the risk of silent failure. For example, if a software process fails in a way that does not trigger an explicit failure indication, like a failure return code or processor exception, then the failure should be detected via failure of a periodic keep alive/heartbeat mechanism, or via unchanging throughput monitor values, or via other secondary/alternate failure detection mechanisms.

Figure 3.14 illustrates these high availability principles in the context of canonical enterprise application architecture. The example application is built around a load-shared pool of server instances S1, S2, and S3 that offer users access to data stored in

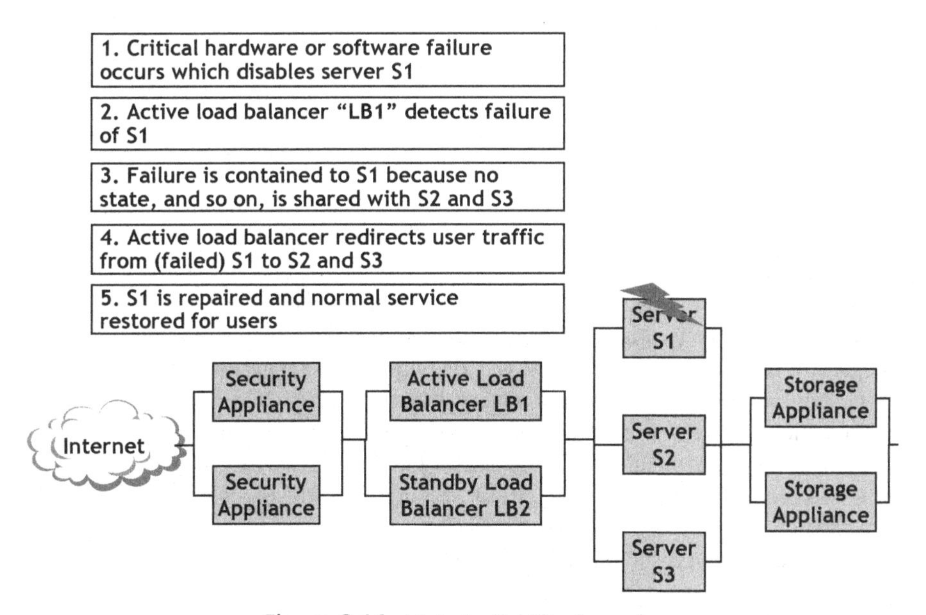

Figure 3.14. High Availability Example.

a highly available storage array subject to some business rules. An active–standby pair of load balancers distributes user traffic across the available server instances, and a pair of security appliances protects the elements from external attack. The example scenario illustrated in Figure 3.14 is as follows:

1. Critical failure occurs on server instance S1; assume that this prevents S1 from responding to requests from end users on the Internet.
2. Active load balancer LB1 observes that server S1 has stopped responding to user requests. Since server instances S2 and S3 remain operational, the load balancer deduces that S1 has failed and raises an alarm to the element management system (EMS).
3. Server application instances were explicitly designed to be independent with no shared information, so failure of S1 does not impact the ability of S2 or S3 to deliver service to their users.
4. Active load balancer stops directing any traffic to S1 and distributes all traffic to S2 and S3. Service is restored for all users.
5. Server S1 is repaired (e.g., by replacing failed hardware or repairing and restarting failed software) and made available. Active load balancer detects recovery of S1 and resumes distributing traffic to S1. Service is restored to full redundancy, and LB1 clears the alarm it raised to the EMS.

[Hamilton] offered the following practical test for the effectiveness of a system's high availability architecture and implementation: "*is the operations team willing and able to bring down any server in the service at any time without draining the work load first?*"

3.7 HIGH AVAILABILITY AND DISASTER RECOVERY

While high availability systems are designed to withstand any single failure, occasionally, force majeure or disaster events cause multiple systems to fail simultaneously. For example, a fire, flood, or roof collapse in a data center is likely to impact both the primary and redundant instances of multiple critical components. As these events overwhelm high availability mechanisms, an additional tier of business continuity planning and disaster recovery is often deployed to protect critical services. Disaster recovery strategies for critical services generally rely on both geographically separated redundant data center facilities and disaster recovery processes and mechanisms to promptly recover critical services to alternate data centers following a disaster.

Disaster recovery planning focuses on two key objectives: recovery time objective (RTO) and recovery point objective (RPO). Figure 3.15 illustrates these objectives in the context of a canonical disaster recovery flow. A system is operating normally when a disaster event occurs, such as an earthquake, fire, building collapse, or other catastrophic event; this catastrophic event causes service to become unavailable. Typically, enterprise staff will first see to the safety of all staff and visitors at the site and then

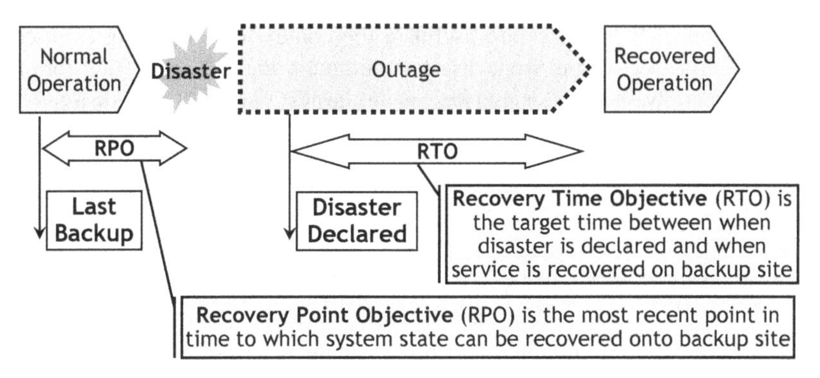

Figure 3.15. Disaster Recovery Objectives.

assess the damage. If the damage is light, then the enterprise may opt to promptly recover service on the equipment in the impacted data center; if the damage is more substantial, then the enterprise will declare a disaster and activate their disaster recovery plan. The disaster recovery plan will typically involve recovering service to equipment in a geographically distant data center that was not impacted by the disaster event.

The disaster RTO is the target time to recover service onto the geographically remote site from the time the disaster recovery plan was activated via formal disaster declaration, to the time application service is recovered to the remote site. Note that as data centers often support many end users, and each user will recover service to the georedundant site at a somewhat different time, one typically deems the recovery to be complete when some percentage of impacted users are fully recovered to the redundant site, such as when 90% of users impacted by the disaster are able to successfully access service from an alternate data center. Disaster RTOs often range from a few hours to a few days.

The RPO is the most recent point in time at which system state can be restored onto the recovery site. Typically, the geographically redundant system is recovered from the last data backup, meaning that any data changes or updates that completed after the last data backup will be lost. The RPO is the most recent point in time that system state can be recovered following a disaster. For example, with daily backups, the maximum disaster RPO should be 24 hours. As all data changes executed in less than the RPO time before a disaster are expected to be lost when service is restored following a disaster, enterprises should carefully set their RPO and engineer their systems accordingly. Operationally, data changes executed less than the RPO period before a disaster are likely to be lost and will either have to be reentered or accepted by the business as a permanent data loss. Daily data backups may offer an unacceptably long RPO for some enterprise data, so information may be replicated to a remote site to shorten RPO. Highly critical data can even be mirrored to assure that the data is securely stored on multiple sites before completing a transaction, thus assuring that no critical data will be lost due to a disaster. Data replication and related technologies can shorten RPOs to hours or minutes, and data mirroring or synchronous write technologies can shorten RPOs to seconds or less.

The RTO and RPO metrics are useful in the context of recovering from critical failures, as well as recovering from disasters, although the RTO and RPO for critical failures are generally seconds or minutes while the RTO and RPO for disaster events are often hours or days.

Disaster recovery and geographic redundancy is considered in detail in Chapter 9.

3.8 STREAMING SERVICES

Transaction-style services nominally return a single response per request, and thus service latency and service reliability is relatively straightforward to measure by considering the linkage between service requests and their corresponding responses. Streaming services, like voice and video calling and conferencing and voice playback, are fundamentally different from transaction style services because a single logical request (e.g., request to play a video or make a call) can result in a huge number of data packets being sent in response. These differences result in somewhat different service quality, reliability, and availability risks.

This section begins by differentiating the logical data plane, which carries streaming content from the logical control plane, which controls the flow of content. Streaming service quality metrics are then discussed. A key difference between transaction-oriented control traffic and streaming data traffic is the expectation of isochrony; isochronal data is covered in Section 3.8.3, followed by a discussion of streaming quality impairments.

3.8.1 Control and Data Planes

Streaming services like voice or video have two fundamental components: session control and user data.

- Session *control* covers operations to create, manipulate, and terminate streaming service sessions. IETF's Session Initiation Protocol (SIP) is a protocol used to control voice calls and video sessions.
- User *data* carry digitally encoded audio/voice or video content to the user, typically via IETF's real-time protocol (RTP).

It is often convenient to view control and data as two logical planes of network traffic; a small number of largely asynchronous control plane messages "control" a much larger volume of synchronous data plane traffic. Control plane traffic is generally transaction oriented so that each request to establish a session produces a finite and well-defined protocol exchange. While the session setup requires only a handful of control plane messages to be exchanged, the audio and/or video content of the call/session will typically be carried in dozens of RTP packets per second containing digitized audio and/or video for the duration of the session.

3.8.2 Service Quality Metrics

While it is theoretically possible to measure the service reliability of a media stream (e.g., the number of packets per million sent that are not correctly received within the maximum acceptable latency), that measure is not particularly useful because digital decoders are designed to mask occasional lost or late packets via concealment algorithms (e.g., replaying the last audio packet received or continuing to display the last video image). Instead of traditional service reliability metrics, service quality of streaming sessions is often characterized via several of the following metrics:

- *Mean Opinion Score (*MOS*).* The quantitative 1 (worst) thru 5 (best) mean opinion score is a standard way (e.g., [P.800], [BT.500]) to characterize multimedia quality of service. Standard definitions of MOS values are given in Table 3.2.
- *Session setup latency* is the time it takes to establish a new session (e.g., voice or video call) or start rendering requested contents (e.g., begin playing a prerecorded video or switch to a different television channel).
- *Impacted or Severely Impacted Seconds or Units of Streaming Service.* Different streaming services have somewhat different units of impact. While video impact may be measured in impacted frames, audio (and perhaps online gaming) impact is measured in milliseconds or seconds of impact or loss of data.
- *Lip Sync.* For streams where both audio and video are provided, it is important for visual images of moving mouths and other sound+producing actions to remain in sync with the audio sound track. If the so-called lip sync is off by more than about 50 milliseconds, then users' quality of experience will be impacted.
- *Session Retention or Retainability.* Captures the probability that the stream remains operational until normal termination (e.g., reaching the end of the prerecorded material or deliberate termination of a call by one of the participants).

TABLE 3.2. Mean Opinion Scores

Mean Opinion Score	[P.800] Quality Rating	[BT.500] Impairment Rating	[P.800] "Effort Required to Understand the Meaning of Sentences"
5	Excellent	Imperceptible	Complete relaxation possible; no effort required
4	Good	Perceptible, but not annoying	Attention necessary; no appreciable effort required
3	Fair	Slightly annoying	Moderate effort required
2	Poor	Annoying	Considerable effort required
1	Bad	Very annoying	No meaning understood with any feasible effort

3.8.3 Isochronal Data

The dictionary [Webster] defines *isochronal* as "uniform in time; having equal duration; recurring at regular intervals." Streaming data for real-time communications is inherently isochronal to assure that the audio and/or video content that was encoded by the sender's device is promptly transported across the network and available for timely decoding into an analog representation for the receiving party to enjoy. While modern encoding standards for audio and video streams aggressively compress redundant data (e.g., suppress "silent" audio periods and unchanging video images to reduce network usage), the data remains largely isochronal.

If the communications is not isochronal, then the receiver has two undesirable options:

1. Vary the pace of rendering the data to track with the arrival rate, so voice/video may be compressed (e.g., higher pitch audio) when traffic is received faster and slower (e.g., lower audio pitch) when congestion or other factors delays network transmission
2. Maintain an isochronal rendering schedule, and if appropriate data isn't available when required then attempt to conceal the missing data (e.g., by filling the "dead" spot by replaying previous data) to minimize the user service impact.

After all, the speakers and displays that render digitized audio and video content are fundamentally isochronal: every few milliseconds, they must be presented with audio or video data to render, or the listener will hear silence and the viewer will see a frozen image, jerky video, pixilation, or other visual impairments. To assure the highest quality of rendered audio and video streams, modern systems maintain an isochronal rendering schedule and receiving systems use de-jitter buffers to compensate for inevitable packet by packet variations in transmission latency across IP networks.

3.8.4 Latency Expectations

Streaming sessions are fundamentally both unidirectional and noninteractive (e.g., playing prerecorded audio or video content) or bidirectional, interactive, or conversational (e.g., a voice or video call). Noninteractive streams have modest bearer latency expectations: the content should begin rendering fairly promptly, but few users will notice if rendered content actually took hundreds of milliseconds or seconds to be streamed from the server, traverse the network, decompress, and be rendered. Interactive or conversational audio and video streams have strict latency expectations so that conversations can maintain a familiar and comfortable dialog between participants that is similar to traditional face-to-face communications.

The International Telecommunications Union (ITU) modeled the service quality perception of users to varying mouth-to-ear delays (i.e., the latency from the time one party speaks into a telephone and the time the other party hears their words); the results are shown in Figure 3.16. When mouth-to-ear latency is below 200 milliseconds, users are very satisfied; when latency doubles to 400 milliseconds, some users

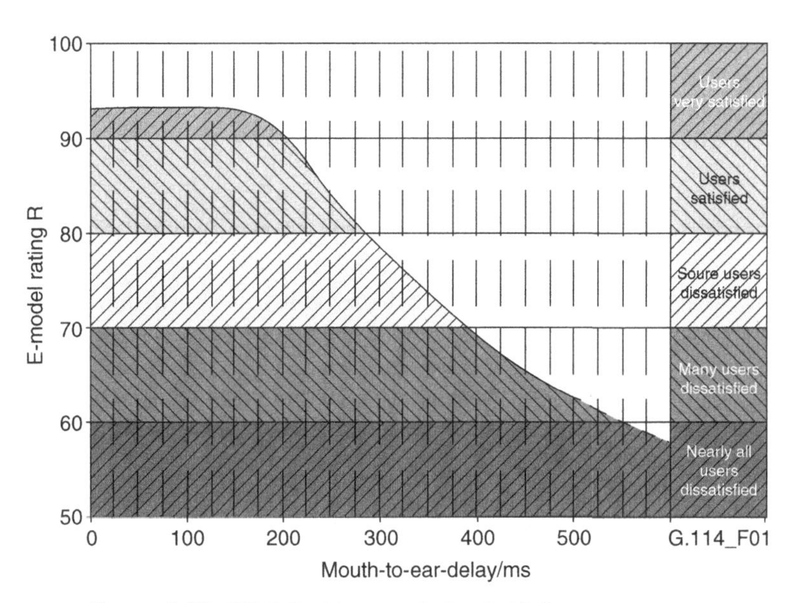

Figure 3.16. ITU-T G.114 Bearer Delay Guideline.
Source: International Telecommunications Union [ITU-T G.114].

are dissatisfied; as latency increases, further more users become dissatisfied. This result should track with readers' personal experience with older wireless phones: when mouth-to-ear latency gets too long, people inadvertently speak over each other because the natural conversational rhythms that they are accustomed to do not work when latency increases past a certain point. Forcing users to explicitly alter their conversational style to use a higher latency communications solution decreases their satisfaction.

Thus, minimizing one-way latency is very important for interactive or conversational streaming services, so encoders, de-jitter buffers, decoders, and other elements in the path of bearer data must be designed and tuned for low latency.

3.8.5 Streaming Quality Impairments

In addition to end-to-end latency, the quality of streaming services is impacted by the following:

- *Packet Loss.* IP networks occasionally lose packets, and thus, IP protocols, including those carrying audio and video streams across the Internet, must be prepared to conceal lost or late IP packets. Two common lost packet concealment strategies are to replay previous data or to fill with benign content like silence.
- *Jitter* is the variation in packet arrival rates. Receivers typically implement de-jitter buffers that introduce latency to give packets more time to traverse the IP network and be resequenced, thereby increasing the probability that data packets will be available at the moment that the decoder needs them to render media to the user. While larger de-jitter buffers—meaning greater latency nominally

consumed by de-jitter buffer to collect and resequence late IP packets—reduces the risk of dropping late packets, it directly increases end-to-end media latency. Less jitter permits smaller de-jitter buffers to be used without reducing service quality, thus shorting end-to-end media latency.

3.9 RELIABILITY AND AVAILABILITY RISKS OF CLOUD COMPUTING

Chapter 4, "Analyzing Cloud Reliability and Availability," considers how the essential and common characteristics of cloud computing introduce new risks to service reliability and service availability. The remaining chapters in Part II, "Analysis," explore these risks in detail. Chapter 10, "Applications, Solutions and Accountability," considers how cloud service models change accountability and measurement contexts for service reliability and availability impairments, and the remainder of Part III, "Recommendations," discusses how to mitigate the risks to service reliability and service availability of cloud computing.

II

ANALYSIS

4

ANALYZING CLOUD RELIABILITY AND AVAILABILITY

The technical challenge this book considers is how one can assure that the benefits of cloud computing (see Section 1.7) are achieved without diminishing service reliability and service availability to levels below those achieved by traditional application deployment models. While the specific reliability and availability risks of a particular application are determined by the architecture, deployment, and operational details of the deployed application, one can consider and usefully analyze the reliability and availability risks inherent to cloud computing.

This part of the book (Part II, "Analysis") analyzes the risks to service reliability and availability of cloud computing; Part III, "Recommendations," discusses techniques for minimizing the service reliability and reliability risks of cloud computing. This chapter frames the general expectations for service reliability and service availability of cloud computing, and gives an overview of how the essential and common characteristics of cloud computing and cloud service and deployment models can impact those expectations. Subsequent chapters in Part II, "Analysis," of this book consider these risks in detail.

4.1 EXPECTATIONS FOR SERVICE RELIABILITY AND AVAILABILITY

Users' baseline expectations for service reliability and service availability are largely determined by the behavior of the service they have historically received. For example,

Reliability and Availability of Cloud Computing, First Edition. Eric Bauer and Randee Adams.

users will expect the service reliability and service availability of long-term evolution (LTE) fourth-generation wireless to be at least as good as the second- or third-generation wireless service (e.g., Universal Mobile Telecommunications System (UMTS), Global System for Mobile Communications (GSM), Code Division Multiplex Access (CDMA), Wideband Code Division Multiplex Access [WCDMA]) they received previously, and subscribers' expectations for Internet Protocol Television (IPTV) service are set in the context of the terrestrial free to air, cable, or satellite service they currently receive. Thus, the initial expectation for applications delivered via cloud computing is likely to be that service reliability and service availability is equivalent to the reliability and availability offered via a traditional, native application deployment architecture. Concretely, an end user accessing a service on the Internet from their browser or client application should not experience lower service reliability or service availability because the application is deployed on a computing cloud rather than natively in a traditional data center.

This expectation cascades to individual applications themselves; if an enterprise ("consumer" in cloud parlance) expects an application to achieve "five 9's" service availability when traditionally deployed on native hardware, then they are likely to expect that application to achieve five 9's service availability when deployed on a virtualized platform (e.g., as a virtual appliance). Users will expect the same service or transactional reliability from cloud-based deployments as well. This means that the rate of failed transactions (e.g., failed calls and unsuccessful or hung web page loads) should be no higher for cloud deployment than for traditional deployment. Therefore, the analysis and recommendations of this book will consider the feasible and likely service reliability and service availability of virtualized and cloud-based applications compared with the baseline of native deployment.

4.2 RISKS OF ESSENTIAL CLOUD CHARACTERISTICS

This section considers the reliability and availability risks of each of the essential characteristics of cloud computing (see Section 1.1): on-demand self-service, broad network access, resource pooling, elastic growth, and measured service.

4.2.1 On-Demand Self-Service

On-demand self-service transforms service provisioning from a rare special case to a key function of cloud-based applications. In particular, on-demand self-service is key for rapid elasticity by enabling cloud consumers to order more resource capacity on the fly as offered load increases. For certain applications and certain customers, self-service provisioning functionality might be so important that loss of self-service provisioning functionality is deemed a partial service outage and is prorated appropriately. Thus, service reliability and service latency of self-service provisioning operations become key quality indicators of cloud computing. Specification of key quality indicators is discussed in Section 8.2.2, "Service Reliability and Availability Measurements."

4.2.2 Broad Network Access

Wireless and wireline access networks, in conjunction with wide area networking, will typically connect users of cloud computing services to the cloud data center hosting the servers executing the cloud consumer's application software. Thus, IP networking service availability, reliability, and latency directly impact users' quality of experience. While modern wireline networks generally offer consistently high service reliability with low latency, wireless networking—especially mobile wireless networking—is inherently subject to additional factors that can adversely impact quality, reliability, latency, and availability of IP networking service. Service quality, reliability, and availability of wireless and wireline IP networking is beyond the scope of this book.

4.2.3 Resource Pooling

Resource pooling enables service providers to boost resource utilization and thereby drive down costs. Resource pooling raises several potential service risks:

- *Virtualization Risks to Service Reliability and Service Availability.* Virtualization technology makes it practical to pool and share compute, memory, networking, and storage resources. Virtualization technology introduces system reliability risks, as well as changing software reliability risks (both covered in Chapter 5, "Reliability Analysis of Virtualization") and hardware reliability risks (covered in Chapter 6, "Hardware Reliability, Virtualization, and Service Availability").
- *Service Latency Jitter Due to Resource Scheduling and Contention.* Discussed in Chapter 7, "Capacity and Elasticity."
- *Service Disruptions Due to Live (Online) Virtual Machine (VM) Migration.* To assure efficient resource utilization, cloud service providers are likely to leverage online and offline migration capabilities supported by virtualization products to maximize resource utilization (e.g., consolidating workloads and taking unneeded capacity temporarily offline) and to complete routine and service transition actions (discussed in Section 11.3, "IT Service Management Considerations"). Online workload migration carries the risk that user service will be delayed or impacted, possibly causing some service requests or transactions to fail (impacting service reliability metrics) or for some operations to be delayed (impacting service latency metrics). Occasionally, some online migration actions may fail to complete successfully and impact all users who were being served by the impacted application instance. These potential risks must be balanced against the benefits of online migration and its potential to greatly reduce downtime that would be realized on more traditional systems.

4.2.4 Rapid Elasticity

Rapid elasticity enables service capacity to expand and contract rapidly while the service is online. The authors will use the term "growth" to refer to expansion of service capacity and "de-growth" to refer to contraction of resource capacity. For example, a traditional hardware "growth" procedure is to add more RAM or another hard disk drive to a server that is installed and nominally in production service. Likewise, a traditional "de-growth" procedure would be to remove RAM or a hard disk drive that is no longer required by one system; a subsequent "growth" operation would presumably install that reclaimed RAM or hard disk drive into another system that could benefit from the additional resources. Growing or de-growing engineered capacity is traditionally a service impacting operation and is often executed offline.

In the context of cloud computing, elastic capacity "degrowth" is the logical opposite of elastic capacity growth. Rapid elasticity contrasts with traditional application architectures where systems are explicitly dimensioned for a particular engineered capacity. For example, the resource capacity allocated for a particular application might elastically grow during a busy period (e.g., the holiday shopping season for a retailer), and then de-grow after the holidays when the additional resource capacity is no longer required. Since cloud consumers pay for resources allocated to them (e.g., per VM per month charge), de-growing capacity that is no longer required reduces the cloud consumer's operating expenses.

Rapid elasticity introduces several general risks:

- *Service Impact of Growth/Degrowth Operations.* Rapid elasticity should have no service impact on active users; this requirement should be explicitly specified and verified.
- *Reliability and Latency of Growth/Degrowth Operations.* Growth and de-growth operations are inherently complex and present a direct risk to availability of engineered capacity to serve offered load. Transactional reliability and transactional latency of online growth and de-growth operations are key metrics. Critical software bugs can cause applications to crash or hang as databases and configuration tables are expanded or contracted, or when executing any of the myriad other online operations necessary to support rapid online elasticity.
- *Elasticity Failure.* If service is not able to grow capacity fast enough to track with increases in offered load before spare online capacity is exhausted, then some users will not be served with acceptable service quality, reliability and latency due to insufficient online capacity to serve offered load. Therefore, overload control mechanisms will still be needed in these cases to manage the increases in offered load before sufficient online capacity has been added.

These risks are considered in Chapter 7, "Capacity and Elasticity."

In addition, the infrastructure as a service (IaaS) service model of cloud computing completely transforms the roles and responsibilities regarding capacity engineering. IaaS suppliers are responsible for providing computing resources to cloud consumers instantly on-demand, and application suppliers and/or cloud consumers are responsible for requesting and assimilating additional resources fast enough so that all offered load is served with acceptable service latency and service reliability. As specific IaaS service providers inevitably have finite resources that are shared across cloud consumers, it is possible that occasionally the IaaS's pool of resources will be fully allocated, and thus additional resource requests may be denied. In this case, the application may burst out of the cloud to engage available resources offered by another cloud. Someone (e.g., cloud consumer or IaaS service provider) is responsible for orchestrating the cloud burst to assure successful elasticity to an alternate cloud and appropriate resource release when those resources are no longer required.

4.2.4.1 Policy Considerations. Since cloud elasticity provides the promise of infinite capacity for an application, from a business point of view, this is often tempered by affordability. Operational policies define guidelines for the capacity, availability, reliability, security, data, privacy, and quality of service requirements that must be met. Those policies are the basis for determining the cost of support and help to determine how much the system can grow and stay within that budget. Policies can also define system boundaries, that is, where the servers and data must be located to meet regulatory standards or quality of service requirements. Policies are thus used to set up the configuration and to monitor adherence to the agreed upon requirements. If the requirements change (e.g., the capacity increases), then additional resources will need to be purchased, and the policies are changed accordingly. The reverse is true if the requirements change per a decrease in expectations (e.g., decrease in needed capacity support). Policies will be discussed in later sections concerning their role in service orchestration (i.e., in Section 8.2, "Policy-Based Management").

4.2.5 Measured Service

Rapid elasticity coupled with a pay-as-you-go pricing model means that it is important for cloud support systems to carefully track resource usage for each application over time. As cloud consumers will generally be charged for resources used, there is an incentive for consumers to release unneeded resources, promptly to minimize their operating expenses. De-growing resource usage has different reliability and availability risks than resource growth has; de-growth risks are also considered in Chapter 7, "Capacity and Elasticity."

The foundation of measured service is obviously the usage data itself. Beyond the simple risk of data unavailability or loss, there are several measurement reliability and integrity risks:

- *Data Accuracy.* Maximum rate of inaccurate usage data (e.g., maximum defective records per million or DPM).
- *Data Completeness.* Maximum rate of missing or damaged usage records (e.g., missing, damaged or corrupted usage records per million ideal records).
- *Timestamp Accuracy.* The maximum rate of usage records (e.g., defects per million [DPM]) for which timestamp accuracy is incorrect by more than a specified number of milliseconds or seconds.

Service measurements are discussed in Section 8.2.2, "Service Reliability and Availability Measurements."

4.3 IMPACTS OF COMMON CLOUD CHARACTERISTICS

This section considers the reliability and availability impacts of six common characteristics of cloud (see Section 1.2).

4.3.1 Virtualization

Virtualization technology decouples application software from the underlying hardware, thereby increasing deployment flexibility and enabling applications workloads to be deployed to computing clouds. Reliability risks of virtualization technology and virtualized applications are discussed in Chapter 5, "Reliability Analysis of Virtualization."

Virtualization itself does not impact customers' expectations for service reliability and service availability, so if an enterprise expected five 9's service availability from the natively deployed application, then they are likely to expect five 9's when the application is deployed as a virtual appliance or on a virtualized platform. Application availability is managed by factoring downtime into appropriate categories, assigning downtime budgets by category, and managing each downtime category to meet its budget. Traditionally application service downtime is factored into software-attributable service downtime and hardware-attributable service downtime. Software downtime of virtual applications is discussed in Chapter 5, "Reliability Analysis of Virtualization," and hardware downtime is discussed in Chapter 6, "Hardware Reliability, Virtualization, and Service Availability." The evolution of downtime budgets as traditional applications migrate to the cloud is considered in Section 10.3, "System Downtime Budgets."

4.3.2 Geographic Distribution

Distributing applications to data centers physically close to end users can both reduce service latency and improve service reliability and availability by minimizing the IP networking equipment and facilities between the serving data center and the end user. Geographic distribution also facilitates distributing application functionality across several data centers, such as pushing/caching contents in content distribution network (CDN) elements close to users to reduce transport latency, and thus improve users' quality of experience.

Geographic distribution is a necessary, but not sufficient, condition for geographic redundancy, and hence for disaster recovery. Chapter 9, "Geographic Distribution, Georedundancy, and Disaster Recovery," explains how geographic distribution relates to geographic redundancy, and how geographic redundancy supports disaster recovery.

4.3.3 Resilient Computing

Mechanisms and architectures that improve the robustness and resiliency of cloud computing platforms and the applications hosted on those platforms will improve service reliability and availability of cloud-based applications. Resilient and high availability computing mechanisms are discussed in Chapter 5, "Reliability Analysis of Virtualization," and in Chapter 11, "Recommendations for Architecting a Reliable System."

4.3.4 Advanced Security

Advanced security is essential in protecting services from denial of service and other security attacks that can adversely impact service availability; this is discussed in Section 7.8, "Security and Service Availability."

4.3.5 Massive Scale

Massive scale systems will require a more complete and thorough set of service management processes supported by service orchestration and automation to ensure the complexity of the large system can be well managed to mitigate risks to service reliability and availability. IT Service Management risks are discussed in Section 4.5.

4.3.6 Homogeneity

Limiting the range of different hardware and software platforms supported to achieve homogeneity should reduce the risk of service provider errors and failures because common policies and procedures can effectively be deployed in homogeneous environments. Homogeneity inherently reduces the number of different types of procedures to automate and/or that maintenance staff execute, and increase the frequency that a smaller set of procedures are executed. As staff gains more experience and expertise with that smaller set of products due to familiarity and frequent execution, the probability of successful execution should increase and the execution time should decrease. Both of these should somewhat decrease overall service downtime. Increased automation should reduce human involvement—thereby eliminating the risk of human error—and maximize consistency and reproducibility; both of these factors should reduce failure rates of IT service management actions. Human errors should be less frequent in homogeneous environments because staff gain more experience executing policies and procedures in a homogeneous environment compared with the alternative of operating in a less consistent heterogeneous environment. It is possible, however, that some customers, particularly large customers, may want custom environments that could include custom tools, policies, and procedures. Customization may attenuate some of the reliability benefits of homogeneity

4.4 RISKS OF SERVICE MODELS

Decomposing traditional IS/IT deployment into application consumers and infrastructure-, platform, or software-as-a-service providers is fundamental to cloud computing (see Section 1.4, "Service Models"). To methodically characterize the roles and responsibilities of both cloud service providers and cloud consumers, we will apply the 8i + 2d model to cloud computing and consider the implications for cloud consumers.

4.4.1 Traditional Accountability

Traditionally, "five 9's" claims and expectations consider only product-attributable impairments (see Section 3.3.6, "Outage Attributability") of individual systems (i.e., application software running on hardware). This was fair for both system suppliers and the suppliers' customers operating the system because the supplier took responsibility for what they directly controlled, and the customer (enterprise) retained responsibility for both customer-attributable and external-attributable (e.g., force majeure) outages. This allocation of responsibility is visualized in Figure 4.1 by crudely overlaying TL 9000 outage accountability from Section 3.3.6, "Outage Attributability," onto the 8i + 2d framework visualized in Figure 3.3 from Section 3.2, "Eight-Ingredient Framework." Suppliers have responsibility for their hardware and software, as well as interworking with other networked elements via application protocols. Customers

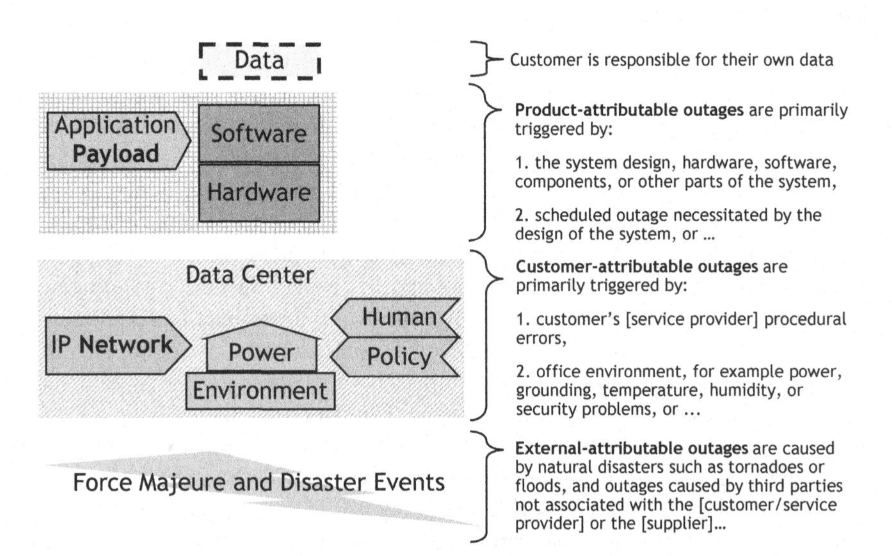

Figure 4.1. TL 9000 Outage Attributability Overlaid on Augmented 8i + 2d Framework.

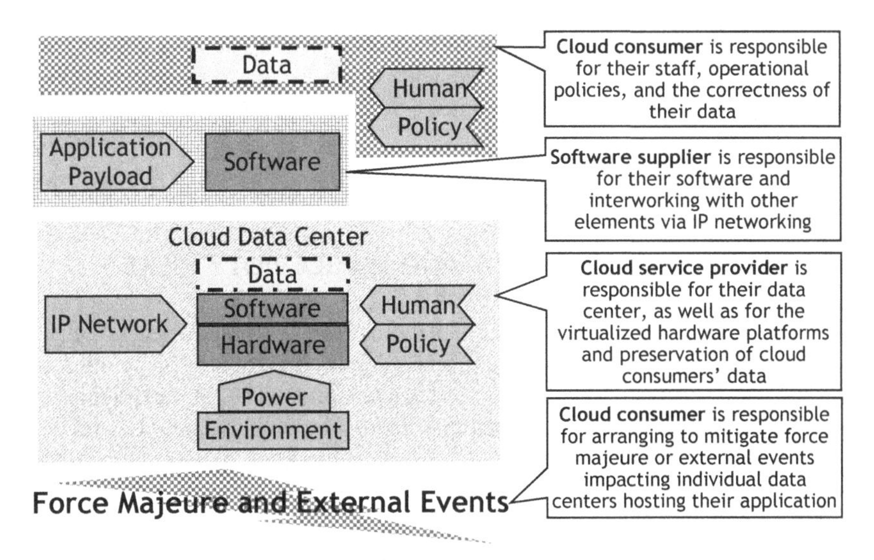

Figure 4.2. Outage Responsibilities Overlaid on Cloud 8i + 2d Framework.

retain responsibility for creation, integrity, maintenance, and other aspects of their data. Note that power, environment, IP networking, human, and policy ingredients have been logically aggregated together as "data center," and this captures the common sense notion that systems from suppliers are installed in data centers maintained by customers/enterprises. Force majeure and external-attributable outages are always a risk to data centers, and this risk must be managed by the customer/enterprise.

4.4.2 Cloud-Based Application Accountability

Figure 4.2 visualizes how outage responsibilities map onto the 8i + 2d cloud framework of Figure 4.1. The cloud consumer is responsible for the creation and integrity of their data and configuration information, as well as the policies and maintenance staff that (remotely) operate, administer, maintain, and provision their application software and data. A system integrator or software supplier is responsible for assuring correct operation of application software, including interworking with other network elements via application protocols. The cloud service provider is responsible for all data center resources (environment, power, network connectivity, operational policies, and maintenance staff), IP networking infrastructure, as well as hardware resources and the supporting platform software. In addition to maintaining configuration data for all virtualized resources and data center facilities, the cloud service provider is also responsible for assuring that cloud consumers' data written to virtualized storage are protected and available on demand. Note that while cloud service providers often operate multiple geographically redundant data centers, it is the responsibility of the cloud consumer to make suitable arrangements to assure that service can be rapidly recovered following force majeure or other external event that renders a single data center unavailable.

4.5 IT SERVICE MANAGEMENT AND AVAILABILITY RISKS

Sophisticated enterprises recognize that while individual components may offer a high theoretical availability, the observed service availability in real-world operation rarely approaches the theoretically maximum service availability. The FAA explicitly distinguishes these notions as inherent availability and operational availability, and defines them as follows:

- Inherent availability (A_i)—"the maximum availability theoretically within the capabilities of the system or constituent piece. . . . Scheduled downtime is not included in the Inherent Availability measure. A_i is an inherent design characteristic of a system that is independent of how the system is actually operated and maintained in a real world environment" [FAA-HDBK-006A].
- Operational availability (A_{op})—"the availability including all sources of downtime, both scheduled and unscheduled. A_{op} is an operational measure for deployed systems that is monitored by NAPRS" [FAA-HDBK-006A].

The difference between the inherent availability (A_i) and operational availability (A_{op}) is largely determined by IT service management (ITSM). IT service management is the implementation and operation of information technology systems to meet the needs of the enterprise. IT service management covers service design, release, delivery, control, and resolution processes for IS/IT services, and perhaps even more in broader definitions of the term. There are a number of IT service management standards and frameworks, including ISO/IEC 20000, Control Objectives for Information and Related Technology (COBIT), Microsoft's operations framework [MOF], and the Information Technology Infrastructure Library (often known simply by its acronym "ITIL"). As ISO/IEC 20000 is based on ITIL service management processes, the authors will analyze IT service management in the context of the 2011 edition of ITIL. This section considers how IT service management impacts operational service availability.

4.5.1 ITIL Overview

ITIL factors IT service management into five categories: service strategy, service design, service transition, service operation, and continual service improvement as shown in Figure 4.3.

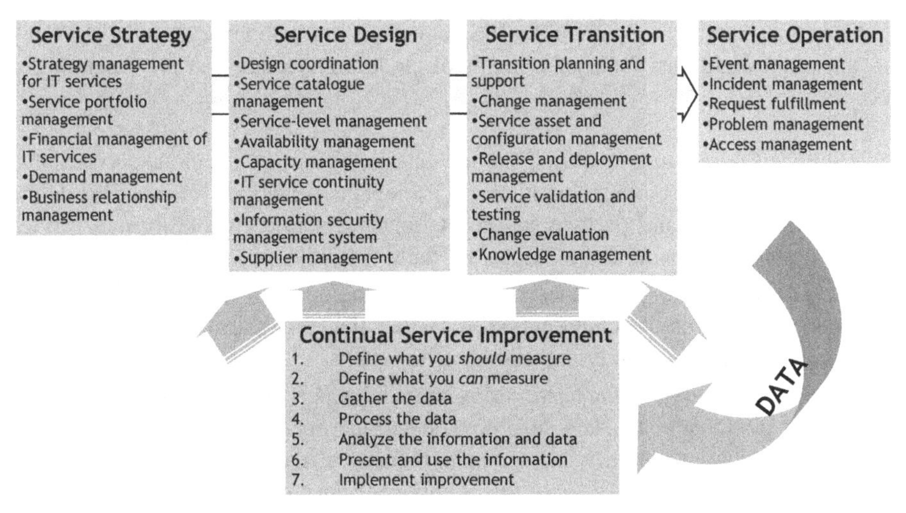

Figure 4.3. ITIL Service Management Visualization.

The following sections review each of these five areas and consider the risks to service reliability and service availability for successful and unsuccessful service management actions. A summary of IT service management risks is also given.

4.5.2 Service Strategy

ITIL Service Strategy [ITILv3SS] essentially defines a plan for delivering and managing information services that meet an enterprise's business needs. The service strategy includes five processes:

- strategy management for IT services;
- service portfolio management;
- financial management of IT services;
- demand management; and
- business relationship management.

While these general processes are crucial to business success, for the most part, they do not have a direct impact on delivered service reliability or service availability. However, specific aspects of the service strategy, especially the use of automation, can have a large impact on service reliability and service availability. Service automation in particular—a major component of "service orchestration" in the context of cloud computing—can improve service reliability by reducing risks associated with complexity and human errors.

Note that Appendix C of [ITILv3SS] considers service strategy and the cloud. This appendix highlights four components of cloud architecture that are essential for success:

- *Service Catalog and Portal.* Communicates to customers the cloud services that are available for purchase and the service level expectations for each. While the service catalog and portal sets expectations for service reliability and service availability, it has no direct impact on them.
- *Service Governance.* Defines and automates the operational policies, standards, and practices of the cloud service provider that facilitate but do not directly impact service reliability and service availability.
- *Service Delivery Management.* Monitors and reports usage by cloud consumers of cloud services. This also covers: service operations processes, capacity management, availability management, security management, and business continuity management. As such, this component does directly impact the service reliability and service availability of cloud-based solutions.
- *Infrastructure and Service Delivery.* Obviously, the service provider's XaaS infrastructure and service delivery mechanisms directly impact the service reliability and service availability experienced by cloud consumers.

4.5.3 Service Design

Service design strives to create IT systems that achieve the goals of the service strategy and require minimal changes after initial deployment. ITIL service design [ITILv3SD] covers the processes of:

- design coordination;
- service catalog management;
- service level management;
- availability management;
- capacity management;
- IT service continuity management;
- information security management system; and
- supplier management.

Of these activities, service-level management and availability management have direct and explicit linkages to the reliability and availability of the offered service. Service-level management essentially sets the service level requirements for reliability and availability, and availability management ensures the feasibility and likelihood of meeting those targets. Chapter 12, "Design for Reliability of Virtualized Applications," and Chapter 13, "Design for Reliability of Cloud Solutions," as well as traditional design for reliability works like [Bauer10], consider these topics from a reliability engineering perspective.

Rapid elasticity is an essential characteristic of cloud computing that enables capacity management to follow a very different paradigm; Chapter 7, "Capacity and Elasticity," discusses this topic in detail. Automation of capacity management in the context of cloud computing is often called service automation; Chapter 8, "Service Orchestration Analysis," discusses this topic in detail.

IT service continuity management (ITSCM) focuses on mitigation of risks and recovery of the service following a critical failure or a disaster. ITSCM of critical services traditionally focused on geographic redundancy to assure prompt recovery time and recovery point objectives could be achieved. With appropriate engineering, geographic distribution of cloud data centers can be leveraged to both support IT service continuity management as well as improve users' quality of experience in normal (i.e., nondisaster) periods; this topic is considered in detail in Chapter 9, "Geographic Distribution, Georedundancy, and Disaster Recovery."

4.5.4 Service Transition

Service transition strives to assure that new service introduction and subsequent changes are efficiently executed with minimal impact to service users. ITIL service transition [ITILv3ST] covers the processes of:

- transition planning and support;
- change management;
- service asset and configuration management;
- release and deployment management;
- service validation and testing;
- change evaluation; and
- knowledge management.

IT service transition focuses on mitigation of risks associated with the introduction of new or changed services. All of the processes contribute to this focus through careful management of assets, version control, and validation. Of particular importance to service reliability and service availability is release and deployment management, which focuses on the successful development, testing, and delivery of a new or changed service in accordance with customer requirements. Activities, such as software upgrade and patch, are facilitated through virtualization and cloud mechanisms, which support nonimpact to user service during these activities. Mitigation of risks associated with service transition activities is discussed in Section 11.3, "IT Service Management Considerations."

4.5.5 Service Operation

ITIL service operation [ITILv3SO] covers the processes of:

- event management;
- incident management;

- request fulfillment;
- problem management; and
- access management.

These service operation processes are covered by five core functions:

- service desk;
- technical management;
- IT operations management;
- Application management; and
- monitoring and control.

Efficiency and effectiveness of a provider's service operations have a major impact on overall service availability. Event and incident management processes directly impact the duration of service disruptions for failures that are not automatically detected and recovered properly. Best in class event and incident management processes strive to proactively detect events before they cascade into service disruptions and resolve them with minimal service disruption. Request fulfillment ensures that requests for changes or for information are handled properly. Problem management processes should ensure that in addition to resolving problems promptly, the root cause of the problem is identified and corrected to minimize the risk of reoccurrence.

While access management technically grants or denies a user's availability to IS/IT services or data, each access management error generally affects a very small portion of IS/IT users—often just a single user—so those events are not generally classified as outages, and thus downtime or unavailability metrics are not impacted.

IT service monitoring is crucial to assure that all failures are promptly detected and mitigated via event and incident management processes. Operations and application management functions must be correctly executed to assure user service is not impacted. Properly trained staff that are able to make excellent decisions assessing and managing risks to service operation is crucial.

4.5.6 Continual Service Improvement

ITIL continual service improvement [ITILv3CSI] strives to improve service quality and effectiveness of IS/IT. ITIL recommends a seven-step continual service process of:

1. Define what you *should* measure.
2. Define what you *can* measure.
3. Gather the data.
4. Process the data, including scrubbing for accuracy.
5. Analyze the data to determine if targets were met, understand trends, and relationships among the data, and propose corrective actions.

6. Present and use the information and corrective actions.
7. Implement corrective actions and improvements.

Continual service improvement also covers regular reporting of key service performance indicators to keep leaders aware of observed service quality, reliability, and availability.

4.5.7 IT Service Management Summary

Figure 4.4 illustrates the IT service management processes and topics that have the most impact on service reliability and availability.

IT Service Management provides well-defined processes to direct all aspects of customer service from strategy to deployment onto monitoring and improvement. Each process in some way supports and enables a highly reliable service, but the processes included in Figure 4.4 have the most direct impact on defining and maintaining service to ensure that it meets the customers' requirements for availability and reliability.

4.5.8 Risks of Service Orchestration

As defined in Section 8.1, "Service Orchestration Definition," service orchestration entails the linking together of architecture, tasks, and tools necessary to initiate and dynamically manage a service. Service orchestration provides an infrastructure for automating the configuration and management of a cloud-based service that conforms to associated operational policies. In the cloud computing environment, this entails not only configuring software to hardware, but also determining efficient work flows,

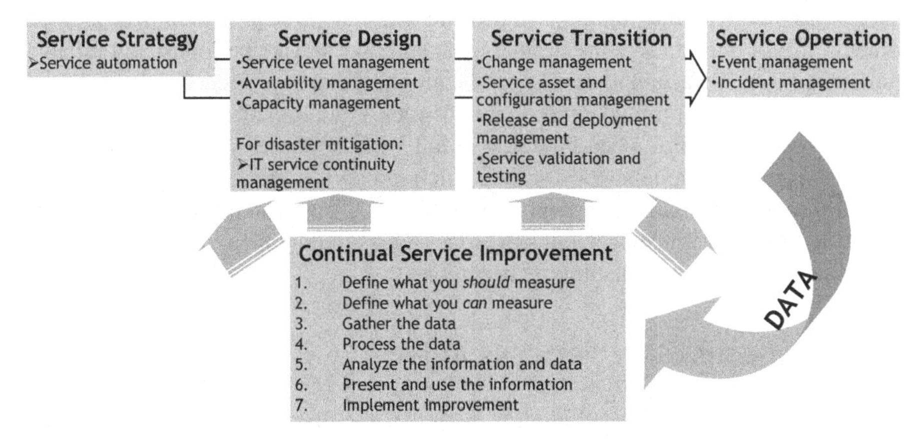

Figure 4.4. IT Service Management Activities to Minimize Service Availability Risk.

adherence to service agreements and standards, billing, and monitoring all in an automated way. Automation is a benefit for service orchestration, but it can also introduce some risks:

- If there is a bug associated with a particular application instance, the automatic creation of additional instances of that application may further spread this issue.
- Service orchestration may introduce a level of complexity that makes the system more error prone.
- Service orchestration coupled with rapid elasticity may result in the automatic allocation of resources that the customer is not prepared to pay for. A more manual approach might be requested by some customers.

The effectiveness of the service orchestration may be measured based on service availability, service reliability (e.g., number of defective transactions against the number attempted), and service latency. The reliability implications of service orchestration will be analyzed in Chapter 8, "Service Orchestration Analysis."

4.5.9 IT Service Management Risks

Service management of cloud-based applications is inherently more complex than traditional application deployments because the organizational split between the cloud consumer and the cloud service provider introduces the risk of confusion over roles and responsibilities for aspects of IT service management. Even when roles and responsibilities are clear, the split between consumer and provider introduces one more organizational boundary, which adds both latency and the risk of errors across the interface.

A more subtle service management risk may arise from historic assumptions about the reliability and availability requirements for systems that support IT service management. [Oppenheimer] observed that some enterprises have historically focused on improving service availability for end users without bothering to improve the resilience, robustness, and redundancy of the systems that support IT service management. This strategy was driven by the notion that since end users are powerless to mitigate service problems, it was important to the business that end user service be robust and reliable. In contrast, IT staff was empowered to work around failures of their support systems, so it was less important to make those systems robust and reliable. While this strategy may have been successful when IT staff had full visibility, access, and control of their traditional systems, fragile support systems may prove too brittle and inadequate for more elastic and dynamic cloud deployment models. Thus, enterprises should reconsider if their IT service management support systems are reliable and robust enough to support the user service availability expectations.

4.6 OUTAGE RISKS BY PROCESS AREA

To better manage service downtime, it is useful to map the 8i + 2d ingredients to the process or best practice areas that assure proper operation and availability of those

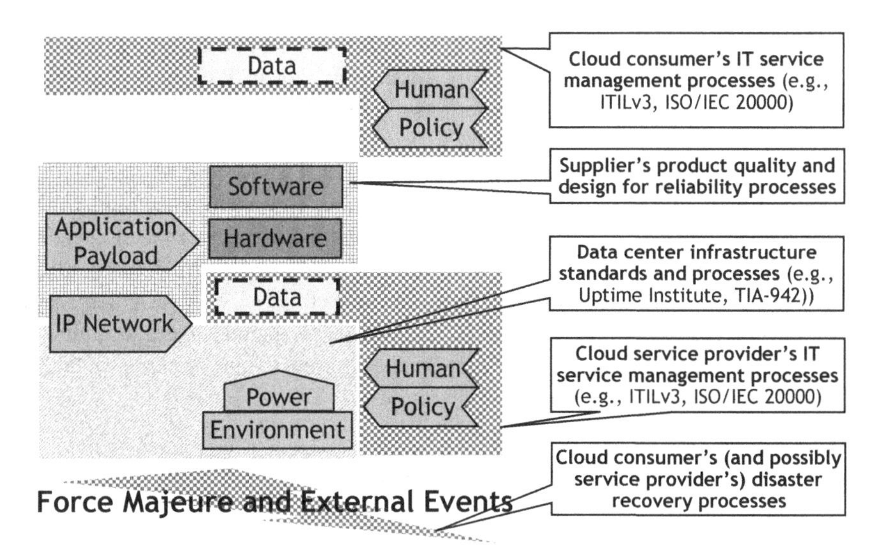

Figure 4.5. 8i + 2d Attributability by Process or Best Practice Areas.

ingredients. Figure 4.5 visualizes how four broad categories of process or best practices cover the 8i + 2d ingredients:

- *Design for reliability best practices* like [Bauer10] for software, hardware, and application payload ingredients (including software hardware and payload portions of IP networking equipment) assure highly reliable and available operation of these ingredients. These best practices, including "Design for Reliability of Virtualized Applications" in Chapter 12, methodically manage the product-attributable service downtime of individual applications.
- *Data center infrastructure standards and best practices* like [Uptime] and [TIA942] cover power, environment, and aspects of networking infrastructure. Standard data center tiers set expectations for service availability of data center power, environment, and networking infrastructure.
- *IT service management*, like ITIL best practices, COBIT (for Control OBjectives for Information and related Technology), and ISO/IEC 20000 families of standards, covers human, policy, and data ingredients. Specifics of the cloud service model will determine the exact split of responsibilities for IT service management activities and processes between the cloud service provider and the cloud consumer.
- *Business continuity planning* covers disaster recovery from force majeure and external events. While comprehensive IT service management frameworks (e.g., ITIL) explicitly reference business continuity planning (e.g., IT service

continuity management or ITSCM, in the context of service design [ITILv3SD]), it is sufficiently distinct from routine operations that it is generally considered separately. Service impact expectations for business continuity plans are routinely specified as recovery time objectives (RTO) and recovery point objectives (RPO), which will be described in Section 3.7, "High Availability and Disaster Recovery." Chapter 9, "Geographic Distribution, Georedundancy, and Disaster Recovery," considers the topic of disaster recovery. Specifics of the cloud service model will determine the exact split of responsibilities for business continuity planning activities and processes between the cloud service provider and the cloud consumer. Note that it is not customary to include average annualized downtime estimates due to force majeure or external events because, thankfully, these events are very rare.

4.6.1 Validating Outage Attributability

To both validate the outage risk by process area factorization and to offer more concrete examples of outage causes, this section maps standard Federal Communications Commission (FCC) outage causes against the four cloud-oriented process risk areas. Service outages in the real world have a direct or triggering cause, and often have one or more contributing causes. Mitigating each of these direct and contributing causes is often addressed by at least one of the four general process areas enumerated above. This hypothesis is casually validated by considering how a standard set of real-world outage causes would nominally be addressed by each of these areas. The U.S. FCC mandates usage of a formal network outage reporting system (NORS) for recording severe communication disruption events. The formal outage reports explicitly identify the root cause, direct cause, and contributing factors for each outage event; the standard set of these causes is given in [NORS]. One can illustrate and validate the outage attributability by process areas by mapping each of these standard outage categories into broad process areas. While one can quibble that some items are actually covered by multiple process areas (e.g., IT service management considers some business continuity planning topics), this exercise does clarify many of the service outage risks each process should be mitigating. For simplicity, several NORS categories that are not directly applicable to cloud computing (e.g., diversity failures of SS7 links) are omitted.

- Supplier's design for reliability processes and diligence should mitigate risk of the following standard outage causes:
  - *Design—Software* causes, such as "*faulty software load*," "*inadequate defensive checks*," and "*ineffective fault recovery or reinitialization action*."
  - *Hardware failure* causes, such as "*memory unit failure*."
  - *Design—firmware* causes, such as "*ineffective fault recovery or reinitialization action*" and "*Insufficient software state indications*."
  - *Design—hardware* causes, such as "*inadequate grounding*" or "*poor backplane or pin arrangement*."

  - *Procedural—system vendor* causes, such as "*ad hoc activities, outside scope of method of procedure (MOP)*," "*insufficient supervision/control*," or "*insufficient training*."
- Data center infrastructure and processes mitigate the risk of the following outage causes:
  - *Environment (internal)* causes, such as "*environmental system failure (heat/humidity)*," "*fire, arcing, smoke damage*," or "*fire suppression (water, chemicals) damage*."
  - *Diversity failure* causes, such as "*power*."
  - *Power failure (commercial and/or backup)* causes, such as "*generator failure*," "*extended commercial power failure*," or "*lack of routine maintenance/testing*."
- IT service management processes and diligence by both cloud consumer and cloud service provider should mitigate risk of the following standard outage causes:
  - *Procedural—service provider or other vendor* causes, such as "*documentation/procedures out-of-date, unusable or impractical*," "*documentation/procedures unavailable/unclear/incomplete*," "*inadequate routine maintenance/memory back-up*," "*insufficient supervision/control*," or "*insufficient training*."
  - *Spare* causes, such as "*not available*" or "*on hand—failed*."
  - *Traffic/System Overload* causes, such as "*Inappropriate/insufficient Network Management control(s)*" or "*Mass calling—focused/diffuse network overload*."
- Business continuity planning and diligence should mitigate risk of the following standard outage causes:
  - *Environment—external* causes, such as "*earthquake*," "*fire*," "*storm—water/ice*," or "*storm—wind/trees*."

4.7 FAILURE DETECTION CONSIDERATIONS

Figure 4.6 visualizes eight traditional product error vectors from [Bauer10]. Figure 4.7 shows the subset of traditional product error vectors that are likely to be primarily the responsibility of the IaaS provider, and Figure 4.8 shows the traditional error vectors that are likely to be primarily the responsibility of software suppliers or the organization responsible for the software, such as the software as a service (SaaS) service provider. Note that additional virtualization-related errors will be discussed in Section 12.6, "Robustness Testing," and several cloud related errors will be discussed in Section 13.6, "Solution Testing and Validation." This section considers responsibility for failure detection and mitigation when applications are deployed to a cloud.

4.7.1 Hardware Failures

The IaaS service provider has primary responsibility for detection and mitigation of hardware failures. Chapter 6, "Hardware Reliability, Virtualization, and Service Availability," explicitly considers this error vector.

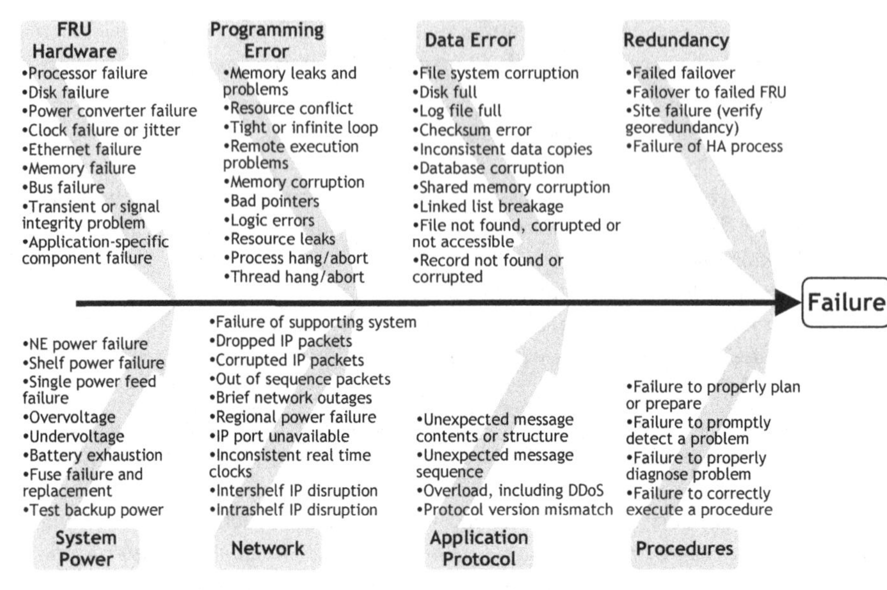

Figure 4.6. Traditional Error Vectors (from [Bauer10]).

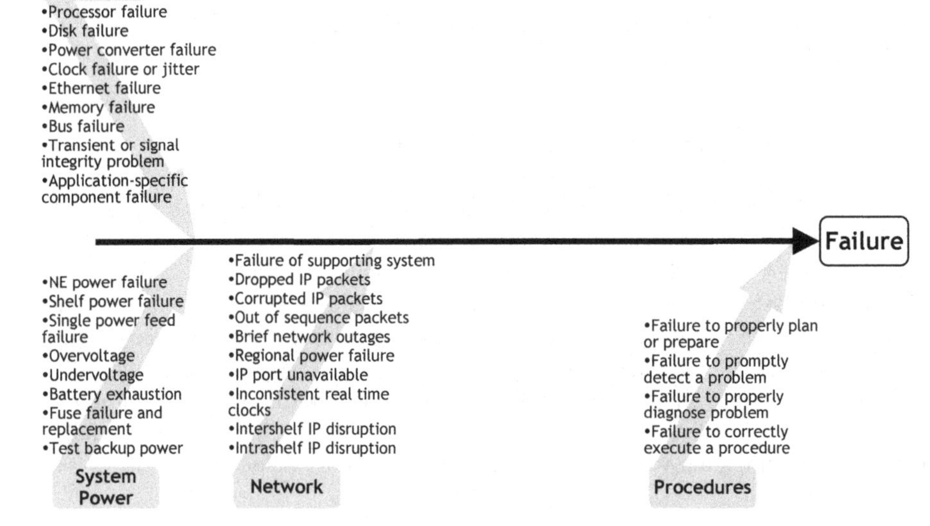

Figure 4.7. IaaS Provider Responsibilities for Traditional Error Vectors.

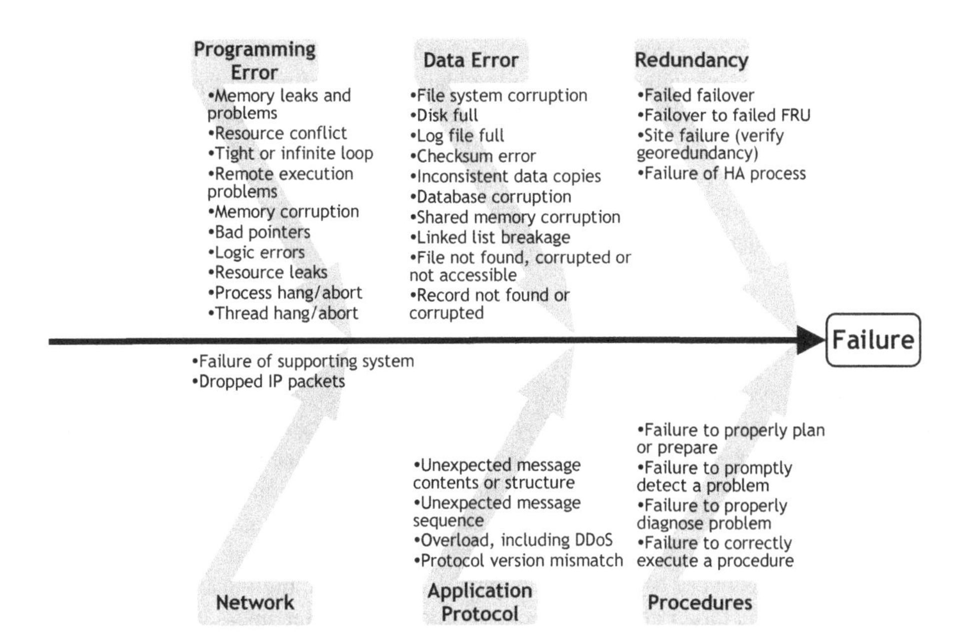

Figure 4.8. Software Supplier (and SaaS) Responsibilities for Traditional Error Vectors.

4.7.2 Programming Errors

Programming errors in application and platform software should be detected in virtualized environments by the application's high availability infrastructure just as they are in native execution environments. Programming errors in host operating system, VM monitor, or hypervisor are likely to appear to applications as a catastrophic (virtual machine) hardware failure.

Application suppliers (and SaaS service providers) have primary responsibility for detection and mitigation of application software; responsibility for failures of platform software and guest OS software is often responsibility of platform as a service (PaaS) suppliers; responsibility for hypervisor failures generally rests with IaaS supplier. Chapter 5, "Reliability Analysis of Virtualization," considers mitigation and detection of errors in this category.

4.7.3 Data Inconsistency and Errors

Data inconsistencies and errors appear and must be detected by application and platform software in virtualized environments just as they are in native environments. One potential advantage of practical virtualized environments is that they may offer additional processing and I/O resources so that an application's data audit routines can be run more often and/or with more elaborate checking logic, thus reducing the risk that a data inconsistency or error will be first encountered by production code and thus cascade into a software failure. For example, database integrity audit programs could be executed in a separate VM instance so that resources consumed for the integrity check would have minimal impact on the running application. It is important to clarify responsibility for detection and recovery from data-related failures. For example, if a file system is corrupted, is the file system repair the responsibility of the XaaS service provider or the cloud consumer?

4.7.4 Redundancy Errors

Virtualization invariably changes the specifics of at least some application redundancy arrangements by mapping some redundancy into virtual machines rather than native hardware devices. Virtualization enabled redundancy mechanisms are discussed in Section 5.4.2, "Virtualized Recovery Options." As virtualization can affect operation and monitoring of both traditional redundancy arrangements (e.g., active/active and active/warm standby) and virtualization-related redundancy arrangements (discussed in Section 5.4, "Recovery Models"), one must assure that the high availability mechanism supporting the target application reliably monitors the readiness of all redundant instances and promptly detects failed switchovers or other redundancy failures. Thus, software suppliers must assure that the application and platform software reliably detect the true operational status of both virtualized and native redundant instances.

4.7.5 System Power Failures

IaaS service providers have primary responsibility for detecting and mitigating power failures and impairments.

4.7.6 Network Errors

Physical and link layer networking issues (e.g., packet collisions, checksum failures, and buffer overrun/under run) are likely to be detected by network adapter hardware and firmware, and thus are likely to be addressed beneath the virtual machine that the application software inhabits. Network layer issues (e.g., IP packets out of sequence, packet jitter) are likely to be passed straight through to the VM and must be detected and mitigated by application and platform software.

4.7.7 Application Protocol Errors

Application protocol errors must be detected and mitigated by application and platform software in the virtualized environment just as they are in native environments. For example, the application software logic that detects and mitigates application protocol failures like *unexpected message contents or structure* or *unexpected message sequence* should behave identically when executing on both virtualized and native deployments.

4.8 RISKS OF DEPLOYMENT MODELS

Reliability and availability risks are fundamentally the same for private, community, public, and hybrid clouds, and thus the same mitigations and metrics are generally appropriate although the associated roles and responsibilities will differ. Cloud bursting is when a workload is dynamically shifted from one deployment strategy to another, such as when migrating some workload from a private or public cloud that is approaching saturation to another provider's public cloud. Cloud bursting has the same primary reliability risks as rapid elasticity (reliability and latency of engaging additional service capacity), plus additional risks associated with the identification, authentication, authorization, and commercial relationships necessary to rapidly shift workloads between distinct enterprises. These additional risks will contribute additional latency to provisioning resources and may cause an individual cloud burst attempt to fail. The risks of cloud bursting are considered in Chapter 7, "Capacity and Elasticity." Cloud bursting also has some architectural challenges such as:

- The virtualization environments must be the same or it must be possible to migrate to the new environment.
- The load balancer (or whichever element is responsible for directing user traffic) must be aware of and able to access the extended cloud environment.
- VM instances must be able to be transferred and instantiated with necessary resources in the extended cloud environment.
- If communication is necessary between cloud environments, the networking layer must support it.
- Sufficient security must be set up to protect the application in the extended cloud environment.

These challenges may make it impractical for most systems to implement cloud bursting.

4.9 EXPECTATIONS OF IaaS DATA CENTERS

The Open Data Center Alliance [ODCA] defines four classes of IaaS data centers: bronze, silver, gold, and platinum. These classes offer a recognized baseline for IaaS performance levels. Table 4.1 from [ODCA-SUoM] gives general characteristics of each class of service; Table 4.2 characterizes service availability expectations. ODCA frames expectations for a variety of data center performance areas, including elasticity (discussed in Chapter 7, "Capacity and Elasticity") and recoverability (discussed in Chapter 9, "Geographic Distribution, Georedundancy, and Disaster Recovery").

TABLE 4.1. ODCA's Data Center Classification

	Bronze	Silver	Gold	Platinum
	Basic	Enterprise Equivalent	Critical Market or Business Sector Equivalent	Military or Safety-Critical Equivalent
Outline	Representing the lower-end corporate requirement, possibly equating to a reasonably high level for a small to medium business customer	Representing a tradeoff more toward cost than service level within the SLA range	Representing a preference for more cost to deliver a higher quality of service within the SLA range	Representing the maximum contemplated corporate requirement, stretching toward the lower end of military or safety-critical needs
Price levels	€ Lowest, commodity	€ €	€ € €	€ € € € Premium
Measures likely to be taken	Standard out-of-the-box components	Standby or reassignable elements		Full duplication with load-balancing or failover, no SPoFs
Performance assurances	Component inputs	Component outputs	Degrees of contention experienced	User applications experience
Scope of assurances	Components	Subsystems	Full systems	End-to-end, including all dependent elements
Security in-built	Basic	Enterprise	Financial	Military
Commitment measurement periods	Averaged over weeks or months	Daily	Hourly	Real time, continuous

Source: Open Data Center Alliance. © 2011 Open Data Center Alliance, Inc. All Rights Reserved.

SLA, service-level agreement.

TABLE 4.2. ODCA's Data Center Service Availability Expectations by Classification

SLA Level	Description
Bronze	Reasonable efforts to attain 99% availability for the IaaS (up to but not including the cloud subscriber's components). Note that the service provider cannot be penalized for any failure of OS or app in the guest VM, except where the failure is clearly the fault of the hypervisor or underlying hardware solution.
Silver	Provisions made to attain 99.9% availability, including increased focus on preventing impact from contention risks.
Gold	Specifically demonstrable additional measures needed to achieve and sustain 99.9% availability and demonstrating resilience to reasonably anticipated fault conditions. Service penalties should apply at this level.
Platinum	Highest possible focus on uptime to achieve 99.99% availability, with the expectation of significantly increased service penalties (beyond Gold tier) if not achieved.

Source: Open Data Center Alliance.

SLA, service-level agreement.

5

RELIABILITY ANALYSIS OF VIRTUALIZATION

This chapter gives a qualitative reliability analysis of virtualization technology. It begins with a review of reliability analysis techniques, and then considers how these techniques apply to the virtualization techniques reviewed in Chapter 2, "Virtualization." There is also an analysis on software failure rates, concluding with a comparison between virtualized and traditional, nonvirtualized applications.

5.1 RELIABILITY ANALYSIS TECHNIQUES

This section reviews several standard reliability analysis techniques:

- reliability block diagrams;
- single point of failure analysis; and
- failure mode and effects analysis (FMEA).

5.1.1 Reliability Block Diagrams

Reliability block diagrams (RBDs) are a simple and powerful tool for visualizing redundancy and analyzing reliability. Reliability block diagrams arrange all service

Reliability and Availability of Cloud Computing, First Edition. Eric Bauer and Randee Adams.

critical elements in a series of connected boxes with elements that are redundant shown in parallel. Figure 5.1 gives an RBD of a hypothetical system with several elements providing critical service: a single critical component A; pairs of components B1, B2 and C1, C2; and three redundant components D1, D2, and D3.

For service to be available across the sample system of Figure 5.1, there must be at least one path through operational ("up") components from one side of the RBD to the other; Figure 5.2 illustrates one such traversal path. For example, either B1 or B2 can fail and service remains available. However, if element A, or both B1 and B2, or both C1 and C2, or D1, D2, and D3 fail, then service will be unavailable.

Individual blocks within RBDs can be aggregated or decomposed to perform the appropriate level of analysis. For example, Figure 5.3 gives an RBD of a canonical simplex (nonredundant) computer-based system: physical computer hardware runs a software platform (e.g., middleware) and operating system (OS) software that hosts a software application. Each of these components is illustrated in a separate box, and for an application service to be operational and available to users, the hardware, OS, software platform, and application must all be operational (aka, "up"). Conversely, if any one of these components is unavailable (aka, "down"), then service is unavailable.

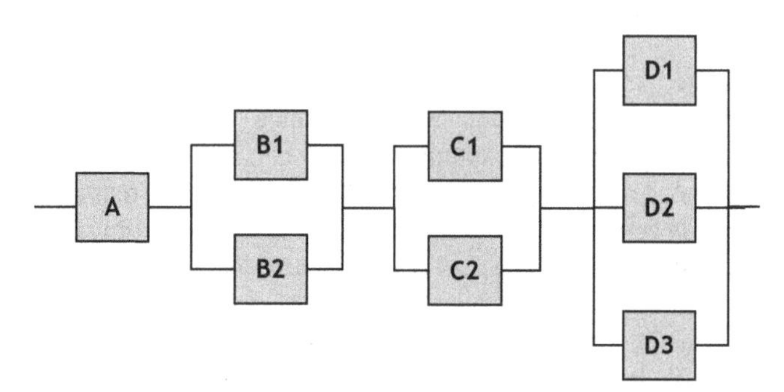

Figure 5.1. Sample Reliability Block Diagram.

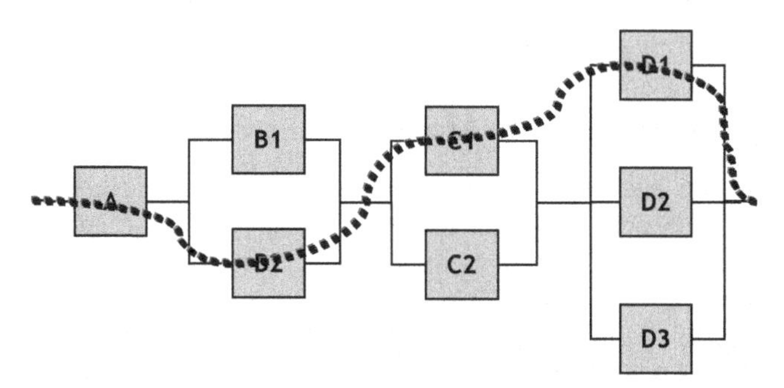

Figure 5.2. Traversal of Sample Reliability Block Diagram.

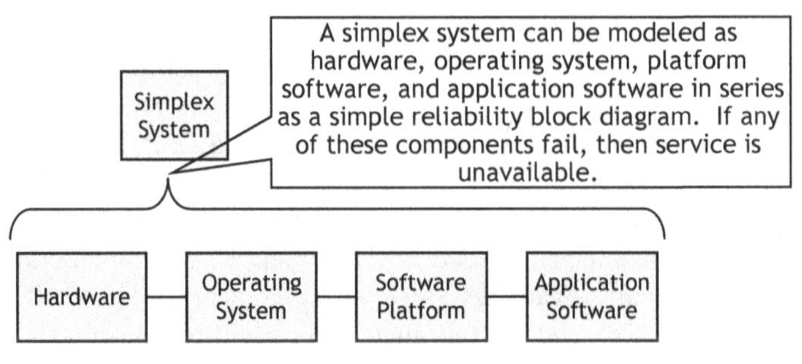

Figure 5.3. Nominal System Reliability Block Diagram.

5.1.2 Single Point of Failure Analysis

Single point of failure (SPOF) is any single component in a system or solution configuration that can fail and cause service to become unavailable. In the RBD example of Figure 5.1, module "A" is a single point of failure. Critical systems must be designed to have no single point of failure. Typically, one reviews a proposed design for single points of failures, and redesigns any components or facilities that are found to be single points of failure.

There is a related architectural concept of a *single point of maintenance* or *single point of repair* that considers the impact of maintenance or repair actions. A single point of maintenance or repair refers to a component that can only be serviced by taking the entire system offline. For example, a basic RAID system may require the unit to be powered off to safely replace a failed hard disk drive, while a more robust RAID configuration of the system protects against a single hard disk failure from being a single point of failure that impacts user service, a simple physical or electrical design makes the hard disk drive a single point of maintenance or repair. The notion of "hot" removal and installation of blades, hard disks, and so on, directly mitigates the risk of single point of maintenance or repair. Redundant hardware can prevent the initial failure from causing a prolonged service outage, but eventually the failed hardware must be repaired or replaced to return the system to full operational redundancy. If that maintenance or repair operation requires the entire element to be depowered or rebooted, then that module represents a single point of maintenance or repair for that element.

5.1.3 Failure Mode Effects Analysis

Failure mode effects analysis is a technique for assuring that a design can contain and recover failures with acceptable service impact. We review the concepts of failure containment and recovery groups as background, and then discuss FMEA analysis methodology.

5.1.3.1 Failure Containment. Containing the impact of a failure is crucial in preventing a cascade of failures that impact more users and functionality. For example, modern OSs and processors rigidly separate memory address spaces of different user processes from each other, thereby enabling individual user processes to experience critical failures (e.g., divide-by-zero exceptions, accessing undefined memory locations, and unaligned memory access) without directly impacting other processes running on the same OS instance. Database transactions are a well-known unit of failure containment; if a transaction fails, then it can be aborted, and any changes executed after the formal beginning of that transaction will be rolled back, thereby containing the impact of the failure to losing that single transaction. Virtualization provides an additional level of rigid failure containment because VM instances are wholly independent of each other, and thus can be started, and restarted, restored, or recovered without directly impacting any other VM instances hosted on the same or different hardware platform. Good hardware designs contain hardware failures to the smallest field replaceable unit. For example, failure of a single hardware blade in bladed system architecture should have no impact on other blades in the chassis.

The ability to rigidly contain failures preventing a failure from cascading into further failures is crucial in high availability systems. Virtualization offers rigid failure containment of software failures at the level of virtual machine (VM). In the simplest scenario, VM containment is equivalent to the containment offered by native server or blade instances; however, the hypervisor is able to better manage a multitenant configuration by more effectively allocating resources and isolating failures experienced by one tenant from the other. Note that application architectures can be tailored to use more VM instances than native server instances, and thus offer tighter failure containment and resource management than with traditional deployment.

5.1.3.2 Recovery Groups. After a failure is contained to a particular hardware or software module, service must be recovered to a redundant or repaired unit; in the case of software failures, service could be recovered to a restarted software instance. Service recovery after failure is inherently more complex if any service state is to be retained, so special attention should be paid to carefully design and thoroughly test those associated software modules to ensure they recover rapidly and reliably in a wide variety of failure scenarios.

Recovery groups are typically arranged into well-known redundancy arrangements like $N + K$ load sharing, active/active, or active/standby, so when a critical failure occurs service is recovered by shifting traffic to a redundant instance, such as to the failed active unit's mate or distributed across other elements in the load sharing pool. The granularity at which service is recovered is the recovery group.

5.1.3.3 FMEA Methodology. Failure mode effects analysis considers the impact on service of individual component, module, or subsystem failure events. FMEAs are generally represented in a tabular form where:

- individual components, modules or subsystems are represented by rows;
- primary services offered by the system are represented by columns; and
- individual cells indicate what, if any, impact there is on a particular primary service when a particular component, module, or subsystem fails and is recovered. Note that the recovery action (e.g., failing over service to a redundant module) could impact users who were not impacted by the original failure.

It is often convenient to add two additional columns to the table:

- Nominal time to detect the failure of the individual component, module, or subsystem; this value is typically represented in seconds.
- Nominal time to complete an automatic switchover or service recovery action after detecting failure of the individual component, module, or subsystem.

Any cells that have unacceptable failure effects or for which failure detection or failure recovery time exceed system requirements can be highlighted; system architects and developers should investigate architectural and design changes that can mitigate the unacceptable effect or behavior. Table 5.1 gives a sample failure mode effects analysis for a virtualized application of Figure 5.1.

TABLE 5.1. Example Failure Mode Effects Analysis

Functional Unit	Redundancy Strategy	Impact on New Sessions	Impact on Stable Sessions	Impact on Transient Sessions	Estimated S/O Latency	Notes
A	Simplex	LOST	LOST	LOST	Not supported	Impact on sessions until "A" is restarted on current or new server
B	Active–standby	LOST	No impact	LOST	10 seconds	Impact on new/transient sessions until failover completed; data replication supports maintaining stable sessions across failover
C	Active–standby	LOST	No impact	LOST	1 second	Impact on new/transient sessions until failover completed; data replication supports maintaining stable sessions across failover
D	$N + K$	No impact	No impact	No impact	1 second	Failure of one unit entails load sharing of traffic across the others

5.2 RELIABILITY ANALYSIS OF VIRTUALIZATION TECHNIQUES

The reliability analysis techniques of the previous section are now applied to full virtualization, OS virtualization, and paravirtualization. The RBDs indicate a single application in the first three sections but the coresidency use case is considered in Section 5.2.4.

5.2.1 Analysis of Full Virtualization

Figure 5.4 illustrates how full virtualization changes the canonical system RBD shown in Figure 5.3 by inserting a virtualization hypervisor and Host OS between the OS and the underlying hardware. Note that the "host" OS can be different from the "guest" OS in full virtualization. For the virtualized deployment of this traditional canonical system to be available, the hardware, "host" OS, hypervisor, "guest" OS, software platform, and application must all be operational.

The application, with its software platform and OS, comprise a VM. Each VM is isolated from the other VMs running on the server and is unaware it is running in a virtual environment. As a result, each VM's failures are contained; VMs can fail and be recovered independently of other VMs running under the same hypervisor. A VM instance can be reset, rebooted, or migrated within its failure recovery group based on the nature of the failure. Since the hypervisor itself is required as the hardware interface for all of the VMs, it becomes the single point of failure along with the hardware for the virtualized system.

5.2.2 Analysis of OS Virtualization

Figure 5.5 illustrates how the canonical system RBD of Figure 5.3 changes when OS virtualization inserts a hypervisor running on a primary "host" instance of the OS. The

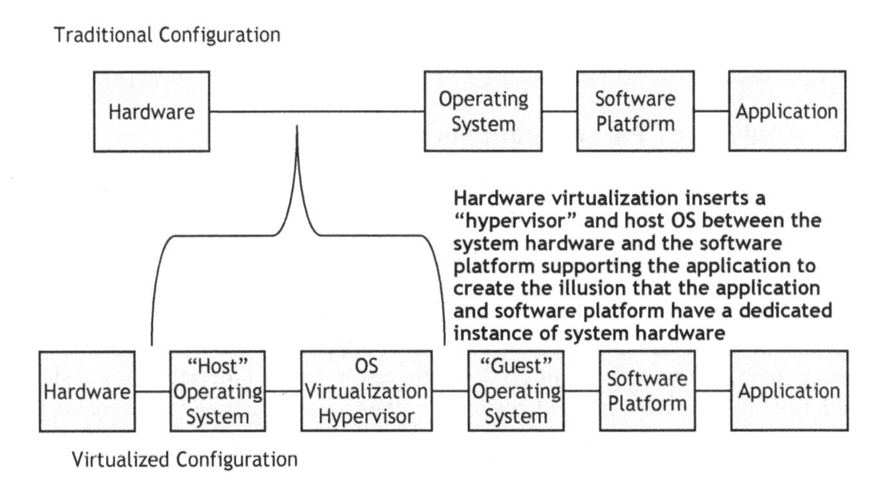

Figure 5.4. Reliability Block Diagram of Full Virtualization.

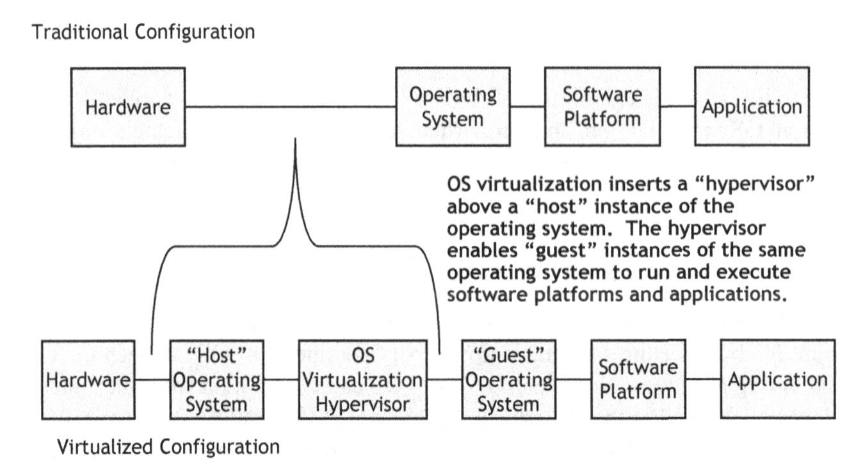

Figure 5.5. Reliability Block Diagram of OS Virtualization.

RBDs for full virtualization and OS virtualization are the same; the difference is that on full virtualization, the guest OSs can be different from the host OS, while with OS virtualization, they must be the same. Note that while both the guest and host OS instances must be the same OS and version, the guest OS instance is separate and can fail (aka, crash) and be recovered independently of the host—or other guest—OS instances.

OS virtualization also provides partitioning of the application instances, along with their software platform and guest OS, into isolated partitions sometimes referred to as virtual environments or containers. This isolation ensures that the failure of one virtual environment does not impact another virtual environment sharing the OS virtualization hypervisor, and that each virtual environment will be included in a separate recovery group than the others.

5.2.3 Analysis of Paravirtualization

Figure 5.6 illustrates how the canonical RBD changes when paravirtualization is used. Like full virtualization, paravirtualization inserts a hypervisor and host OS between the system hardware and the software platform supporting the application to create the illusion that the application and software platform have a dedicated instance of system hardware. The difference is that the guest OS is modified to include integrated device drivers that offer more direct access between the application instance and hardware resources since they provide for direct communication without the need for translation. As with full virtualization, paravirtualization entails the partitioning of the applications into VMs; each VM can run on an OS different from the host OS. As with full virtualization, each VM with paravirtualization is isolated from the other VMs running on the server and is unaware it is running in a virtual environment. As a result, each VM's

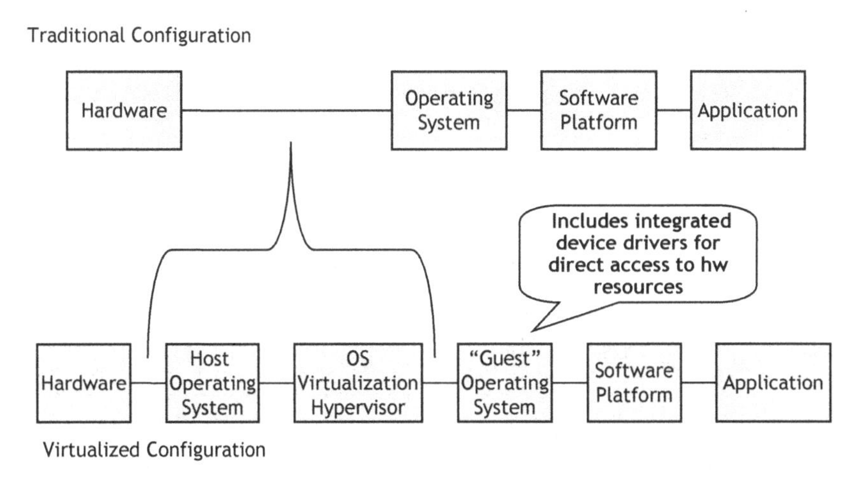

Figure 5.6. Reliability Block Diagram of Paravirtualization.

failures are contained; a single VM can fail independently of other VMs running under the hypervisor and recover within its own recovery group.

Paravirtualization combines advantages of both full virtualization and OS virtualization. OS virtualization offers direct calls from the applications to the OS without translation since the applications all function on the same OS. This is also the main disadvantage of OS virtualization, that is, the host and all of the guests must run on the same OS and version. Paravirtualization provides more direct access for the applications to the hardware resources than full virtualization, but not as much as OS virtualization. Paravirtualization thus offers some of the best of both full virtualization and OS virtualization, with better performance through more direct access to hardware resources and the ability to support VMs on OSs different from that of the host at the cost of needing to alter the guest OSs.

5.2.4 Analysis of VM Coresidency

A popular use of virtualization technology is to enable several VM instances to share hardware resources, thereby improving hardware utilization and efficiency. When it applies to multiple applications, this is referred to as VM coresidency, that is, multiple VM application instances residing on the same server. This VM coresidency feature is leveraged in the popular server consolidation use case. Figure 5.7 illustrates how virtualization enables one to take the applications and software platforms of traditional system deployments "A" and "B" and make them coresident on a single hardware instance via virtualization. Assuming that both applications "A" and "B" are required to be operational for service to be available, then one can easily see how server consolidation replaces excess server hardware from the reliability block diagram with a virtualization layer. Assuming that the virtualization layer is more reliable than the

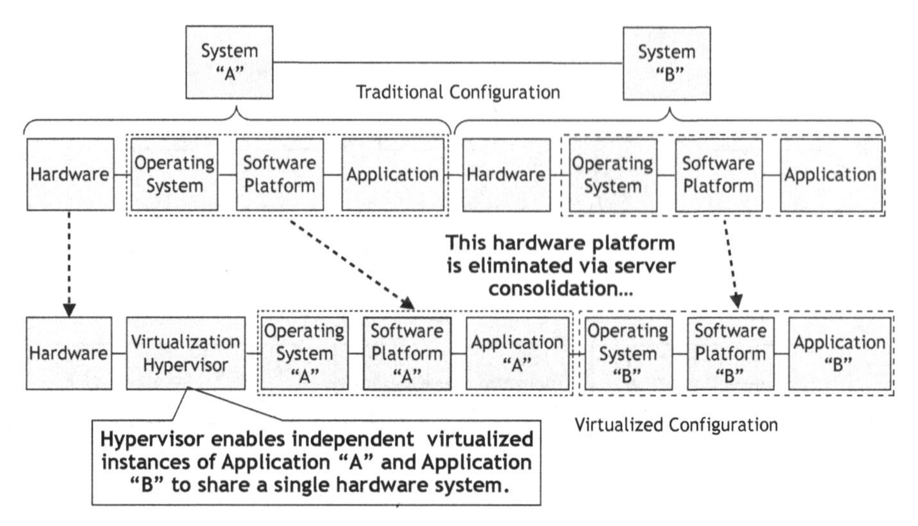

Figure 5.7. Reliability Block Diagram of Coresident Application Deployment.

hardware that is consolidated, service availability is likely to improve because the failure contribution of one hardware platform has been removed from the critical service delivery path.

5.2.4.1 Failure Containment for Coresident Applications. Virtualization also provides the ability to isolate the two applications so that a failure of one of the applications does not impact the other. Because of this isolation, the applications still have their own recovery group as they did in nonconsolidated deployments; if a software failure occurs that triggers the recovery of the VM even to another server, then coresident VMs should not be impacted. Of course, a catastrophic hardware or hypervisor failure will impact all of the coresident VMs; ideally, those affected VM instances will be migrated to alternate virtualized servers that have spare capacity so the service impact can be minimized.

An important characteristic of the different virtualization types is failure containment, isolating the applications in a virtual environment or VMs so that even though the application is sharing a host computer, it is not impacted by failures from another application. All three virtualization types provide failure containment for the VMs managed by the hypervisor, and allow them to be included in separate recovery groups from the other VMs. The hypervisor is a key component responsible for the isolation of the VMs, their access to system resources, and may include an extra layer of high availability software facilitating the recovery of the VM. It could also be a single point of failure for the coresident applications, since it manages the use of most if not all of the hardware resources.

A failure mode effects analysis of the simplex functional entities included in the RBD in Figure 5.7 are included Table 5.2. Failures experienced by a single VM or its components (application, software platform, and guest OS) are isolated to that VM instance and can be resolved by rebooting or resetting the failed VM instance. This

TABLE 5.2. Failure Mode Effect Analysis Figure for Coresident Applications

Functional Entity	Impact on App "A"	Impact on App "B"	Recovery Mechanism
Hardware	Service impact	Service impact	Restart/recover hardware
Hypervisor	Service impact	Service impact	Restart/recover hypervisor
Operating system "A"	Service impact	No impact	Restart/recover virtual machine "A"
Software platform "A"	Service impact	No impact	Restart/recover virtual machine "A"
Application "A"	Service impact	No impact	Restart/recover virtual machine "A"
Operating system "B"	No impact	Service impact	Restart/recover virtual machine "B"
Software platform "B"	No impact	Service impact	Restart/recover virtual machine "B"
Application "B"	No impact	Service impact	Restart/recover virtual machine "B"

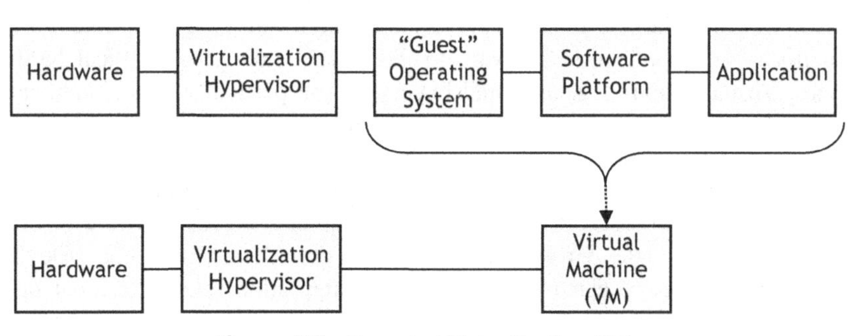

Figure 5.8. Canonical Virtualization RBD.

should result in recovery of the VM instance or recovery to another instance within its failure recovery group when redundancy has been configured. Common elements, such as the server, hypervisor, and the host OS (if it exists) will have a potential impact on all coresident VMs. Failure of the common elements will result in the recovery of those elements, as well as the impacted VMs. Critical failures may result in the recovery of the VMs onto another hardware platform.

5.2.5 Discussion

Since the three virtualization types have such similar characteristics from a failure containment, failure management, and recovery point of view, a generic RBD, such as depicted in Figure 5.8 may be used throughout this book rather than three specific ones.

5.3 SOFTWARE FAILURE RATE ANALYSIS

Critical software failure rate is the most influential parameter when considering software attributed downtime because it drives how often automatic or manual failure detection mechanisms and recovery strategies are activated. This section considers how the application, software platform and guest OS failure rates change when deployed on a hypervisor, as well as considering the software failure rate of the hypervisor itself.

5.3.1 Virtualization and Software Failure Rate

There are residual defects in all software that will result in software failures given the execution of particular scenarios under certain conditions. While a particular piece of application software contains the same residual defects when it is executed on native or virtualized platforms, virtualized deployments should offer a far smaller range of configured operational profiles for the software than native deployments. Virtualization decouples application software from specifics of the underlying physical hardware configuration by presenting a canonical virtualized hardware image to guest OS, platform, and application software, and the virtualization manager carefully addresses the particulars of mapping the canonical hardware image presented to guest OS, platform, and application software to the actual underlying physical hardware resources. Virtualization in general and hypervisors in particular enable applications to execute in the same virtualized configuration regardless of the actual physical configuration of the underlying hardware. Note that virtualization enables software to run on hardware that is nominally faster (e.g., higher clock rate) than the native hardware, so while differences in execution timing that trigger software errors are theoretically possible, the virtualization manager should minimize this risk. Thus, virtualization technology can mask hardware configuration related residual defects in application and platform software from being activated because virtualization technology assures that application and platform software always sees the same canonical hardware image. The software supplier's system test efforts can thoroughly test against that canonical hardware image, and the virtualization supplier assures that all deployments of their virtualization technology should present that same canonical hardware image to guest OS, application, and platform software, regardless of the specifics of the underlying hardware configuration. Overall, virtualized application deployments may experience somewhat lower critical application software failure rates than traditional deployments.

At the time of this writing the authors are unaware of published data comparing the software failure rates of application or platform software on native (nonvirtualized) configurations with execution on virtualized platforms. However, anecdotal data from experts with significant field deployment of both virtualized and non-virtualized instances of their application software indicates that the critical software failure rate of virtualized deployment appears somewhat lower than for native (nonvirtualized) deployments.

5.3.2 Hypervisor Failure Rate

Hypervisors are complex system software modules that are confronted with diverse requests from VM instances and real-time events from underlying hardware elements. The latest release plus stability patch loads of massively deployed hypervisors that support diverse applications on a wide variety of hardware platforms should have few residual defects provided:

- The target application and platform software are architecturally similar to other applications that are broadly and successfully deployed on the hypervisor.
- The guest OS is supported by the hypervisor and has demonstrated reliable operation across a very large set of deployments.

Note that the OS changes required to support paravirtualization increase the risk that modified OS software may have a slightly higher critical failure rate than unmodified guest OSs. This incremental risk is likely to be modest in early releases of paravirtualized OSs, and will likely become negligible as specific paravirtualized implementations mature.

5.3.3 Miscellaneous Software Risks of Virtualization and Cloud

In some cases, the actual application and/or platform software or configuration may be different on virtualized deployment compared with native. For example, software licensing may be implemented differently in virtualized deployment compared with native deployment, and hence the fundamental software failure rate may be different between the configurations. As virtualization platforms strive to faithfully emulate native hardware, any variations should be small and thus the extent of different code—and hence risk of different residual defects—is likely to be small.

In contrast to virtualization-related application software differences that are likely to be minimal compared with native, changes to take advantage of cloud related features like rapid elasticity (aka, "autoscaling") are likely to require at least a moderate amount of moderately complex software, and thus increase the risk of software failure; this risk is considered in detail in Chapter 7, "Capacity and Elasticity."

5.4 RECOVERY MODELS

After a critical failure is detected, a recovery action must be automatically or manually executed to recover user service. This section reviews both traditional recovery options and virtualized recovery options, and considers the implications.

5.4.1 Traditional Recovery Options

Traditional systems can support four general recovery models:

1. *Hardware Repair.* Hardware failures of nonredundant systems render service unavailable until hardware can be repaired and software restarted. The mean

time to repair (MTTR) hardware is primarily a function of the operational policies of the enterprise, such as: whether reserve field replaceable hardware is maintained on-site; whether trained repair staff is available; and what support arrangements are in place with hardware suppliers. Typical hardware repair times range from hours to days.

2. *Software Restart.* Software failures of nonredundant systems are often recovered by either restarting a process, application software, or restarting the entire OS and platform software. Typical software restart times are measured in minutes.
3. *Switchover to Redundant Active Element.* Systems that are deployed with online and "active" redundant elements (e.g., load sharing or active/active configurations) can recover user service by switching (or redirecting) users to an online and active redundant element. Service switchover to an online and active redundant element will have some latency but should be faster than software restart and far faster than hardware repair.
4. *Switchover to Redundant Standby Element.* Some systems are deployed with redundant units that are not online and actively serving users; these redundant units are said to be in "standby" because some actions are necessary to bring them to full readiness to serve users. It inevitably takes time to bring the standby element to active status, plus latency to switchover service from the failed active element to the newly active element. "Hot" standby elements will require time to promote the standby software element to active; "warm" standby elements will also require time to start up the application itself; and "cold" standby elements will require even more time to startup the underlying hardware platform itself. Thus, switchover to "hot" standby elements is often faster than software restart; switchover to "warm" standby elements is often comparable with software restart time; switchover to "cold" standby elements is significantly slower than software restart, but it is much faster than hardware repair.

Figure 5.9 organizes these traditional recovery options on a quasi-logarithmic timeline by service recovery latency, from switchover to redundant active and hot standby elements nominally taking seconds (depending on system architecture and configuration), to hardware repair taking hours or days (depending on sparing strategy and support agreements with hardware suppliers).

5.4.2 Virtualized Recovery Options

Virtualization enables several new redundancy options that are not possible with traditional deployments; these options generally supplement traditional recovery strategies, like process restart. Virtualized recovery options are best understood in the context of the DTMF's virtual system state model, which was described in Section 2.4,

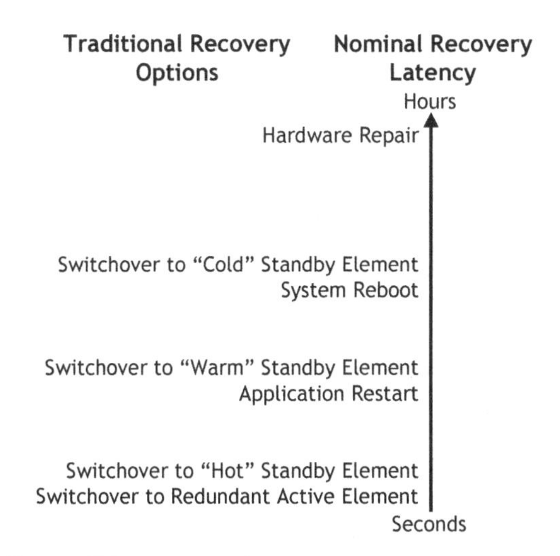

Figure 5.9. Latency of Traditional Recovery Options.

"Virtual Machine Lifecycle," and illustrated in Figure 2.6. These virtualized recovery options are:

- *VM Can Be Moved (Offline) and Restarted on Different Hardware.* Restarting a VM on a different available server can recover service far faster than the time it takes to repair failed hardware. This recovery may be manual or automatic. The likely benefit of this strategy is discussed in Section 6.9, "MTTR of Virtualized Hardware."
- *Virtual System State "Active" Redundancy.* A traditional "active" redundant element can be mapped into a VM so that both the primary and redundant active elements are in separate VMs. This is illustrated in Figure 5.10. Multiple "active" VMs can be configured in standard arrangements, like N + K load sharing, active/standby, and so on. Since it is using the traditional redundancy mechanisms, the resulting system should have essentially the same service availability characteristics as native deployments. Note that the multiple "active" VMs should be deployed on different virtualized servers to prevent underlying hardware from being a single point of failure.
- *VM Instances Can Be Reset or Rebooted.* Virtualization adds the ability to either reset or reboot a VM instance. Figure 5.11 illustrates VM reboot entailing the reboot or recycle of the VM while maintaining their allocated resources. Figure 5.12 illustrates VM reset in which the VM transitions from deactivate to activate without a corresponding deallocation and reallocation of resources.

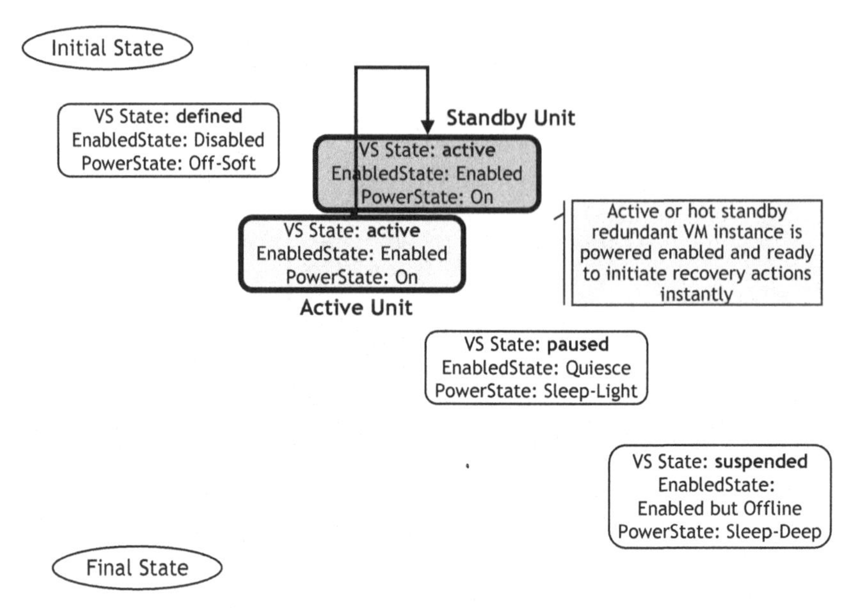

Figure 5.10. Traditional Active-Standby Redundancy via Active VM Virtualization.

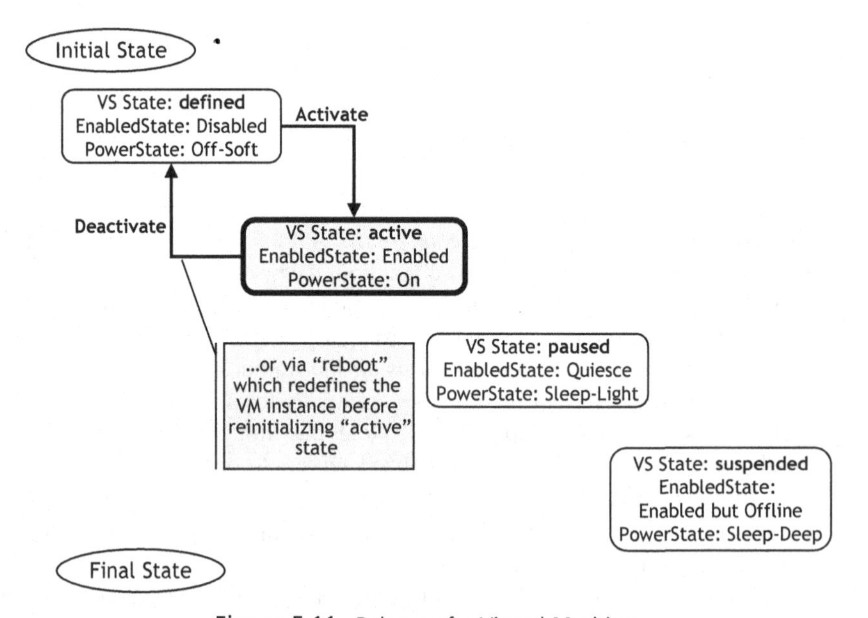

Figure 5.11. Reboot of a Virtual Machine.

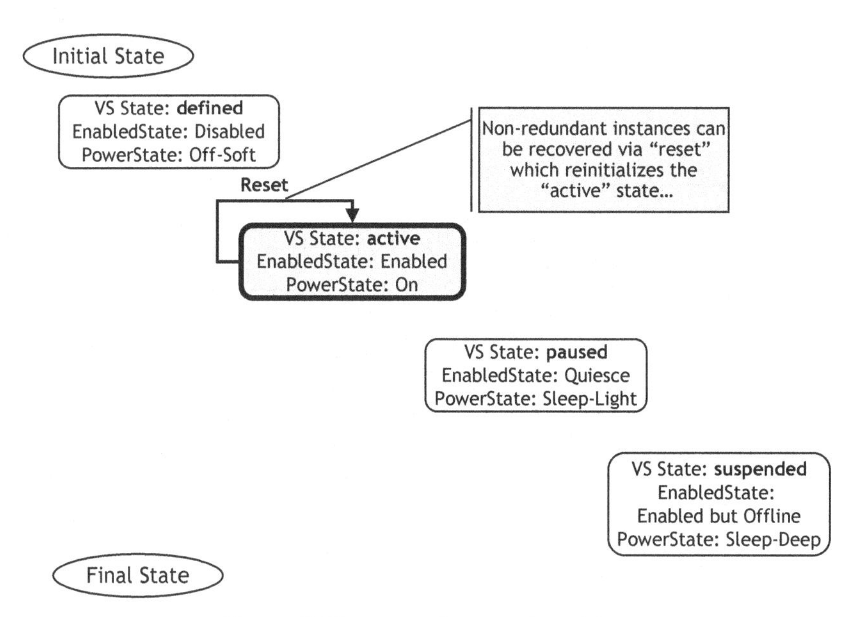

Figure 5.12. Reset of a Virtual Machine.

- *Paused VM Instances Can Be Activated.* Redundant VM instances can be allocated, application software made ready, and then the VM instance can be paused; these paused VM instances can be activated and made available instantly to recover service when necessary. As shown in Figure 5.13, paused VM instances are sleeping "lightly" so there is additional service recovery latency compared with active VM redundancy, but fewer platform (e.g., CPU) resources are consumed to maintain paused VM's nearly online than for online active VM instances. Note that paused redundant VM instances should be hosted on different virtualized server platforms from the active instances to prevent the virtualized server platform from becoming a single point of failure.
- *Suspended or Snapshot VM Instances Can Be Activated.* Redundant VM instances can be allocated, application software made ready, and then the VM can be suspended or snapshot. Redundancy can be provided by suspended or snapshot VM instances that can be activated when required. As shown in Figure 5.14, suspended VM instances are sleeping "deeply" so there is more incremental recovery latency compared with paused VM redundancy, and significantly more latency compared with active VM redundancy, however even fewer virtualized platform resources are consumed to support suspended VM redundancy. Note that suspended redundant VM instances should be hosted on different virtualized server platforms from the active instances to prevent the virtualized server platform from becoming a single point of failure.

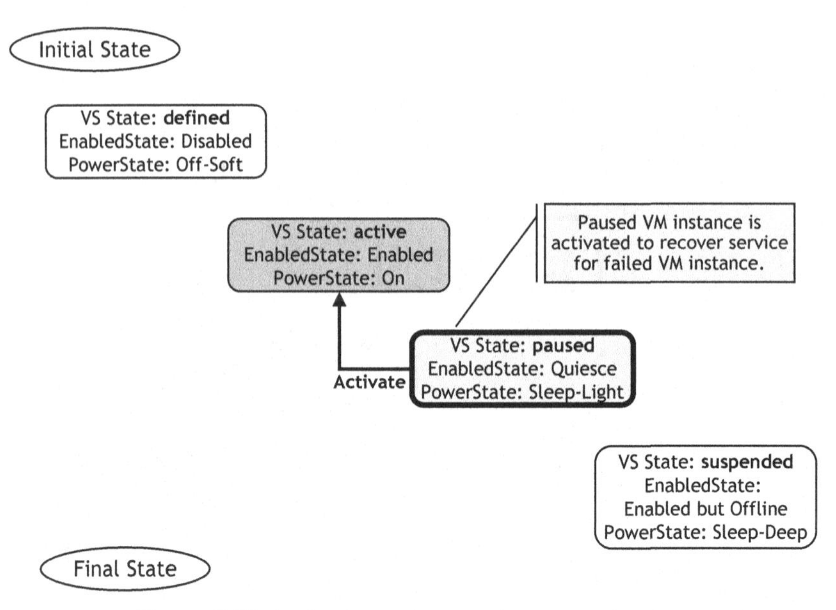

Figure 5.13. Redundancy via Paused VM Virtualization.

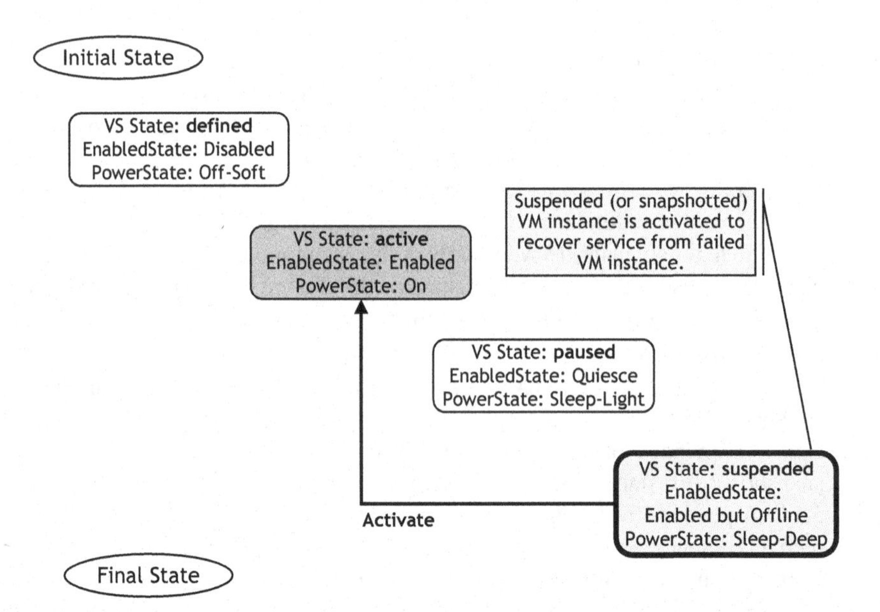

Figure 5.14. Redundancy via Suspended VM Virtualization.

When a paused or suspended VM is activated, the activated VM instance resumes with the same system state that it had at the instant it was paused or suspended. This system state is inevitably different from the state of the VM instance that later crashed, and thus system state must be refreshed, recovered, or rebuilt by either the activated VM instance, or the service clients, or by both VM instance and clients. While some state recovery or rebuilding may be required with either traditional or active VM redundancy, the fact that those redundant instances are nominally executing offers the possibility for those redundant instances to proactively retrieve or recover fresher service status.

5.4.3 Discussion

Figure 5.15 overlays the nominal recovery latency of virtualized recovery options to the right of the traditional recovery options from Figure 5.9. Although this is intended as a qualitative visualization rather than a quantitative analysis, virtualized recovery options do offer somewhat different recovery latencies than what is typically offered by traditional recovery options.

Note that virtualized redundancy may introduce an implicit single point of failure risk not found in traditional deployments if redundant VM instances are served by the same hypervisor instance. Virtualization implicitly breaks the linkage between VM instances and the hardware that supports them, and thus by default, redundant VM instances for a particular application might be mapped to a single physical virtualized server, thereby making that physical server a single point of failure for that particular application at that particular time. It is essential to ensure that redundant VM instances are not hosted on the same hypervisor or physical server. This can be done through affinity rules as discussed in Section 11.1, "Architecting for Virtualization and Cloud."

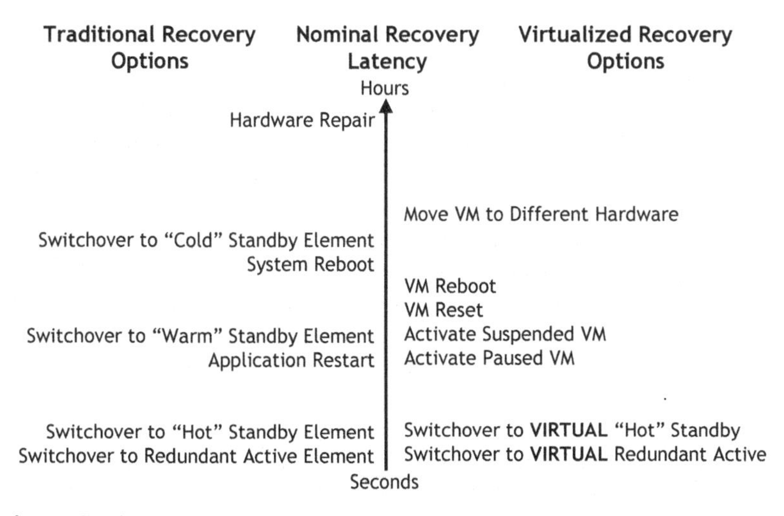

Figure 5.15. Nominal Recovery Latency of Virtualized and Traditional Options.

5.5 APPLICATION ARCHITECTURE STRATEGIES

There are several common high level architectural patterns that are used in some traditional environments but have special meaning when applied to virtualized and cloud-based applications. This section reviews the following common patterns and considers the reliability characteristics of each:

- on-demand single-user model;
- single-user daemon model;
- multiuser server model; and
- consolidated server model.

5.5.1 On-Demand Single-User Model

The simplest and most common architectural pattern is the on-demand single-user model in which an application instance is spawned exclusively for the use of that single user. Once spawned on user request, the application instance runs until the user explicitly terminates (e.g., closes) the application instance. Most applications running on personal computers (PCs) and Smartphones follow this pattern, such as browsers, word processors, games, and so on. Desktop as a service follows this architectural pattern. A virtualized desktop application instance is spawned on user request and remains operational until the user terminates the instance.

In the single-user model, the extent of all failures is limited to the individual user who requested the application, and thus a single failure should never impact more than a single user, meaning that there is no notion of partial capacity outages. Each individual application instance should be independent, that is, a failure of one single user's application should not impact service offered by another user application instance.

Because the application is explicitly started and stopped on user request, an application instance is likely to be offline for the majority of any weekly, monthly, or other measurement period. Thus, traditional service availability metrics (e.g., Availability = Uptime/[Uptime + Downtime]) are not generally useful because they do not adequately account for offline time. Instead, accessibility and retainability metrics should be used to characterize application availability. For an on-demand single user application, accessibility is the probability that a user's request to launch the application will successfully create an operational application instance within a maximum acceptable time (e.g., seconds), and retainability is the probability that an application instance will remain operational until the user explicitly terminates the instance. Accessibility impairments are generally either transient, meaning that a second (or third) attempt to spawn the application will succeed, or persist meaning that the application will not spawn correctly until the root cause of the failure has been corrected. While a statistical measurement that considers the probability of transient accessibility failures is interesting, there is little point in considering the accessibility when the application experiences a persistent failure because accessibility will be nil until the root cause of the failure has been corrected. Operationally, retainability captures the probability that the application will not crash during a session, and this is likely to be a key service metric for on-demand single users.

5.5.2 Single-User Daemon Model

This application pattern assigns a persistent VM instance to a single object or logical user, and that persistent application instance is expected to be continuously available. For example, if control software for a wireless base station is virtualized and moved into the cloud (thereby reducing the hardware resources that are required to be deployed to remote, unstaffed outdoor locations), then the cloud-based control function for each individual base station may become a single-user daemon application. One VM instance is uniquely associated with a particular physical object (a wireless base station in this example); the number of VM instances grows and shrinks with the number of objects of interest. When one VM instance fails, a single object is impacted, and the VM instance should be automatically recovered as quickly as possible. Availability is a primary service metric for single-user daemon deployments. It is easily computed by comparing the cumulative object down-minutes in the measurement period across the population of VMs to the total number of minutes in the measurement period multiplied by the number of VMs. Service reliability and service latency are also useful metrics. Service accessibility and service retainability are not particularly interesting.

5.5.3 Multiuser Server Model

Multiuser server is the canonical architecture model for traditional application servers. Each application server instance can handle a variable but finite workload of work and/or set of active users. Traditional application instances have a fixed maximum engineered capacity, but the cloud's rapid elasticity characteristic means that applications are expected to grow horizontally, vertically, and/or out as load increases. As load shrinks, multiuser server applications are expected to gracefully release unneeded resources.

Service reliability, service latency, service accessibility, and service retainability are good metrics for multi user server. As explained in Section 13.7.1, "Cloud Service Measurements," service availability is sometimes an awkward metric for cloud-based multi-user server applications because it is often hard (or impossible) to know the true extent of service impact—as opposed to duration of service impact—because the number of impacted application users varies across time.

5.5.4 Consolidated Server Model

One of the most common use cases of virtualization is the consolidated server model in which multiple applications or application instances share the same physical server. The applications are referred to as coresident applications and are discussed in Section 5.2.4, "Analysis of VM Coresidency."

The following figure indicates how multiple applications each on its own hardware component can become coresident through virtualization:

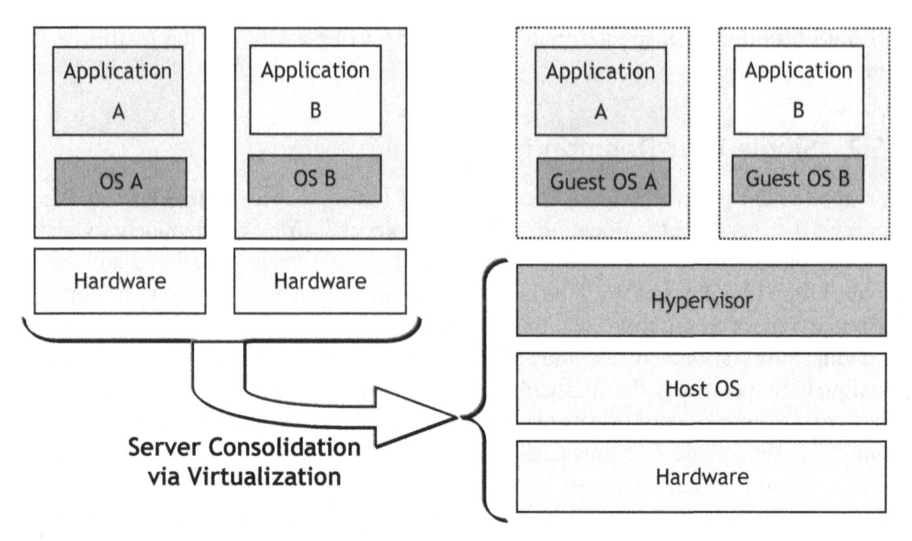

Figure 5.16. Server Consolidation Using Virtualization.

With virtualization (as depicted in Figure 5.16), server consolidation is supported by creating a VM for each application, including its OS. Each VM will act like a separate computer. The hypervisor manages the complexity of the coresident applications by monitoring the Guest OSs and determining how the resources are allocated to the VMs. There are no changes needed to the Guest OSs. The hypervisor will provide all of the hardware interfaces for the VMs.

Since virtualization supports the containment of failures to the individual VMs, if a failure occurs on one VM, the other VMs residing on the same server will not be impacted. However, failures associated with the hardware or virtualization platform will likely impact all VMs on that server.

Traditional service availability calculations are still relevant. Product availability calculations will be based on the VM instances that comprise the product and do not then include other coresident VMs. Service reliability, service latency, service accessibility, and service retainability are good metrics for the consolidated server model.

5.6 AVAILABILITY MODELING OF VIRTUALIZED RECOVERY OPTIONS

At the highest level, systems are deployed either:

- *Simplex* (or standalone) with no redundant instances allocated or configured prior to failure.
- *Redundant* with sufficient resources pre-allocated to promptly recover service following failure.

Mathematical modeling is commonly used to estimate the feasible and likely availability of system architectures. This section considers how service availability of both simplex and redundant system configurations is likely to be impacted by the use of virtualization.

5.6.1 Availability of Virtualized Simplex Architecture

Conventional (vs. highly available) systems are deployed simplex, meaning that a single nonredundant operational unit is available to provide each critical function supporting a service. A single, simplex unit has two well-known operational states, shown in Figure 5.17:

1. *Up*, or working, in which the system is known to be operational.
2. *Down*, or failed, in which the system is known to be nonoperational, and presumably repair or replacement is planned.

As explained in Section 3.3, "Service Availability," service availability is uptime divided by uptime plus downtime (Availability = Uptime/[Uptime + Downtime]), so availability of simplex systems can be improved in two general ways:

1. reducing failure rate; and
2. shortening recovery time.

5.6.2 Availability of Virtualized Redundant Architecture

Highly available systems are designed to automatically detect, isolate, and recover from any single failure. To achieve the fastest possible recovery, service is often restored onto redundant modules that are either active or in standby and ready to rapidly recover service from a failed module. In some cases, the entire element may be duplicated, such as having two identical jet engines on an airplane. High availability systems are built by architecting redundant arrangements of all critical modules so that no single unit failure will necessarily produce a service outage.

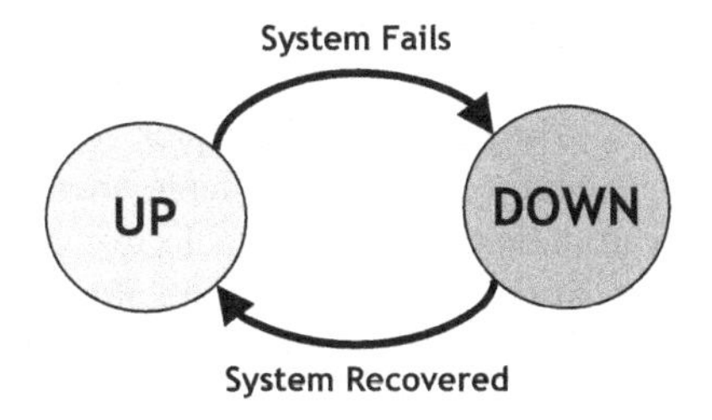

Figure 5.17. Simplified Simplex State Diagram.

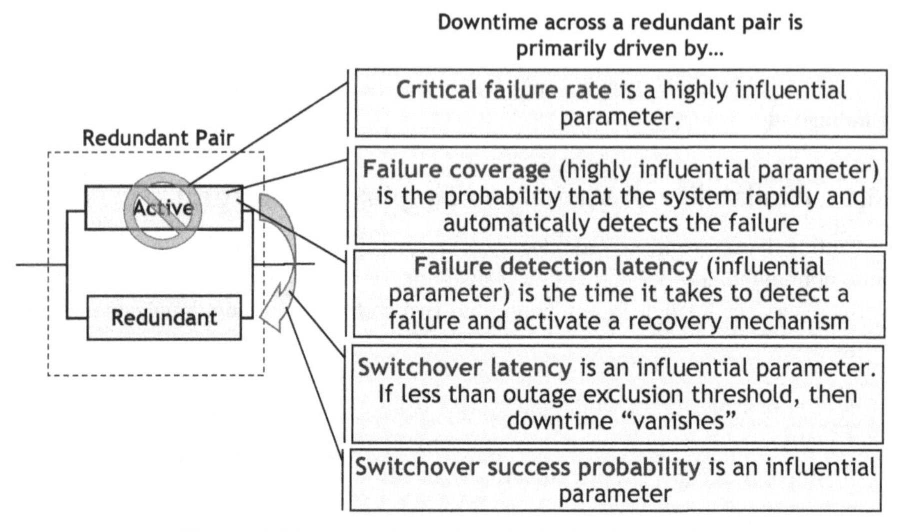

Figure 5.18. Downtime Drivers for Redundancy Pairs.

Fundamentally, redundant pairs of components can either share the load across both elements (called "active–active") or have one unit actively serving load while the redundant unit waits in standby (called "active–standby"). In both of these cases, failure of a unit actively serving load (i.e., "active" unit in "active–standby" arrangements, or either unit in "active–active" arrangements) will impact service availability at least momentarily. The service impact can be simply modeled by considering five factors shown in Figure 5.18: critical failure rate, failure coverage, failure detection latency, switchover latency, and switchover success probability. The following sections consider how virtualization impacts each of these characteristics, and then discusses the overall implications

5.6.3 Critical Failure Rate

Critical failure rate is the rate of service-impacting failure events. As discussed in Section 5.3, "Software Failure," we assume that the critical failure rate of the host OS and application and platform software is roughly the same for virtualized and native deployments. Virtualization introduces a hypervisor software, and perhaps a host OS, that is not present in native deployments; this complex software will inevitably contribute some failures, but the maturity and massive deployment of this software suggests that this incremental software failure rate attributable to the hypervisor and host OS should be significantly less than from the application software. Chapter 6, "Hardware Reliability, Virtualization, and Service Availability," considers hardware reliability in detail and observes that hardware failure rates are likely to be equivalent for both traditional and virtualized application deployments.

5.6.4 Failure Coverage

Failure coverage is the probability that the system will rapidly detect and correctly isolate a critical failure to the appropriate recoverable module. Failure coverage is driven by the efficacy of failure detection, which is detailed in Section 3.6,."Redundancy and High Availability." Software failure coverage for application software, software platform, and guest OS should be the same in native and virtualized deployment. Hardware coverage is primarily the responsibility of the "guest" OS and the application platform in native deployments, and of the host OS and hypervisor in virtualized deployments. Mature and high-quality OS, platform, and hypervisor software should achieve comparable levels of hardware failure coverage.

Rapid and reliable detection of failures is the first step in the activation of high availability mechanisms. Software failures of guest OS, platform, or application software in a virtualized deployment should be detectable via the same mechanisms used for native deployments; however, there may be an increased latency in fault detection and recovery based on resource sharing efficacy. While the hypervisor might explicitly see software-triggered processor exceptions like divide by zero events, these exceptions should be caught and addressed by the guest OS, platform, or application software, so there should be no material difference in the efficacy of software failure detection.

5.6.5 Failure Detection Latency

Failure detection latency is the time it nominally takes the system to automatically detect and correctly isolate a critical failure. Nominal failure detection latency is often related to the failure detection strategy and mechanism. Some failures will be detected synchronously and automatically, such as when the OS raises a processor exception for access to an illegal memory address or throws a signal on death of child process. Other failures are detected asynchronously, such as during periodic integrity scans of data structures, like shared memory. Failure detection latency is influenced by configured parameters, like heartbeat time outs, maximum retry counts, frequency of integrity audits, and so on. Assuming that similar or identical detection-related settings are configured on virtualized deployments as on native deployments, failure detection latency should be similar for both.

5.6.6 Switchover Latency

Switchover latency is the time it takes for the system to recover service onto the redundant unit and is driven by the redundancy strategy used to mitigate the failure. Strategies where the redundant unit is active naturally have shorter switchover latencies because the redundant unit need not take time to bring itself to full operational readiness. Hot standby has shorter switchover latency than warm standby; warm standby has shorter switchover latency than cold standby. Additional latency might be consumed if it is necessary to activate a paused or suspended VM instance, or if VM instances must be created and initialized to complete the recovery.

5.6.7 Switchover Success Probability

Switchover success probability is the probability that an automatic switchover will successfully restore service in the nominal automatic switchover latency. Note that in field deployments (and hopefully in lab testing), automatic switchovers are initiated following a critical failure that has driven part of the system into an ill-defined or undefined state. Thus, automatic switchover must be fast and reliable even when it is recovering service that may have been left in a messy state. Switchover success probability is influenced by the architecture, implementation, and testing of the redundancy strategy and recovery model. Activating paused or suspended VM instances increases the risk of switchover failing by adding a level of complexity to the action.

5.6.8 Modeling and "Fast Failure"

Common software engineering guidance is to "fail fast," meaning that it is often better to initiate automatic recovery actions when one is fairly sure something has failed (e.g., "the preponderance of evidence") rather than waiting longer until one is absolutely sure (e.g., "beyond a shadow of doubt"). After all, a false positive (triggering recovery when system had not actually failed) is generally preferable to a false negative (silent or sleeping failure in which the system is unavailable but no action recovery action is taken because there is no indication there is a failure). In fact, the bulk of predicted downtime comes from these false negative situations, since the failure is not recognized,

TABLE 5.3. Comparison of Nominal Software Availability Parameters

Parameter	How Does Value Nominally Compare for Virtualized and Native Deployments?	Comments
Software failure rate	Same to slightly better	Virtualization should assure a narrower operational profile for production software, thereby reducing the risk of residual defects
Software failure coverage	Same	Same failure detection mechanisms are used in both native and virtualized environments.
Failure detection latency	Same	Same failure detection mechanisms are used in both native and virtualized environments.
Switchover latency	Different	Switchover latency depends on characteristics of selected traditional or virtualized recovery strategy.
Switchover success probability	Different	Switchover success probability depends on characteristics of selected traditional or virtualized recovery strategy.

and thus downtime—and user dissatisfaction—accrues until extraordinary actions (e.g., angry calls from end users) alert the maintenance engineers to the failure so recovery actions can be initiated.

This common-sense advice is modeled via the combination of failure detection latency and failure coverage that together represent the portion of failures that are successfully detected within the nominal detection latency. While it is infeasible to detect 100% of all failures in 0 milliseconds, the "fail fast" guidance reminds engineers to strive for that goal.

5.6.9 Comparison of Native and Virtualized Deployments

Table 5.3 gives a side-by-side comparison of how key reliability parameters change when application software is deployed on a virtualized platform compared with a native deployment. Software failure rate should be no worse for the virtualized platform, and may be slightly better. Software failure detection coverage and failure detection latency should be essentially the same for both deployment scenarios because the same mechanisms are used in both configurations. Switchover latency and switchover success probability are where availability predictions may differ because different virtualized and native recovery strategies can have significantly different performance characteristics.

Thus, virtualization permits new cost-effective redundancy options that enable applications to achieve service availability of traditional architecture and deployment options.

6

HARDWARE RELIABILITY, VIRTUALIZATION, AND SERVICE AVAILABILITY

This chapter assesses how hardware reliability considerations change for applications deployed on a virtualized platform compared with the same application natively deployed on hardware (i.e., no hypervisor). The chapter begins by reviewing hardware downtime expectations and the basics of hardware failures. The chapter then considers the impact of virtualization, especially the server consolidation use case, on hardware failure rates. We then consider limitations on containment and the risk of a cascade of hardware failures in virtualized configurations. Next, a review of the fundamental recovery strategies that can be used to mitigate hardware failure events is presented. The chapter concludes with a discussion summarizing how hardware failures of virtualized platforms impact service availability.

6.1 HARDWARE DOWNTIME EXPECTATIONS

Traditionally, service availability expectations of applications (e.g., "five 9's") took into account downtime attributed to hardware, software and procedural (a.k.a., human) failures. While virtualization decouples application software from the underlying hardware, it does not eliminate the hardware attributed downtime; hardware still fails, and

Reliability and Availability of Cloud Computing, First Edition. Eric Bauer and Randee Adams.

automatic or manual actions must be taken to recover service following failure. To achieve a hardware downtime target, the hardware failure rate must be sufficiently low, and the probability of rapid failure detection and successful service recovery must be sufficiently high so that the long-term, annualized, prorated service downtime across a population of system attributed to hardware does not exceed the target value. As explained in Chapter 5, "Reliability Analysis of Virtualization," mathematical modeling combines quantitative estimates of key parameters to estimate the feasible and likely service downtime.

6.2 HARDWARE FAILURES

Hardware is susceptible to a variety of failure modes, including:

- Random failures from manufacturing defects, such as poor solder joints;
- Time- and temperature-dependent (aka, wear out) failures, such as electro migration that dissolves metal connections into surrounding silicon or dielectric breakdown of gate oxide, or which causes the breakdown or loss of physical properties over time or use.
- Corrosion from gases like H_2S and H_2SO_4.
- Hydrophilic dust that accumulates on hardware components and assemblies, absorbs water and electrically shorts pins.
- Soft (i.e., nonpersistent) bit errors from cosmic rays or alpha particles.
- Electrical or thermal overstress.
- Damage during shipping.

Reliability qualification, electrical and thermal derating of components, robust design-for-manufacturing guidelines, highly accelerated life or design environmental stress testing, diligent factory testing, and other techniques should minimize the rate of hardware faults throughout the hardware's designed service life. Nevertheless, hardware failures will occur, and thus systems must rapidly detect and isolate hardware failures so that system software can activate appropriate recovery actions.

Practical hardware error scenarios to consider are:

- *Processor Failure.* Complex and highly integrated devices like microprocessors, digital signal processors, network processors, field programmable gate arrays, and so on are critical to hardware functionality and are often more susceptible to wear out due to environmental-related effects. For example, weak thermal design and elevated ambient temperatures can cause a high-performance processor to run with excessive junction temperatures, thus accelerating time- and temperature-dependent failure modes, which lead to premature failure.
- *Disk Failure.* Hard disk drives are built around high-performance spinning platters and moving magnetic heads. Over time moving parts (e.g., lubricated bearings) will wear and eventually fail. Although hard disks may have low random

failure rates during their designed service life, their service lifetime is often shorter than the designed lifetime of the system electronics, and thus hard disks may fail and require replacement before the system's electronics has reached the end of its useful service life.

- *Power Converter Failure.* Board-mounted power modules are used to convert voltages provided on the system's backplane to the voltages required by devices on the board itself. As these compact devices inherently dissipate high power and run hot, they tend to have appreciable hardware failure rates; failure of a power converter impacts power delivery and thus renders impacted hardware inoperable.
- *Clock Failure.* Oscillators drive the clocks that are the heartbeat of digital systems. Clock failure will impact (and likely disable) the circuitry served by the clock.
- *Clock Jitter.* In addition to hard (persistent) clock failures, the clock signal produced by an oscillator can jitter or drift. Clocks tend to drift as they age for a variety of reasons, including mechanical changes to crystal connections or movement of debris onto crystal. This jitter or drift can cause circuitry served by one oscillator to lose synchronization with circuitry served by another oscillator, thus causing timing or communications problems between the circuits.
- *Switching/Ethernet Failure.* These devices enable IP traffic to enter and leave the hardware unit, and thus are critical.
- *Memory Device Failure.* Memory devices are typically built with the smallest supported manufacturing line geometries to achieve the highest storage densities. In addition, many systems deploy large numbers of memory devices to support large and complex system software. Dynamic RAM is susceptible to soft bit errors; FLASH memory devices wear out with high write voltages, and over long time periods can lose data.
- *Parallel or Serial Bus Failure.* High-speed parallel and serial busses are very sensitive to electrical factors like capacitance and are vulnerable to crosstalk.
- *Transient Failure or Signal Integrity Issue.* Weak electrical design or circuit layout can lead to stray transient signals, crosstalk, and other impairments of electrical signals. As these issues are transient rather than persistent, they are often difficult to debug.
- *Application-Specific Component Failure.* Application-specific components like optical or radio frequency devices may be more failure prone because of small device geometries, high power densities, and newness of technology or manufacturing process. Components like fans, aluminum electrolytic capacitors, and batteries are also subject to wear out.

All of these fundamental error scenarios are applicable to hardware regardless of whether applications are executing natively on the hardware or if a hypervisor is virtualizing the application's access to the hardware. Thus, the key hardware downtime questions to consider are:

1. Does virtualization impact the rate of hardware errors/failures?
2. Does virtualization impact the latency or effectiveness of hardware failure detection?
3. Does virtualization impact containment of hardware failures?
4. Does virtualization impact the latency and effectiveness of service recovery from hardware failures?
5. Can virtualization itself mitigate the impact of hardware failures?

These questions are addressed in the remainder of this chapter.

6.3 HARDWARE FAILURE RATE

Hardware failure intensities, or rates, follow the so-called "bathtub" curve, which features three phases:

- *infant mortality phase* when weak units fail or due to manufacturing defects;
- *useful service life phase* when random hardware failures occur at a fairly constant rate that is nominally below the predicted hardware failure rate; and
- *wear-out phase* in which failure rates increase until all units in the population eventually fail.

Given this "bathtub" behavior, the hardware failure rate questions are:

1. Are virtualization and cloud operational characteristics likely to shorten the useful service life time of hardware elements and cause wear out failures to begin prematurely?
2. Are virtualization and cloud-related factors likely to increase the random hardware failure rate during the useful service life of hardware elements?

These two questions are visualized in Figure 6.1.

Failures are primarily driven by thermal, voltage, and current stress, as well as mechanical vibration. For example, components that run hotter (e.g., semiconductor devices with higher junction temperatures) tend to wear out faster than devices operated at lower junction temperatures. Hardware failure rates for virtualized and cloud deployments may be somewhat higher, and useful service life somewhat shorter than for traditional, native deployments because of increased stress on hardware components due to:

1. *Increased Hardware Utilization Reduces the Time Components Engage Power Management Mechanisms to Reduce Thermal Stresses.* Server consolidation and increased hardware resource utilization are primary motivations for deploying virtualization, and increased utilization can both increase the duty cycle of hardware components and reduce the opportunities to engage power

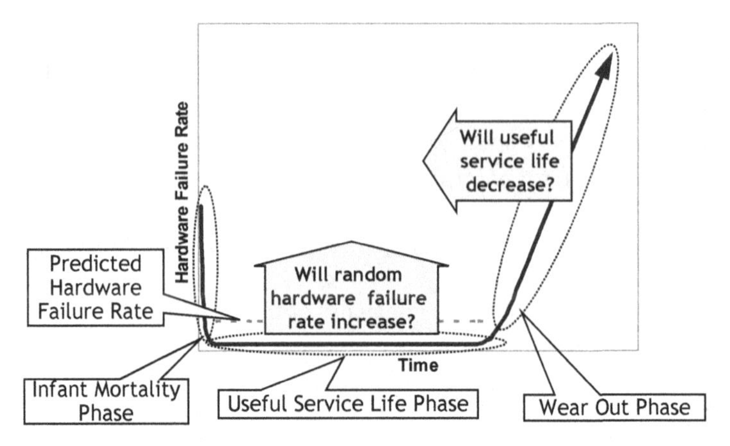

Figure 6.1. Hardware Failure Rate Questions.

management technologies that can reduce electrical and thermal stress on components. For example, some modern processors include advanced power management that slows clock speeds to reduce power consumption—and heat generation, hence thermal stress—when the processor is lightly loaded. While traditional servers are often very lightly loaded so that power management mechanisms can engage to reduce thermal stress on components, server consolidation and cloud deployments are likely to keep systems at much higher workloads to minimize capital expense and operating expense, thereby making it less likely that power management will activate; thus devices are more likely to endure higher thermal stress. Therefore, higher hardware resource utilization via server consolidation may both increase the hardware failure rate during the useful life period, as well as accelerate the onset of the wear-out phase, which reduces the useful service life of hardware elements.

2. *Increased Thermal Stress Due to Elevated Ambient Temperatures in Data Centers.* To reduce operating expense for power, data center operators—including cloud service providers—may increase the ambient temperature in their data centers, which reduces their cooling load and hence power consumption. Higher ambient temperatures increase thermal stress on hardware components, regardless of whether they are virtualized or traditional systems.
3. *Reduced Derating Rules and Design Margins by Hardware Suppliers to Reduce Costs.* Virtualization's ability to decouple software from the underlying hardware combined with the massive scale and homogeneity are common characteristics that tempt many traditional and cloud computing data center owners to deploy less expensive commodity hardware. Performance can sometimes be boosted (or component costs reduced) by reducing design margins, such as operating components closer to or at their maximum rated values for

temperature, voltage, and current. Commodity hardware suppliers may adopt less conservative derating and design rules, driving components closer to their design limits to reduce component costs and/or to boost performance, and thereby increasing stress on components.

4. *Increased Duty Cycle on Hard Disk Drives and Other Components.* Server consolidation in particular and cloud computing in general strive to increase hardware utilization rates by squeezing the maximum value out of the capital investment in hardware resources. Increased duty cycle can accelerate aging of hardware components like hard disk drives. Higher usage of hard disk drives can increase vibration and hence mechanical stress.

Interestingly, a study by Microsoft of hardware reliability in their cloud computing data centers [Vishwanath] reported the surprising observation that "*the age of the server, the configuration of the server, the location of the server within a rack [owing to temperature/humidity gradient within rack we might have expected different failure characteristics], workload run on the machine, none of these were found to be a significant indicator of failures.*" Thus, cloud deployment might not impact hardware failure rates as much as one might expect, but further research is necessary.

6.4 HARDWARE FAILURE DETECTION

Hardware failures are detected either synchronously when the resource is accessed during normal operation (e.g., hardware component returns an error code or raises an error interrupt) or asynchronously during execution of a periodic hardware audit routine. Since the hypervisor and host operating system (OS) (if present) should have access to the same hardware visibility mechanisms and essentially the same software drivers as the guest OS, hardware error and failure detection by the virtualized platform should theoretically be comparably effective to native detection by the native OS in nonvirtualized system configurations.

Theoretically, virtualized platforms might have more effective hardware failure detection capabilities because a hardware failure can be detected when executing any one particular virtual machine (VM) instance running on the physical hardware, and thus the hardware failure may be known to the virtualization platform before other VMs are exposed to the hardware failure, thereby creating opportunities for proactive hardware failure detection. The practical question then becomes: if the guest OS, platform or application software running in a VM instance detects a hardware failure, then is there a mechanism to signal the hypervisor of the hardware failure so the hypervisor can initiate failure mitigation actions?

Note that since virtualization introduces an additional layer of platform software (i.e., the hypervisor), real-time notification of hardware failure may be slightly slower than in native configurations, especially if multiple VM instances are affected and need to be alerted.

6.5 HARDWARE FAILURE CONTAINMENT

Good hardware design assures that hardware failure of one field replaceable unit should not cause hardware failure of another hardware unit. For example, the failure of a processor or electronic component on one rack-mounted server or compute blade should not cascade to cause adjacent or nonadjacent server or blade hardware to fail. However, the fundamental nature of the server consolidation use case means that by default, a single hardware failure will be presented to every VM instance that runs on the hardware until the hardware failure is repaired. Thus, although good hardware design assures that the hardware failure will not cascade to other hardware elements, virtualization inherently presents the risk that a hardware failure will be impressed upon all active VM instances associated with the failed hardware until the hardware failure is mitigated and eventually repaired. Note that this is also an issue with nonvirtualized multitenancy architectures.

6.6 HARDWARE FAILURE MITIGATION

Hardware failures are traditionally mitigated by either switching service from the failed hardware resource to a redundant hardware resource or by replacing or repairing the failed hardware and restarting the software. To maximize service availability a redundant application instance will be online on a redundant hardware resource and will be ready to serve users immediately (e.g., active/active or active/hot standby), but other traditional (e.g., active/warm standby and active/cold standby) and virtualization-related redundancy arrangements (discussed in Section 5.4, "Recovery Models") are possible.

To mitigate further service impact due to the hardware failure, once the virtualization platform/hypervisor detects the underlying hardware failure, it should stop accepting requests to create new VM instances on the failed hardware. Fundamentally, preexisting VM instances associated with hardware that is believed to have failed but which have not yet experienced the failure can be addressed via one of the following strategies:

- *Virtualized platform mitigates hardware failure.* It is theoretically possible for the hypervisor, host OS, or other components supporting VM instances (e.g., RAID) to detect and mitigate some hardware failures, thus masking the underlying hardware failure from the VM instances. For example, it is possible to implement LAN bonding across a pair of physical NICs in the host OS so that the failure of an individual NIC need not be exposed to VM instances across the virtualized NIC interface. This is discussed further in Section 6.7, "Mitigating Hardware Failures via Virtualization."
- *VM instances run to the point of failure* (typically the default behavior). The hardware failure is implicitly or explicitly presented directly to running VM instances where the guest OS, platform, and application software is fully exposed to the hardware failure. At some point, the VM instance software will probably

experience the hardware failure, and service will be impacted. After the VM instance fails, the virtualization platform should start a new VM instance on fully operational hardware.

- *Hypervisor stops vm instances to (implicitly) activate application-level high availability mechanisms.* If the hypervisor detects the hardware failure or is notified of the hardware failure by a VM through an API, then it can decide not to risk running the VM instances to the point of failure. Instead, the hypervisor can stop (e.g., pause) VM instances tied to impacted hardware, thereby mimicking a catastrophic software failure. This action should cause each application's HA mechanism (if implemented) to activate and rapidly recover application service to a redundant application VM instance.
- *Hypervisor stops VM instances and activates virtualization platform/hypervisor high availability mechanisms.* If the hypervisor detects the hardware failure, then it can elect not to risk running the VM instances to the point of failure, and destroy the VM instances associated with the impacted hardware, create new VM instances on fully operational hardware, and boot the applications into the new VM instances.
- *Hypervisor live (online) migrates VM instances* to other virtualized platform hardware instances. For each VM instance, if the hypervisor (1) detects the hardware failure, (2) deems that a VM instance has not been compromised by the failure event, and (3) the hardware is sufficiently operational, then the hypervisor can theoretically attempt live migration of the VM instance to fully operational hardware. This technique presents four fundamental risks:
    1. *Availability of hypervisor may be compromised.* The hardware failure may have compromised the hypervisor's state and/or its ability to correctly execute any recovery actions.
    2. *VM instance may have already been compromised*, so migration risks prolonging the period of service impact since cascaded software failures will be separately detected, isolated, and recovered, inevitably extending the period of service impact. In addition to the service disruption period experienced during live migration, the compromised VM instance image will likely eventually run to failure on the other hypervisor. Thus, it would have been easier and faster to restart the application instance promptly once the hardware is operational rather than migrating the damaged software and then having to restart the application once the failure has occurred after the migration.
    3. *Live migration may be so slow that application-level HA mechanisms activate*, and thus the migrated VM instance will end up fighting with the redundant application instance(s) that are attempting to take over the service for the impacted VM instance. Competing application instances increase the risk of slower or unsuccessful service recovery.
    4. *Live migration of VM instance may be unsuccessful*, and thus the opportunity costs of time and resources of the attempted live migration are wasted. The virtualized platform/hypervisor will still have to create a new VM instance

on fully operational hardware, and application-level HA mechanisms will activate, if available.

Thus, live migration is not generally a feasible option to mitigate hardware failure.

6.7 MITIGATING HARDWARE FAILURES VIA VIRTUALIZATION

The virtualization layer of software (the hypervisor plus the host OS, if used) decouples the VM instance from the physical hardware; this section considers how this layer of software can mitigate the impact of hardware failures on VM instances. The virtualization hypervisor exposes virtual CPU, virtual memory, virtual storage, and virtual network interface cards (NICs) to guest OS, software platform, and application instances. One could even draw an application-centric RBD that explicitly includes these virtualized devices, as in Figure 6.2.

Note that the virtualized application, platform, and guest OS software, and enterprise IS/IT may have a completely different perspective of virtualized devices because device and performance monitoring, redundancy, and high availability mechanisms can be hidden beneath the virtualized device interface that is exposed to guest OSs. For example, one can easily imagine how a virtualization platform would be configured to map virtualized storage operations onto a high availability RAID configuration to offer applications higher availability storage than they might expect from traditional deployments. The specific management of CPU, memory, storage, and network resources is detailed in the next sections.

6.7.1 Virtual CPU

A virtual CPU represents the abstraction of the available physical CPUs or processor cores. VM instances are configured with one or more virtual CPUs. The hypervisor is

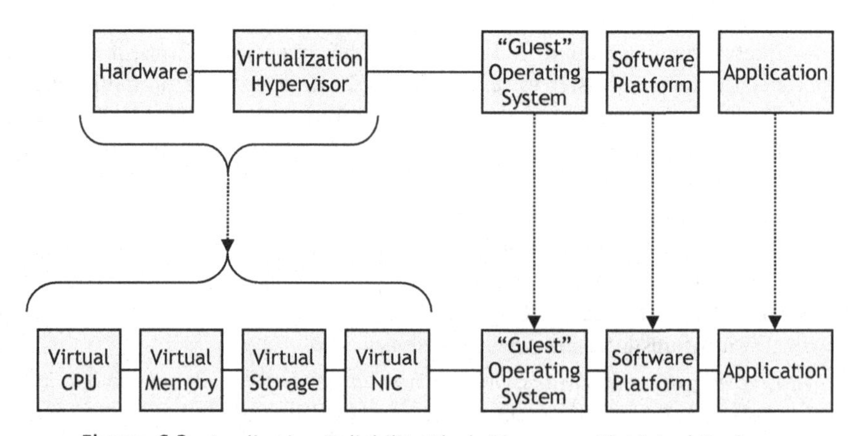

Figure 6.2. Application Reliability Block Diagram with Virtual Devices.

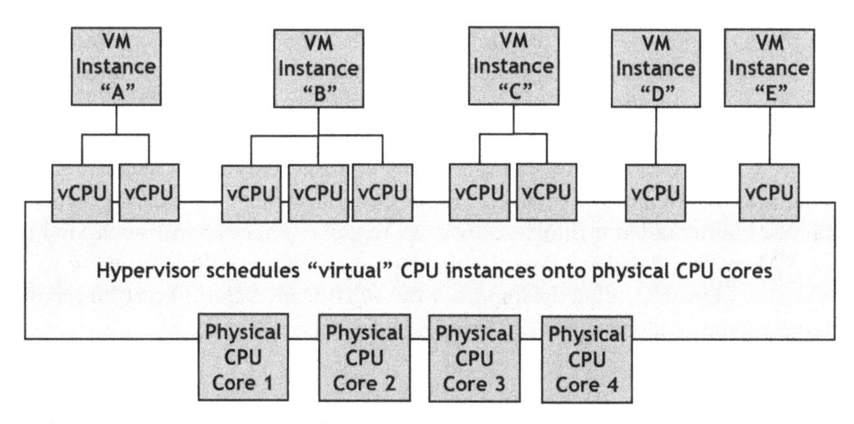

Figure 6.3. Virtual CPU.

responsible for managing the allocation of the physical processing core resources to virtual CPUs to meet the needs of all of the VM instances within its control. The benefit of virtual CPU is the ability to share and more fully utilize the physical CPU resources by allocating them to multiple VMs based on their requirements. Figure 6.3 depicts the assignment of one or more virtual CPUs to each VM instance and the hypervisor in turn mapping the virtual CPUs to physical cores.

Virtual CPUs could theoretically be used by the hypervisor to mitigate the impact of a single physical CPU failure if it is able to detect a failure of a physical CPU, quarantine the failed physical CPU, reallocate physical CPU resources from the other nonimpacted physical CPUs to the affected virtual CPUs, and restart the VMs. Since the VMs are unaware of the physical CPU resources, the hypervisor could be used to provide a quicker recovery of the impacted VM's. The following risks exist when attempting to mitigate the impact of a single physical CPU failure:

1. The CPU failure may have compromised the hypervisor, host OS, or VM instance, so it is no longer capable of executing properly, and thus must be restarted.
2. Sometimes, it is not possible to recover a single physical CPU; all of the physical CPUs on the server may have to be recovered as well. In this case, all impacted VMs will have to be migrated to another server before the recovery is attempted.
3. Failure of a physical CPU could escalate into a more serious OS problem that cannot be resolved by just quarantining the CPU.
4. The remaining nonfailed physical CPUs may not have enough resources to allocate to the VMs.

6.7.2 Virtual Memory

Virtual memory management by the hypervisor provides a means of efficiently allocating and deallocating memory to the VMs in a way that masks the fact that they are sharing physical memory. This maximizes the utilization of the memory and helps to ensure that each VM has enough memory to meet its needs yet not interfere with memory access by another VM.

The hypervisor could use various techniques for virtual memory management, such as memory compression and swapping memory to disk, to avoid shortages, thus mitigating failures due to memory shortage. In the case of memory failures, the hypervisor could block out the failed memory sectors and remap memory allocation around those sectors. VM instances that are directly impacted by memory failure events (e.g., their active memory image is directly impacted by a hardware failure) must be recovered via high availability mechanisms.

6.7.3 Virtual Storage

Virtual storage abstracts the logical storage from physical storage. The virtualized storage may be local to the processing resources (e.g., collocated on the processing blade or in the rack mounted server) or may be networked such as via a storage area network (SAN) or network-attached storage (NAS). Storage virtualization can be classified into two general types:

- *Block virtualization* can be managed at the server level, storage device level, or network level. At the server level, a local volume manager can intercept all attempts to access the disk and provide the proper mapping to the physical resources. At the storage device level, a software controller (such as used by RAID mechanisms) manages access and replication across the disk arrays. At the network level, the SAN provides the mapping between the applications' storage requests and the storage controllers managing the physical resources. The Internet Small Computer System Interface (iSCSI) standard supports the transport of data across the IP network. Network File System protocol supports access to files on storage devices across the network.
- *File virtualization* provides a mapping of access requests to the actual directory and file level in order to mask the physical resources from the applications. This provides more flexibility for the storage and management of the files. Logical storage can be abstracted from those physical storage pools.

Storage virtualization offers a means of moving or redirecting file access to a different device due to a failure or exhaustion of existing resources with no user impact. Multiple layers of virtualization may be supported. Mapping of virtual to physical resources can be implemented using mapping tables (sometimes referred to as meta-data) or more dynamically via algorithms that calculate the location. Providing multiple paths to the storage resources with failover capabilities can be configured to provide additional robustness if one of the paths is unavailable.

Readers will be familiar with the well-known ability of properly configured RAID configurations to successfully detect and mitigate hard disk failures. Mapping virtualized storage onto high availability RAID storage can mask hard disk failures from virtualized applications. Failure containment depends upon the type of underlying storage mechanisms that are used and the configuration of the paths to the storage, but storage virtualization can provide robust access to the data across failures of individual devices or migration to different storage devices. For example, Amazon Web Services reports that their fault tolerant Elastic Block Store (EBS) reduces the 4% annual failure rate of the commodity hard disk drives that underlie EBS to a 0.1–0.5% annual failure rate observed by EBS users [AWSFT].

6.8 VIRTUALIZED NETWORKS

Network virtualization entails a combination of hardware and software network resources and network functionality providing a single, software-based administrative entity, that is, a virtual network. Networking is virtualized at several levels:

- virtual network interface cards
- virtual local area networks;
- virtual IP addresses; and
- virtual private networks.

Each of these is considered separately.

6.8.1 Virtual Network Interface Cards

A network interface card (NIC) is a hardware component that connects the host computer to the external network. A virtual NIC provides an abstraction of that physical component to a user residing on the host (i.e., guest OS) by mapping to a physical NIC or to a virtual network. In the case of a virtual network, the network may be contained within the server such as between the coresident VMs. In the case of an internal virtualized network a virtual network interface card (VNIC) is a type of interface managed by the hypervisor to provide communication between the VMs within its control. VMs on the same host can share resources and exchange data using the VNIC and virtual switch without needing to use the external network.

A server can have multiple physical NICs. Each physical NIC can be partitioned into several virtual NICs. The virtual NICs can then be assigned to the VMs residing on the server. Figure 6.4 indicates the flow of packets from an external LAN to the appropriate VM by way of the physical NIC, hypervisor, to the configured VNIC on that VM. VMs communicate internally (within the server) and externally via the virtual NICs. In this way, the hypervisor takes care of managing the network I/O activities for their VMs. To improve service availability, VMs can be configured to multiple physical NIC's via their Virtual NICs.

Failures associated with a particular physical NIC can be mitigated using bonding (sometimes referred to as NIC teaming) to aggregate links associated with multiple physical NICs and mapping this bonded interface to a virtual NIC. If there is a failure of one of the physical NICs traffic to and from the VMs will be moved to the other physical NIC. Risks associated with this bonding mechanism include:

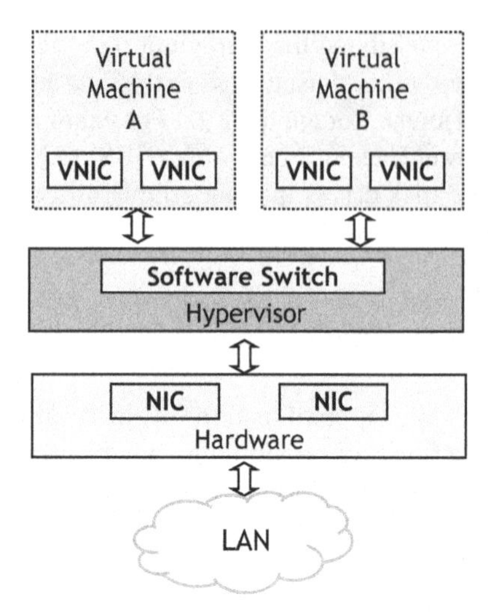

Figure 6.4. Virtual NIC.

1. The virtualization platform must support this bonding mechanism, as well as mechanisms to detect network interface failures and failover/failback to another network connection within its aggregated set. Proper configuration is required for the bonding of the physical NICs and the mapping to the Virtual NICs configured to the VMs.
2. If the physical NICs are attached to the same physical networking IC, a failure of the IC will represent a single point of failure and impact both NICs.
3. The host OS is often bypassed for performance reasons so bonding may not occur.

6.8.2 Virtual Local Area Networks

Virtual local area networks (VLANs) provide a means of grouping together a collection of nodes into a single broadcast domain regardless of whether they are in the same physical location. LANs are thus configured with software rather than with physical cables. One of the purposes of VLANs is to provide network separation (e.g., separation of network management traffic from user traffic). Even if multiple VLANs share an Ethernet switch, they cannot communicate with each other directly on that switch; a router would be required for communication between the VLANs. A VNIC can only be associated with a single VLAN. Multiple VLANs can be instantiated to provide network robustness. Each VLAN can be configured to a different physical switch. Each physical NIC can be connected to multiple physical switches using separate ports. The physical NICs can then be connected to a virtual switch for communication with the VMs. Virtual LANs are not specific to virtualized environments, and are included here for completeness, but will not be analyzed further in this book.

6.8.3 Virtual IP Addresses

A virtual IP address (VIP) is an IP address that is not associated with a particular physical network interface. It can be mapped to physical network interfaces on multiple servers or associated with multiple domain names. Many redundancy mechanisms make use of VIP addresses so that interfaces to a particular server or component only have to know one IP address. In an active/standby redundancy situation, the active component assumes the virtual IP address so that all traffic directed to that VIP is managed by the active component. If the active component fails, then the standby component will assume the virtual IP address as part of its activation procedure and will then receive all of the traffic directed to that VIP. Virtual IP addresses are mentioned here for completeness since they provide a means to mitigate the impact of a server failure by supporting the activation of a redundant mate without interfacing components having to keep track of that change. However, since VIPs are a commonly used mechanism implemented on nonvirtualized platforms as well as virtualized platforms, they will not be analyzed further in this book.

6.8.4 Virtual Private Networks

The main purpose of a virtual private network (VPN) is to provide a secure, reliable connection through encryption to a private local area network even via a remote public network. VPNs were created to save costs by remotely connecting a private Intranet by eliminating the need to lease physical facilities. VPNs are not particular to cloud-based solutions and are thus outside the scope of this book.

6.9 MTTR OF VIRTUALIZED HARDWARE

While highly available systems can mitigate hardware failures by switching service to a redundant hardware unit, hardware failures of simplex systems render service unavailable until the underlying hardware can be repaired. Hardware mean time to repair (MTTR) estimates the time required to troubleshoot the hardware failure, complete the hardware repair, and return the system instance to normal operation. For simplex (non-redundant) systems, MTTR is the same as the mean time to restore service (MTTRS); for systems with redundancy, the MTTRS is far shorter than the MTTR, an underlying hardware failure.

Since virtualization decouples application software from the underlying hardware, hardware failures of virtualized platforms can effectively be mitigated by promptly recovering application software to another hypervisor. This enables MTTRS for hardware failures for even simplex (nonredundant) systems to be decoupled from physical MTTR provided another server is available to support the application. Equation 6.1 (repeat of Equation 3.4, for convenience) gives the simple linkage between failure rate (the mathematical reciprocal of MTBF), repair time (MTTR), and service availability.

TABLE 6.1. Example of Hardware Availability as a Function of MTTR/MTTRS

Predicted Hardware Availability as a Function of MTTR/MTTRS			
Hardware MTBF (Hours) = 100,000			
MTTR/MTTRS (Hours)	MTTR/MTTRS (Minutes)	Availability (%)	Annualized Down Minutes
0.1	6	99.9999	0.5
0.25	15	99.9998	1.3
0.5	30	99.9995	2.6
1	60	99.9990	5.3
2	120	99.9980	10.5
4	240	99.9960	21.0
8	480	99.9920	42.1
24	1440	99.9760	126.2
48	2880	99.9520	252.3

$$\text{Availability} = \frac{\text{MTBF}}{\text{MTBF} + \text{MTTR}}.$$

Equation 6.1. Availability as a Function of MTBF/MTTR

Table 6.1 solves this equation for hardware for a canonical hardware MTBF of 100,000 hours and a range of MTTR values from 6 minutes to 2 days; results are expressed both as an availability percentage as well as annualized downtime minutes. Note that this estimate considers only hardware-attributed failures covered by the 100,000 hour MTBF estimate; estimating system downtime would require one to consider software attributed failures and downtime as well.

Table 6.1 shows that with the assumed hardware MTBF and an aggressive native hardware repair time assumption of 4 hours, over the long term, the system would accrue 21 annualized minutes of hardware-attributed service downtime; less aggressive hardware repair time assumptions would accrue more downtime. Assuming reserve virtualized resources are available (i.e., online or near line) and appropriate operational policies are in place, it might be reasonable to assume that the virtualization platform and/or data center staff could restart an application on an alternate hypervisor following hardware failure in minutes. If the hardware failure rate is constant at 100,000 hour MTBF, then a 30-minute (0.5-hour) MTTR yields less than 3 minutes of annualized hardware-attributable downtime compared with more than 20 minutes of annualized downtime for 4 hour MTTR values. Therefore, when robust data center operational policies are coupled with virtualization, the hardware attributed service downtime of simplex applications can be dramatically reduced. If software is recovered by activating snapshots of VM instances of partially or fully booted applications, then software recovery time (and presumably hardware recovery time, as well) can be reduced compared with native deployment, further boosting service availability of simplex system configurations.

6.10 DISCUSSION

Although the server consolidation and cloud use cases of virtualization may increase hardware failure rates somewhat compared with traditional hardware use scenarios, virtualization can be used to mitigate some of the impact of inevitable hardware failures. Virtualization also offers the potential of drastically reducing the effective hardware MTTR for standalone systems to dramatically reduce the hardware attributed downtime of simplex (nonredundant) applications. If the hardware failure is not explicitly detected and/or mitigated by the host OS, hypervisor, or guest OS, then the application software and/or platform must be prepared to detect and recover from uncovered hardware failure.

7

CAPACITY AND ELASTICITY

Rapid elasticity is an essential characteristic of cloud computing that is radically different from both traditional deployment models and from redundancy. Redundancy is designed to rapidly provide resources to recover the prefailure service capacity, and redundancy is typically expected to recover the impacted service load in seconds or minutes. Elasticity is designed to increase (or decrease) the capacity available to serve offered load, and elasticity is typically expected to alter capacity in hours, rather than weeks or months for traditional deployments.

This chapter begins by reviewing system load basics, overload, and traditional capacity planning, and then discusses how rapid elasticity in cloud computing changes traditional capacity planning assumptions. Capacity-related service risks, as well as security risks, are discussed.

7.1 SYSTEM LOAD BASICS

Many applications have usage patterns that vary based on hour of the day, day of the week, and time of year. Figure 7.1 illustrates the day/night usage pattern of a sample application with most usage during business and evening hours, and light usage when

Reliability and Availability of Cloud Computing, First Edition. Eric Bauer and Randee Adams.

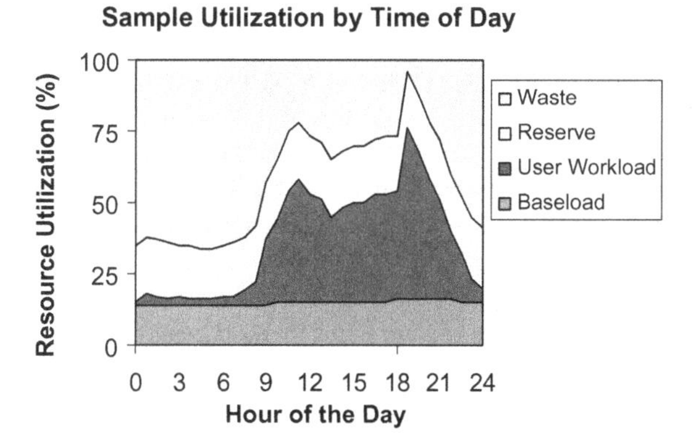

Figure 7.1. Sample Application Resource Utilization by Time of Day.

most people are sleeping. Many applications show day of the week patterns, such as heavier usage on either weekdays or weekends. Some applications exhibit other seasonality patterns, such as retailing applications experiencing heavy volumes before Christmas, and financial applications experiencing peak usage when quarterly and annual financial results are prepared. For traditional applications to meet these expected peak loads, the enterprise would have to purchase sufficient resource to serve that projected peak load, as well as some reserve capacity to mitigate failures that might occur during peak usage periods. When offered load is below the peak engineered load, the excess resource capacity (beyond necessary reserve online capacity) is unused and hence wasted.

Deeper examination of the Figure 7.1 example reveals insights into how resources are actually used. There is a constant base processing load to support application monitoring, management, visibility, and controllability. The resource utilization to serve user traffic is highly variable, with minimal traffic when most people are sleeping and peak at the end of the business day. The figure clearly shows how the system explicitly maintains reserve (or redundant) capacity to rapidly recover user service on failure of a component actively serving users. Note that sufficient reserve capacity is maintained to mitigate the failure of a component serving baseload even when offered user workload is very light, such as in the middle of the night. The figure also clearly shows that capacity not required for base processing, user workload, or reserve capacity is wasted in this traditional configuration. Statically sizing resource capacity to serve the peak load in the early evening means that significant resource capacity is unused (i.e., wasted) by this application in the middle of the night.

Server consolidation with complementary applications is one way to increase resource utilization. For example, one could imagine running batch jobs—like processing usage records to generate bills for customers—on the platform from midnight to 6 a.m. local time to utilize some of the resource capacity that would otherwise be wasted. The resource pooling essential characteristic of cloud computing, coupled with virtualization, can enable this intelligent resource sharing.

7.1.1 Extraordinary Event Considerations

The offered service load on some applications is highly correlated with natural disasters, concert ticket sales, reality show voting, or some hard-to-predict events of regional, national, or commercial significance. For example, within minutes of an earthquake or terrorist attack, there is likely to be a spike in traffic load both related to emergency response by governmental and to other organizations involved or impacted by the event. As news of the event reaches the general population, then there may be a traffic spike as citizens seek to assure that their family and friends are ok and learn more about the event or as emergency responders send updates on their progress. The 1989 Loma Prieta earthquake [LomaPrieta] was unexpectedly broadcast live to a national audience watching game 3 of baseball's 1989 World Series, and this prompted many to call family and friends in the San Francisco Bay Area. Figure 7.2 gives an example of a traffic spike due to an extraordinary event. One can see normal daily and weekly traffic patterns with an extraordinary event causing traffic to spike far above normal traffic volumes. Obviously, enterprises strive to have their applications always deliver acceptable service quality and reliability to all users, even during periods of unusually high demand.

7.1.2 Slashdot Effect

The Slashdot effect [Slashdot] occurs when a larger website creates a link to a smaller website that produces a huge boost in traffic to the smaller website. For example, moments after a popular website showcases a little known website, the highlighted website might observe a huge spike in traffic; Figure 7.3 illustrates a moderate example

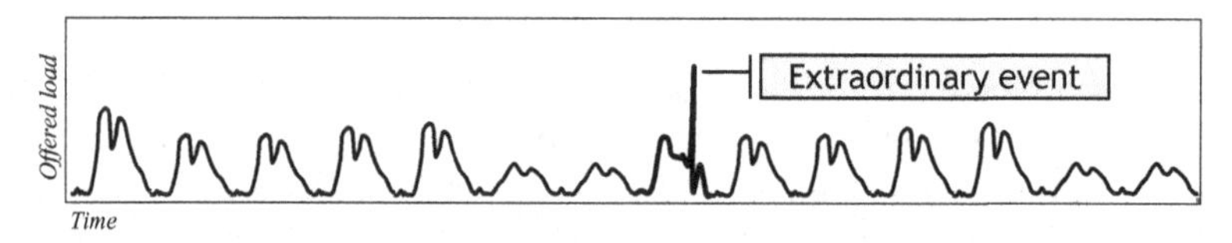

Figure 7.2. Example of Extraordinary Event Traffic Spike.

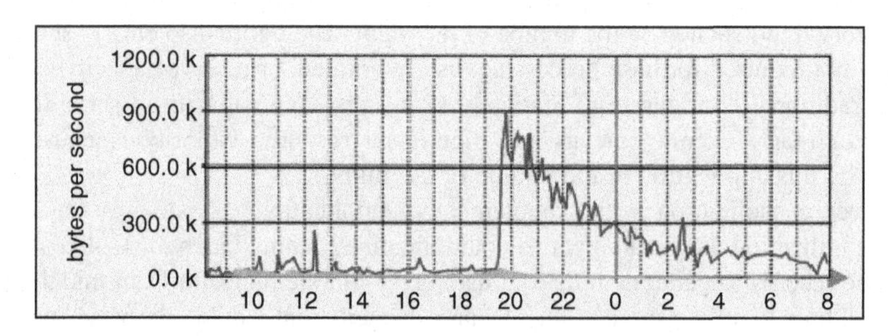

Figure 7.3. The Slashdot Effect: Traffic Load Over Time (in Hours).
Source: Wikipedia.org, at http://en.wikipedia.org/wiki/File:SlashdotEffectGraph.svg.

of the Slashdot effect. Slashdot effect traffic is hard to predict in advance because the effect is fundamentally driven by actions of popular websites controlled by others, and the reactions of users to those actions. Traditional capacity planning strategies do not generally address Slashdot effect events well because it is difficult to predict both the maximum offered load during the Slashdot event and the timing of that event.

7.2 OVERLOAD, SERVICE RELIABILITY, AND SERVICE AVAILABILITY

Applications ultimately depend on sufficient physical processing, storage, networking and other resources being promptly available to serve the offered load. If sufficient resources are not available to meet the offered load, then either load is shed gracefully or service performance (latency) and ultimately reliability and service availability can be impacted. There are three canonical offered load operating regions, as shown in Figure 7.4:

- *Offered load is at or below configured capacity*, so the system operates normally, and all requests should be served with acceptable service reliability and service latency.
- *Offered load is greater than configured capacity but below the maximum overload capacity*. If the system attempts to serve an offered traffic load above its configured capacity limit, then service latencies are likely to increase as work queues fill faster than the queues. Well-engineered traditional systems will implement

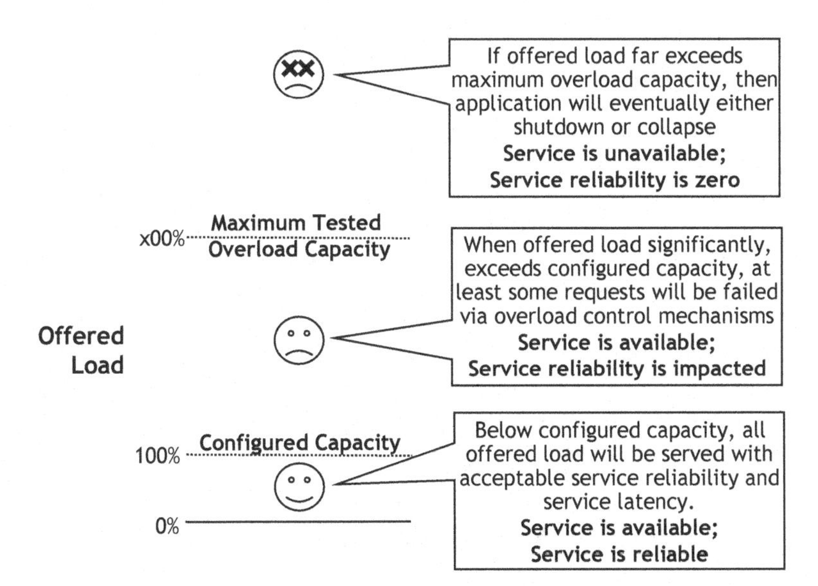

Figure 7.4. Offered Load, Service Reliability, and Service Availability of a Traditional System.

overload control mechanisms that detect when the offered load exceeds the capacity of the resources available to serve load, and take proactive steps to shape the load so that acceptable service is delivered to the maximum number of users or those with priority (e.g., emergency calls). A common overload control action is to explicitly queue the request for brief bursts of traffic (e.g., "*all agents are busy; please remain on the line and your call will be served by the next available agent*") or to return a "too busy" indication to some service requests for sustained overload (e.g., "*all circuits are busy; try your call again later*"). While the overloaded system remains available and in control of the load, requests from at least some users do not complete successfully and thus may be counted as failed or defective transactions and thereby impact service reliability metrics. As a system enters overload, many requests appear to be unacceptably served from the user's perspective because being told to try again later or wait is not the same as correctly serving the request. Users will often consider requests that were not served correctly from their perspective as not being reliable, and thus messages rejected due to successful operation of overload mechanisms are often counted as impairments against service reliability metrics, even if the overload responses are returned within the specified latency targets (and hence do not count as service latency impairments). Note that these impacted transactions are attributed to the enterprise or service provider because they failed to engineer sufficient capacity to serve the load rather than to the supplier because the application is managing the additional load for a period of time and/or correctly responding to overload. As application overload controls cannot deal with an infinite offered load, the nominal overload control capacity is generally specified as a multiple of the engineered capacity of a particular configuration (often 2–10 times), and overload control testing will verify that the system can endure sustained load at this maximum overload capacity and automatically revert to normal operation shortly after the offered load falls to or below the engineered capacity of the system.

- *Offered load exceeds maximum overload capacity*. When offered load far exceeds available processing, storage, and/or networking resources (e.g., during a distributed denial of service [DoS] attack) the system must take dramatic actions, such as flushing and discarding all (or virtually all) network traffic, or risk failing catastrophically under the crushing traffic load. When an application discards all, or virtually all, user requests it is not available for user service, and thus is not generally considered available. When an application stops responding to user requests in a last-ditch attempt to avoid collapse (or because of collapse/catastrophic failure), service availability is nil because no traffic is served. While the application may continue to execute and respond to management commands, since no more than a tiny portion of the offered load of user traffic is served, the application is effectively unavailable for user service.

7.3 TRADITIONAL CAPACITY PLANNING

Reconfiguring physical hardware (e.g., adding more RAM or processors to a server) is typically an activity that requires the server to be powered off, thereby potentially impacting all users served by applications hosted on that server. Thus, growing (or degrowing) a traditional server's hardware configuration is typically a service-impacting action with significant operational expense for the following activities:

- Migrating traffic served by applications hosted on the server to be reconfigured;
- Executing the hardware growth (or degrowth) procedure, which typically includes shutting down the server before executing the change and powering the hardware on after the change is completed;
- Reconfiguring the operating system, platform, and application software to use the expanded (or contracted) hardware resources;
- Restarting application software;
- Gracefully migrating user traffic back to the expanded (or off the contracted) application instance.

These activities often require direct human involvement and carry a nontrivial risk of failure; failure of any task during a growth or degrowth operation could increase the duration of service impact outage and increase operating expense to address the failure.

Thus, growth or degrowth of hardware resources supporting traditional applications deployed directly on physical hardware is an expensive and time-consuming activity that carries a risk of failure that could produce a service outage. Some enterprises find the expense and effort of upgrading computer hardware to be so onerous that it is more cost-effective to simply deploy new servers rather than upgrading systems that are deployed and in service. To minimize the opex and service availability risk of hardware resource growth, enterprises would generally engineer their hardware configurations to serve the largest expected busy hour, minute, or second of offered load. The assumption was that the larger capital expense investment for higher capacity up front would eliminate at least some of the opex and service availability risk of hardware resource growth of a production system. In addition, the opex and service availability impact of hardware resource degrowth coupled with the difficulty in redeploying reclaimed/salvaged hardware resources meant that many underutilized hardware resources were simply left in place because it was more cost-effective to leave them in place than to undertake the expense and risk of resource salvage and redeployment.

7.4 CLOUD AND CAPACITY

Rapid elasticity is an essential characteristic of cloud computing that enables additional resources to be applied as needed to support application load, and reclaimed later when they are no longer needed. The measured service characteristic of cloud assures that cloud consumers are charged only for the resources they actually use, and thus consumers have a financial motivation to use resources wisely. Cloud enables three types of compute capacity growth (and degrowth):

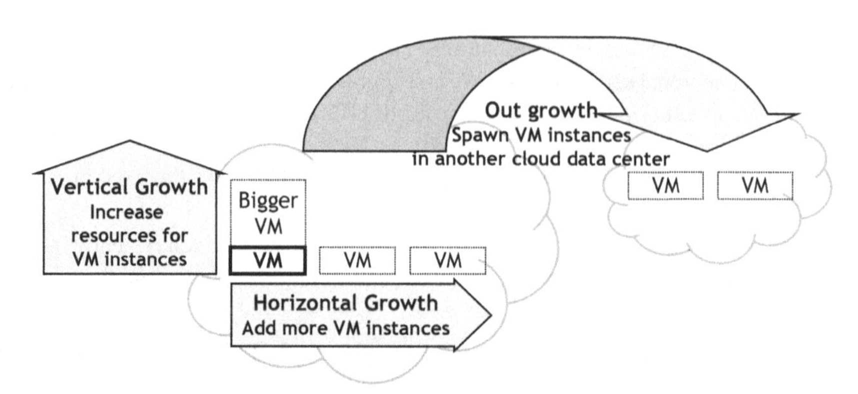

Figure 7.5. Visualizing VM Growth Scenarios.

- *Vertical Growth.* Individual virtual machine (VM) instances can be allocated more CPU, memory, and network resources.
- *Horizontal Growth.* More VM instances can be spawned to enable an application to meet the offered load.
- *Outgrowth.* Additional independent instances of the application can be run in different data centers, typically leveraging the common cloud characteristic of geographic distribution. When supported, this outgrowth can even be to alternate cloud service providers in what is called cloud bursting.

Note that applications must be engineered to support one, two, or all three growth strategies. These three growth scenarios are visualized in Figure 7.5.

Nonvolatile data storage (i.e., disk) grows (and degrows) vertically based on storage volumes or file systems, which can grow (or shrink) in capacity. New storage volumes or file systems can also be created or destroyed to meet application capacity needs providing horizontal growth and degrowth. If necessary, nonvolatile data storage can also experience outgrowth by sharing storage volumes or file systems in another cloud data center.

7.4.1 Nominal Cloud Capacity Model

Figure 7.6 visualizes nominal usage of a pool of virtualized application instances in a computing cloud. A pool of online application server instances is available to instantaneously serve a maximum finite load with acceptable service quality. Under normal circumstances, a portion of that capacity is engaged serving the offered load, and the remainder of online capacity is spare. Cloud service providers will strive to configure resource pooling so that resources (e.g., processing) that are not required by one application might be used by another, similar to how an operating system schedules runnable processes onto available processors. The offered load varies over time, and the application should automatically engage spare capacity as load increases, potentially

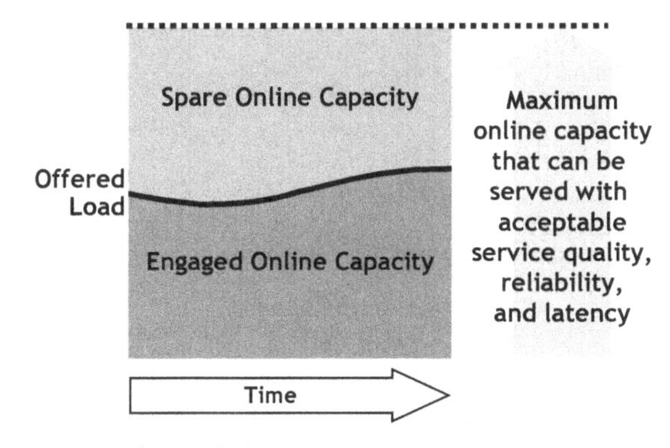

Figure 7.6. Nominal Capacity Model.

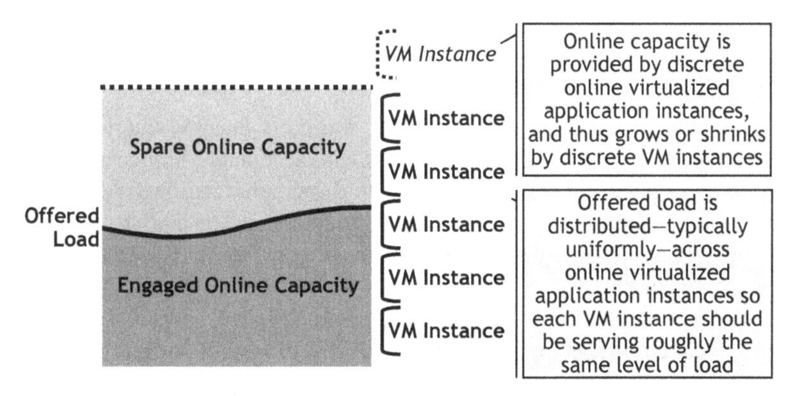

Figure 7.7. Implementation Architecture of Compute Capacity Model.

preempting another lower priority application, and engage less capacity as load decreases (implicitly increasing spare capacity), as the time varying offered load line shows in Figure 7.6. CPU, memory and network consumption are likely to vary directly with offered load, while storage consumption are often somewhat coupled with offered load.

Behind the high-level visualization of Figure 7.6 are a set of online VM instances each hosting a portion of the application's service capacity, as shown in Figure 7.7. This visualization nominally represents the capacity available to a single application instance in a single data center, but a similar visualization could be used to represent aggregate capacity across a pool of application instances in several data centers. Nominally, load will be distributed uniformly across the online VM instances, such as via DNS, a load balancer, or application distribution controller (ADC). Each VM instance may be running on different hypervisor instances on different server hardware to minimize the risk of a single hardware or hypervisor failure impacting an unacceptably large portion of service capacity. Ideally, users would be served by a VM instance in a data

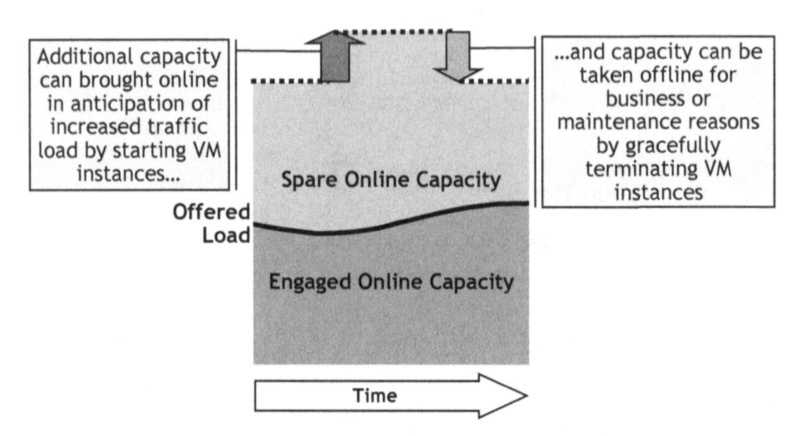

Figure 7.8. Orderly Reconfiguration of the Capacity Model.

center physically close to them to minimize transport latency, but well-engineered networks will often be capable of serving users with acceptable service quality, latency, and reliability from several data centers in their geographic region.

Figure 7.8 illustrates how cloud service providers can instantiate additional VM instances (e.g., horizontal growth) to increase online capacity, and release VM instances to reduce online capacity.

Note that we use the term "spare online capacity" to refer to service capacity that is instantly available to the application when needed but is not currently engaged. One can logically divide this spare capacity into:

- *Reserve online capacity*, which is maintained to cover both: (1) spikes in offered load until rapid elasticity mechanisms can bring additional service capacity online and (2) as redundant capacity for high availability mechanisms to instantly mitigate the impact of inevitable failure events.
- *Wasted capacity*: Capacity that is beyond what is required to serve the offered load (aka, engaged online capacity) and the level of reserve online capacity dictated by enterprise operational policy is effectively unneeded and thus deemed waste. Inevitably, there will be some nominally "wasted" spare capacity beyond the level of reserve capacity dictated by the application's operational policies due to quantization of resources (e.g., you can't allocate or deallocate half a VM). However, when this level of wasted online capacity exceeds the infrastructure as a service (IaaS) provider's allocation/deallocation unit size, then one should consider deallocating the unneeded resources.

Conversely, when offered load increases, some spare capacity will become engaged, thereby logically shrinking the pool of online spare capacity. When the level of spare online capacity falls below a minimum reserve threshold, then the application can elastically grow to rebuild spare capacity, subject to the consumer's operational constraints (e.g., budget and software license limits). If insufficient spare online capacity is available to serve the offered load with acceptable service quality and comply with operational policies for minimum acceptable online reserve capacity, then the application should activate overload controls.

7.4.2 Elasticity Expectations

Elastically growing or degrowing the resources available to an application requires:

- the cloud service provider to locate and allocate requested resources for the cloud consumer; and
- the application software to reconfigure itself to use those newly available resources.

Elastic degrowth is logically the reverse: a running application instance must release some used resources and the cloud service provider reclaims those resources to make them available for other cloud consumers.

Slew rate in electronics refers to the maximum ability of a circuit, especially an amplifier, to drive the output to track with changes in input. The classic example of slew rate is illustrated in Figure 7.9 as the output of an amplifier tracks with a square wave input. While the input is assumed to be capable of "instantly" changing signal levels from low to high, the amplifier takes finite time to drive the output level from low to high for fundamental physical reasons (e.g., capacitive load).

As with physical amplifiers, application capacity cannot be grown infinitely fast to track with changes of offered load. A decision must first be made to add application capacity, and then additional resources must be requested from the cloud service provider (nominally via the on-demand self-service essential characteristic of cloud computing). The cloud service provider must locate suitable available resources and allocate them to the cloud consumer. Then the application must:

1. start up the host operating system, platform, and application software in the newly allocated VM instance; and
2. integrate this new service capacity with the preexisting independent application instance before the service capacity is fully operational and available to serve offered load.

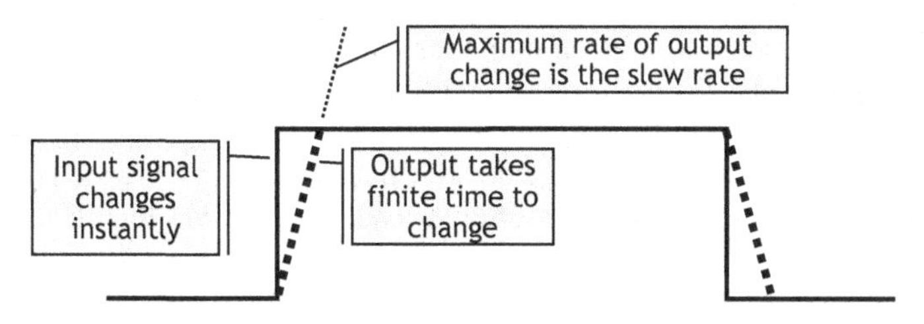

Figure 7.9. Slew Rate of Square Wave Amplification.

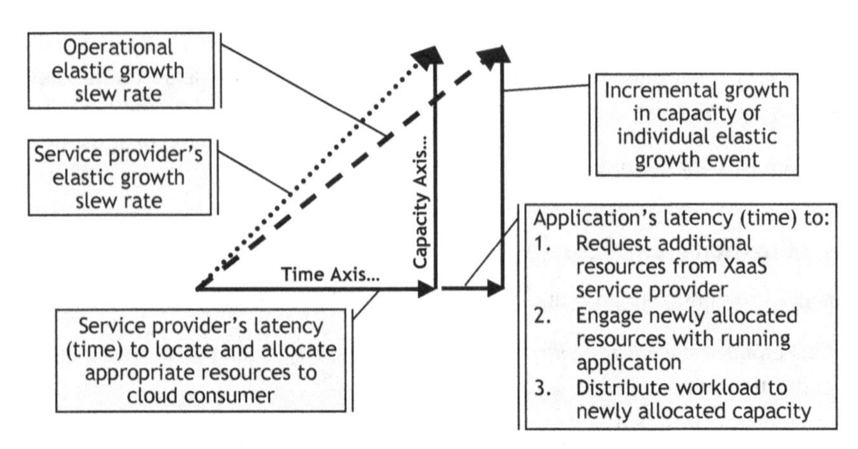

Figure 7.10. Slew Rate of Rapid Elasticity.

Thus, the (growth) slew rate is the increment of additional capacity divided by the latency to decide that additional capacity is needed, to allocate suitable capacity from the cloud provider, and to engage that newly allocated resource capacity with a running application instance. Note that the time for the XaaS platform to allocate new resources is a critical contributor to elastic growth latency, but additional application latency is required to make the newly allocated capacity available to serve end users. This is illustrated in Figure 7.10.

Elastic growth slew rate can be improved by:

- increasing the capacity growth per elastic growth event (increasing Δ capacity or *dy* from calculus); or
- reducing the capacity growth latency (decreasing Δ latency or *dx* from calculus).

It is likely that horizontal, vertical, and outgrowth events will each offer different increments of Δ capacity, and the Δ latency may also vary as both the cloud service provider(s), and application instances must do different work to support each of these three growth options. Likewise, there will be practical and design limits to the resource capacity increments that are offered by the cloud (e.g., no 1 THz virtual CPUs are available) and architectural limits on individual application instances (e.g., supporting a maximum number of VMs per application instance). Thus, highly elastic applications will support multiple options for horizontal, vertical, and/or out growth. Given the finite growth slew rate, applications must decide how much excess spare capacity to maintain online to follow normal variations in offered load, and at what load level to engage more capacity to minimize the risk of elastic failure.

Allocating and reclaiming resources is neither trivial nor instantaneous. Table 7.1 gives the Open Data Center Alliance's (ODCA's) expectations for the rate of elastic

TABLE 7.1. ODCA IaaS Elasticity Objectives [ODCA-SUoM]

SLA Level	Description
Bronze	Reasonable efforts to provide ability to grow by 10% above current usage within 24 hours, or 25% within a month.
Silver	Provisions made to provide ability to grow by 10% within 2 hours, 25% within 24 hours, and 100% within a month.
Gold	Significantly additional demonstrable steps taken to be able to respond very quickly to increase or decrease in needs; 25% within 2 hours, 50% within 24 hours, and 300% within a month. Penalties to be applied if this capacity is not available to these scale points when requested.
Platinum	Highest capacity possible to flex up and down by 100% within 2 hours, 1000% within a month, with major penalties if not available at any time as needed.

Source: Open Data Center Alliance.

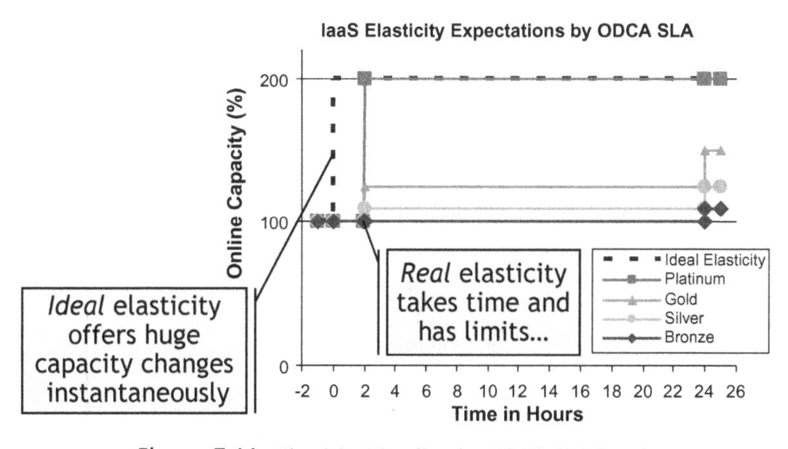

Figure 7.11. Elasticity Timeline by ODCA SLA Level.

growth and degrowth by service-level agreement (SLA) level. Figure 7.11 visualizes these expectations in a plot of expected elastic capacity growth by time compared with the theoretical scenario of nominally doubling resource allocation instantly. The key insight from Table 7.1 and Figure 7.11 is that elastic growth is not fast enough to simply replace overload controls. Instead, elastic growth should enable applications to gracefully grow their online capacity ahead of offered load so that overload controls need not activate as frequently as might be necessary with static application capacity arrangements that cannot easily change on a timely basis. Likewise, application capacity can be gracefully reduced over time to reduce a cloud consumer's operating expense.

As Figure 7.11 shows, the committed elastic growth increment (e.g., plus 10% of allocated capacity, plus 25% or plus 100%) of application capacity is likely to be limited by the elasticity SLA of the cloud service provider. The delay in making that additional application capacity available to end users is likely to be at least a couple of hours, and perhaps a day or more. Thus, applications should maintain sufficient spare online capacity and activate overload controls to manage traffic beyond that capacity so that they can likely serve a growing traffic load with acceptable service quality until elastically grown application capacity can be brought online. Likewise, elastic growth should replenish the pool of online spare resource capacity that will be consumed prior to the elastically grown capacity coming online so that additional growth in user workload—or service recovery on failure—can be served with acceptable service quality.

7.5 MANAGING ONLINE CAPACITY

The on-demand self service essential characteristic of cloud computing integrates with the rapid elasticity characteristic to permit close to real-time management of an application's online resource usage. The process of online capacity management has several high-level steps that are visualized in Figure 7.12 and discussed below.

- *Monitor offered load and resource usage* to accumulate absolute and trend data to support business decisions on appropriateness of current resource allocation
- *Decide.* Business logic is applied to the monitored data, along with historic data, trend analysis, and heuristic considerations to decide either:

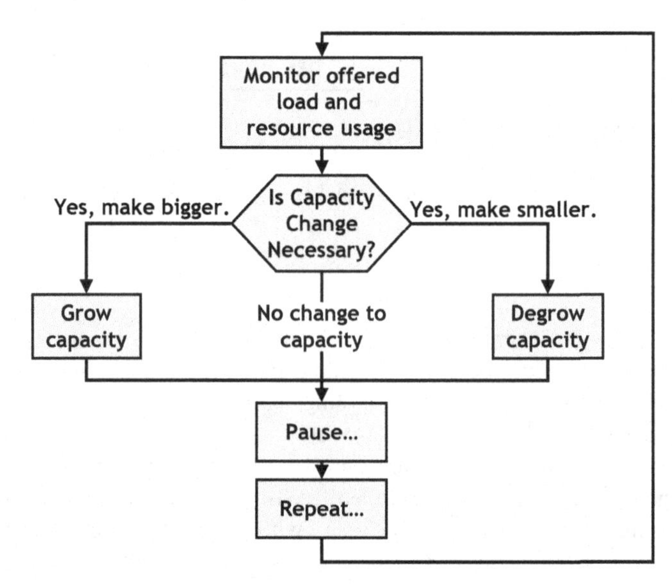

Figure 7.12. Capacity Management Process.

1. Insufficient spare capacity is online to assure an acceptably high probability of serving anticipated load with acceptable service quality, reliability and availability, and thus resource allocation should be grown.
2. Excess spare capacity is allocated, and thus resource allocation should be decreased (degrown)
3. Current spare online capacity offers an acceptably high probability of serving anticipated load with acceptable service quality, reliability, and availability without unacceptable waste, so no capacity change is needed at this time.

- *Grow Capacity.* If resource growth is decided, then a request for additional resources must be passed to the IaaS supplier. The IaaS supplier should promptly allocate and furnish the requested resources to the application. The application then engages the new resources (e.g., initializes application software in the newly allocated VM instance) to bring the new application capacity online to serve users.
- *Degrow Capacity.* If resource degrowth is decided, then the application must decide exactly which resource(s) (e.g., VM instance and block[s] of storage) will be disengaged of the targeted resource from service (e.g., drain traffic from the VM instance or move active data from the data block[s]). After the targeted resource is successfully disengaged, the application requests the IaaS service provider to deallocate the resource (e.g., by gracefully shutting down a VM instance or releasing storage block[s]).
- *Pause and Repeat.* Sufficient spare capacity should be maintained online so that capacity management decisions can be made periodically (e.g., every 15 minutes) rather than continuously (e.g., every microsecond), because it will inevitably take some time for each capacity management change to be completed and for traffic to properly engage on the reconfigured resource allocation. The repetition frequency for capacity management decisions can be statically configured (e.g., every 15 minutes) and/or triggered by threshold crossing alarms (e.g., resource high or low water marks crossed) and/or by heuristics (e.g., x minutes after a growth or degrowth operation).

7.5.1 Capacity Planning Assumptions of Cloud Computing

The cloud service model enables the cloud service provider to focus on assuring that adequate resource capacity is available to meet consumers' demand, and thus the cloud consumers can focus on engaging and releasing resources to meet the needs of their users. This is the heart of the utility computing vision of cloud computing. Traditional application capacity planning assumptions are shattered by the shift to cloud computing. Consider how the often implicit assumptions of traditional capacity planning no longer apply to applications deployed on computing clouds.

1. *Future traffic demands must be carefully anticipated because acquiring additional hardware resources to serve offered load requires a long lead time.* The rapid elasticity offered by IaaS cloud providers assures that

additional hardware capacity will be available in nearly real time (i.e., rapidly). The on-demand self-service essential characteristic assures that the actual procedure for requesting additional resources is streamlined and simple to use. The measured service characteristic assures that consumers are charged only for the resources consumed, so more careful resource usage lowers cloud consumers' operating expenses.

2. *Releasing unused/unneeded hardware capacity is pointless because those resources cannot be cost effectively reclaimed/reused by other applications.* IaaS cloud service providers focus on efficiently managing huge pools of hardware resources, so any hardware resources released can be effectively reused by other applications. In addition, the measured service essential characteristic of cloud assures that cloud consumers will not be charged for unused resources that are released and reclaimed by the cloud service provider.
3. *Capacity change events are inherently risky and expensive, and thus the number of growth (or degrowth) events should be minimized.* On-demand self service reduces the incremental opex of capacity change events, and the automation behind the cloud's on-demand self service interface(s) enables more rigorous automated checking compared with traditional processes, thus reducing the risk of error. The measured service nature of cloud computing assures keeping resource usage closer to offered load (e.g., via frequent small capacity changes) can reduce opex compared to traditional (e.g., occasional large capacity changes). Thus, the closer a cloud consumer tracks their resource usage to the offered load via frequent capacity management events, the lower the consumer's cloud service operating expense.

Having discarded the traditional assumptions of capacity planning, one is free to reconsider the fundamentals of capacity planning based on the essential and common characteristics of cloud computing. The authors suggest the following capacity planning goals for cloud based applications:

1. *Cloud hosted applications should support rapid online resource growth in modest cloud-oriented resource allocation units (e.g., individual VM instances and storage blocks).*
2. *Applications should support graceful service migration of users* (i.e., draining traffic) from target VM instances to another VM instance so excess service capacity can be released.
3. *Applications should support release of online but unneeded resources without disrupting user service* (i.e., after user traffic has been drained from the target VM instance).
4. *Applications should support multiple independent application instances running simultaneously—often in geographically distributed data centers—and possibly hosted by different cloud service providers (i.e., cloud bursting).*
5. *Applications should support balancing offered traffic load across multiple independent application instances.*

6. *Applications should support graceful (even if service impacting) migration of user traffic from one independent application instance to another.* This simplifies draining an independent application instance that was spawned to meet peaks in service demand after offered load has returned to normal levels and the additional service capacity is no longer required.

7.6 CAPACITY-RELATED SERVICE RISKS

Overload is the primary capacity-related risk that traditional applications are vulnerable to when offered load exceeds engineered capacity. Applications that support rapid elasticity are *theoretically* not vulnerable to traditional overload because the engineered capacity can nominally increase to perpetually stay above the offered load. Note that for practical architectural and design reasons, individual application instances can only expand to a finite physical limit, and thus elastic applications should support the creation of an arbitrary number of application instances to serve large offered loads, and mechanisms must be available to efficiently distribute offered load across an arbitrary large aggregate pool of application instances.

If the offered load grows faster than the cloud can allocate additional resources and the application can bring additional capacity online, then an elasticity failure occurs, which impacts at least some offered load. In addition to elasticity failures, cloud deployments are subject to increased service latency risks due to multitenancy and other factors. Critical failure of an application VM instance can also present the risk of partial service capacity loss outage. This section considers elasticity failures, service reliability, and latency impairments and partial capacity loss failures.

7.6.1 Elasticity and Elasticity Failure

Figure 7.13 illustrates how successful elasticity addresses increases in offered load: as offered load increases and less spare online capacity is available, additional resources can be allocated and additional VM instances created to serve the load and maintain an acceptable cushion of spare capacity that is available to serve new traffic. Note that online capacity is increased in discrete steps as individual VM instances come online and are available to serve offered load. Likewise, online capacity can be released one VM instance at a time as offered load declines and excess spare online capacity is available.

Since instantiating application VM instances involves starting VM instances, loading images, and initializing virtualization and application software, it takes a finite amount of time to bring new service capacity online. Thus, there is a risk that if offered load increases so fast that new capacity cannot be brought online before all spare online capacity is consumed, then some offered load will not be served with acceptable service quality, latency, and reliability, as shown in Figure 7.14. The exact behavior experienced by users when offered load exceeds online capacity will be determined by the overload control mechanisms and policies implemented by the application and the application

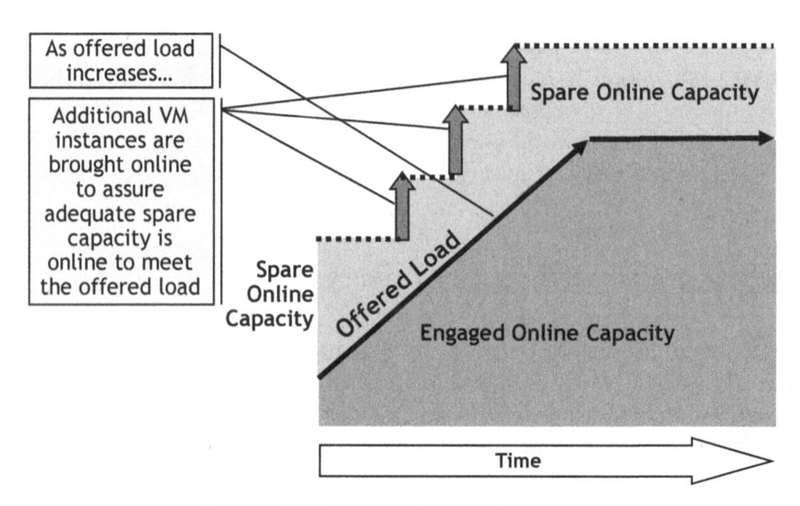

Figure 7.13. Successful Cloud Elasticity.

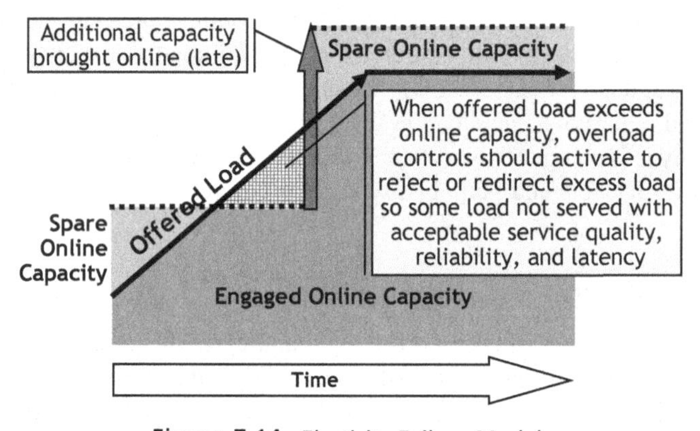

Figure 7.14. Elasticity Failure Model.

itself, but at least some users are likely to experience one of the following unacceptable responses to service requests:

- Explicit overload failure indications, such as SIP's 503 Service Unavailable response (meaning "*The server is temporarily unable to process the request due to a temporary overloading or maintenance of the server*" [RFC3261]).
- Increased service latency, possibly even service latencies beyond the maximum acceptable service latency requirement.

- Time-out expiration because a message request was discarded by the cloud platform or application without sending a reply.

The application's overload control policy can determine how user service is impacted when offered load exceeds online capacity, and this influences whether elasticity failures (i.e., overload situations) appear as service latency impact (e.g., service unacceptably slow), service reliability impact (e.g., server busy), service availability impact (e.g., response latency exceeds maximum acceptable service latency and server appears down), or a combination. Note that overload control policies can explicitly treat individual users differently. For example,

- new session/logon requests might be rejected from some classes of users to shed offered load (e.g., requests from "silver" and "bronze" users are rejected while "gold" users' requests are served);
- resource-intensive requests that consume more scarce online capacity might be rejected during overload periods;
- every "*N*th" request from some classes of users might be rejected in a round-robin fashion to shed offered load;
- active sessions, pending, or queued service requests might be cancelled for some classes of users; or
- combinations of these or other policies might be executed.

As discussed in Chapter 3, "Service Reliability and Service Availability," the service impact of elasticity failures must be considered for individual users:

- Users who experience significantly increased service latency will deem service quality to have degraded. Service quality impact can be quantified by the absolute number of users who experience significantly increased service latency or normalized by the percentage of active users who experience significantly increased service latency.
- Users who have individual and isolated service requests that fail or are unacceptably delayed will deem service reliability to be impacted. Service reliability impact is quantified by the absolute number of impacted service requests or normalized by the number of impacted service requests per million transactions, especially during the overload period.
- Users who are unable to obtain service in less than the maximum acceptable service disruption period (e.g., 10 seconds) will consider the service unavailable.

7.6.2 Partial Capacity Failure

Critical software failures of virtualized applications will typically impact a single VM instance. The critical failure event will impact whatever offered load was actively being served by the application instance experiencing the critical failure, and will reduce the

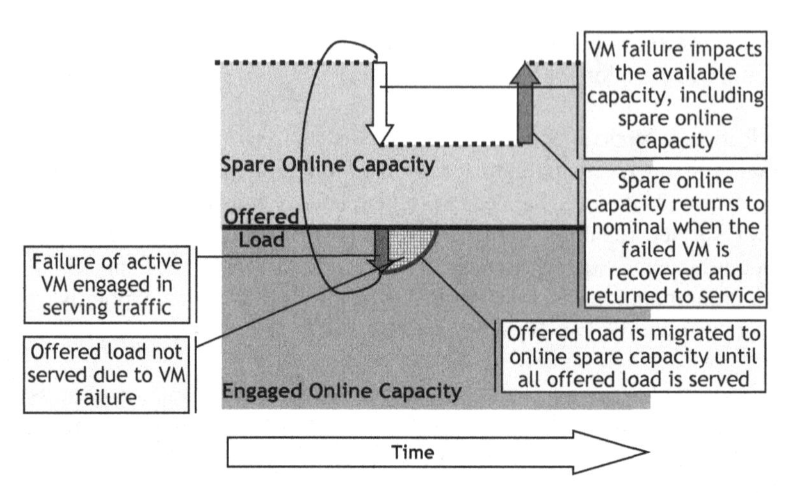

Figure 7.15. Virtualized Application Instance Failure Model.

spare online capacity, as shown in Figure 7.15. The users who had pending transactions with the failed application VM instance are likely to experience service reliability impact if pending transactions and/or perhaps volatile information in the failing VM instance is lost; users whose sessions were served by the failed application instance are also likely to experience some visible service impact on their next service request to the failed application instance. Assuming there is sufficient spare online capacity to serve the impacted traffic load and the application is properly designed, service requests from impacted users will migrate to other VM instances; the exact service recovery time depends on the application protocol options, the application architecture, behavior of the user's client application and other factors. High availability mechanisms should instantiate a new VM instance to restore online service capacity to its prefailure level. Note also that highly available systems are configured so that no single failure will decrease online capacity below the engineered capacity, and that the impact of a VM failure may be similar to the failure of a single blade or server in traditional deployment architecture. Thus, for high availability systems, the service impact is limited to the time it takes to migrate impacted users from the failed instance to "spare" online capacity. In contrast, systems that were not engineered with adequate redundancy may experience capacity loss until the failed instance can be recovered.

The service impact of partial capacity failures is quantified by considering the number of users who are impacted by the failure event and the duration of service impact while the failure is detected, and they are migrated to spare online application instances. This absolute number of users or impacted user minutes value gives a tangible metric for actual service impact.

7.6.3 Service Latency Risk

Virtualization explicitly decouples application software from the underlying hardware, and the essential cloud characteristic of multitenancy increases the risk that resource contention may increase latency for an application to utilize a shared resource like CPU, networking, or storage, especially when one or more applications are under stress. The incremental latency comes from several sources:

- *Resource Contention Latency.* Any resource sharing increases the risk of resource contention that is typically addressed via some form of queuing and queuing entails a wait period, thus accruing latency. Resource contention is more likely to occur as the traffic load on the system increases. As increasing resource utilization is a primary goal of many virtualization deployments (e.g., server consolidation), increasing resource contention is a likely side affect of virtualized deployments. Carefully tuned queuing/scheduling strategies are essential to assure that application instances receive timely access to resources so they can deliver acceptable service latency to users.
- *Real-Time Notification Latency.* While access to physical resources like compute cycles, disk storage, or networking are likely to be randomly distributed across time, real-time clock interrupt notifications are inherently synchronized. If multiple application instances request notification for the same real-time clock interrupt event, then some executions may be serialized and thus implicitly shift the real time understood by the applications that are notified last. If the application requires periodic or synchronous real time notification, such as for streaming media, then any variations in execution timing of application instances can introduce notification jitter. While the virtualized application may or may not be aware of any notification jitter, end users will directly experience this jitter; if this jitter is severe enough, then the users' quality of experience will degrade. Real-time notification latency risk increases as more applications execute processing triggered by clock interrupts.
- *Virtualization Overhead Latency.* System calls made by application software may pass through the additional layer of hypervisor software in virtualized deployment to access hardware resources, and the additional layer of software may add some latency. Note that the virtualization overhead latency is dramatically shortened as virtualization enabled processors are deployed and hypervisors evolve to fully leverage hardware support for virtualization.

These service latency risks are fundamentally the same as those that confront applications running on time shared operating systems; however, since virtualization enables more applications to share hardware resources and achieves higher hardware utilization levels than are typically experienced on traditional architectures, the risk may be even greater. Compared with native deployment of an application instance executing on a traditional operating system platform running directly on physical hardware, virtualization slightly increases service latency due to execution overhead of the virtualization manager. Application service latency of a virtualized configuration is likely to increase somewhat compared with native as the service load approaches the full engineered capacity of the physical hardware configuration due to the scheduling overhead of the virtualization manager. Theoretically, appropriate system configurations and well engineered virtualization configurations can potentially produce better overall service by optimally sharing more powerful hardware platforms compared with more modest hardware that might have been used for standard application deployment.

Virtualization may slightly increase both baseline service latency and as well as service latency for some requests, especially when the virtualized platform is under load. This impact is likely to be manifest in changes to the distribution of service latency for application requests compared with baseline (nominal, or 50–60% of engineered load) performance. If the application and virtualized platform are well designed, engineered, and configured, then the shape of the service latency distribution should remain the same at full engineered load as it was at nominal load, although the latency for 50th percentile at full engineered load may be slightly higher than at nominal load, and the 95th[1] percentile may be proportionally larger at full load than it was at nominal load. A particular risk is if service latency, especially the 95th percentile tail, increases dramatically as the system reaches full engineered capacity, then more of the latency tail will fall beyond the maximum acceptable service latency time, and thus impact service reliability metrics. For example, if the 95th percentile is twice the 50th percentile at nominal (50% capacity) load, but the 95th percentile jumps to five times the somewhat higher 50th percentile latency at full engineered load, then more requests in the distribution tail are likely to have unacceptably long service latency, and thus be counted as service reliability impairments resulting in disappointed users. The key performance metric is to characterize how much the 95th percentile service latency increases as the traffic load increases to full platform capacity, and to verify that even at that increased service latency, the rate of transactions having greater than maximum acceptable service latency is low enough to meet business needs.

Thus, applications might want to monitor that the cloud provider is actually delivering the resource capacity expected (e.g., that another cloud tenant is not compromising the target application's access to resources). If availability of allocated resources is degraded (e.g., due to increased latency), then it may be expedient for the application to request additional resource capacity on a different hypervisor with the expectation that additional capacity will not be burdened with the same exogenous processing load that is degrading the target VM instance.

7.6.4 Capacity Impairments and Service Reliability

Service reliability impairments (e.g., failed or defective transactions) accrue due to the following factors:

- *Critical failures*, which cause pending transactions to be lost and cause service requests to fail until failure detection and recovery is completed.

[1] The authors use 95th percentile as a reasonable point in the latency tail (along with 50th percentile) for analysis; readers can use whatever tail point is customary for their industry or application (e.g., 99th percentile, 99.5th percentile, and 99.9th percentile).

- *Activation of subcritical software defects*—occasionally, residual software defects will cause some requests to execute incorrectly, but will not cause a critical failure that requires explicit recovery action.
- *Transient or one-shot events* like lost or corrupted IP packets or queue overruns due to a random, momentary traffic spike.
- *Failure of a supporting system*—complex applications often rely on other systems, such as database servers, authentication/security servers, payment systems, and so on. When that supporting system experiences transient, brief or prolonged service disruption, then the target system may be forced to fail some or all service requests, and thus service reliability metrics for the target system will be impacted.
- *Excessive service latency* as discussed in Section 3.5, "Service Latency," and Section 7.6.3, "Service Latency Risk."
- *Live migration of VM instances*, which may cause some requests to be lost and/or delays in responses to be so long as to be deemed failed transactions, and thus count as service reliability impairments.

Note that although service availability impairments (aka, outages) are generally isolated to the actual system instance experiencing the failure to enable accurate system-specific service availability metrics (especially for contractual liability for service availability attached to SLAs), brief transient events are often not tracked to the primary root cause. Instead, service reliability impairments are often broadly bucketed as either chronic impairments or acute impairments. Acute impairments are often correlated with critical failures of the target system or a supporting system, or some network event (e.g., router failure), or some application overload event. Chronic impairments are not generally traced to or correlated with specific failures or events; activation of subcritical software failures and recurring transient events (e.g., lightning, occasional buffer overflows) generally fall into the chronic impairment bucket.

Excessive service latency and transient or one-shot failures can be minimized by appropriate system architecture and configuration (e.g., automatic protocol retries for lost messages) and adequate testing to validate and baseline service performance. Residual critical and subcritical software defects are removed prior to commercial service startup via appropriate quality processes, especially adequate system verification testing. Critical hardware failures are minimized via appropriate system architectures, robust hardware design and reliability diligence, and high-quality component sourcing and manufacturing processes. If live migration has any service impact on user service offered by an application, then the number of live migration events should be minimized to reduce the overall service reliability impact to users.

7.7 CAPACITY MANAGEMENT RISKS

This section considers the reliability risks to the generic capacity management process that was described in Section 7.5, "Managing Online Capacity." Section 11.1.5 will

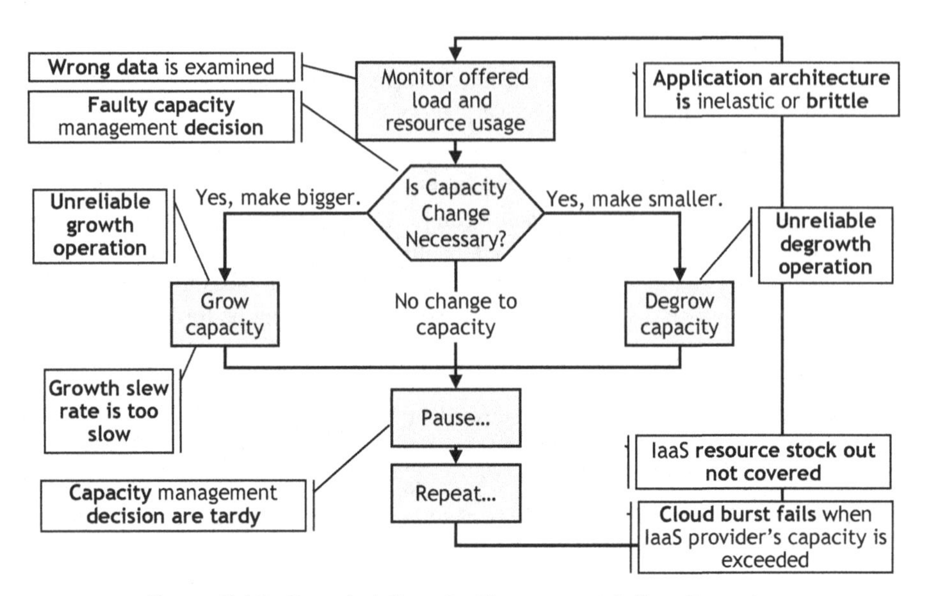

Figure 7.16. Canonical Capacity Management Failure Scenarios.

discuss ways to mitigate the impact of some of the risks. Figure 7.16 overlays the canonical capacity management failure scenarios onto the canonical capacity management model shown in Figure 7.12. Each of these failure scenarios are detailed in a subsequent section.

7.7.1 Brittle Application Architecture

Cloud applications should be architected to be highly elastic to gracefully serve whatever traffic load is presented by expanding and contracting the resources used by the application rather than being brittle and constrained. Individual application instances should efficiently scale from small to moderate to large capacity, and it should be possible to create additional application instances that can cooperate to serve a larger traffic load. Individual application instances should share no components to eliminate the risk of a single component failure impacting more than one application instance. It should be possible to geographically distribute individual application instances both to improve service quality offered to users by serving them from application instances that are geographically close to the users (and thus have less transmission latency), as well as to support georedundancy for disaster recovery.

Elasticity of application architectures can be constrained, or brittle, on several levels:

1. the number of users served by a single application instance; and
2. the number of application instances that can be federated together to serve a larger pool of users.

If the application does not support some form of graceful federation of application instances, then the maximum number of users served by a single application instance is a brittle limit of the architecture. If federation of application instances is supported, then architecture brittleness will appear at or below the product of the maximum number of federated application instances and the maximum number of users per application instance.

7.7.2 Faulty or Inadequate Monitoring Data

Capacity management decisions are based on data about offered load and spare online capacity. Design flaws or failures that compromise this data can cause capacity management processes to break down. Likewise, if the solution or application architecture does not permit visibility to key system load parameters that actually characterize the true workload on a VM instance (e.g., depth of work queues and triggers when work queues overflow), then efficacy of capacity management processes will be impacted.

7.7.3 Faulty Capacity Decisions

Capacity management is fundamentally about making decisions to proactively manage the resources allocated to an application so users are served with acceptable service quality, reliability, and availability at acceptable cost to the cloud consumer. Even when provided with correct input data or rules to manage and fulfill business policies, faulty capacity management decisions can be made if:

- business policies for growth and degrowth are flawed (e.g., inappropriate capacity growth and degrowth triggering criteria thresholds are used); or
- residual software defects (aka, bugs) in decision logic cause business policies to be executed incorrectly by automatic software mechanisms.

7.7.4 Unreliable Capacity Growth

Capacity growth involves two steps, each of which can fail:

1. *Application requests additional resource from IaaS service provider but the IaaS service is unable to serve the request.* If the request fails with a transient error, then the application should retry the allocation request. If the request fails repeatedly or with a persistent error or insufficient resources are provided by the IaaS service provider, then the application should burst to instantiate another application instance in another data center or cloud to grow capacity. The scenario of not bursting on resource allocation failure is covered in Section 7.7.8, "Resource Stock Out Not Covered." The case of cloud burst failing is covered in Section 7.7.9, "Cloud Burst Fails."
2. *Application engages newly allocated resource.* Application initializes the allocated resource, synchronizes/integrates the resource with the preexisting

application instance, and begins engaging the resource to serve users. Failures in resource initialization, synchronization, or engagement should be automatically detected, alarmed, and recovered. Some transient failures may be mitigated simply by retrying the operation; other failures will require more elaborate recovery strategies.

7.7.5 Unreliable Capacity Degrowth

Once a capacity decision has been made to release a particular resource, a multistep degrowth process must be executed to prevent the associated hazards:

1. *Application stops directing new requests to targeted resource.* Traffic that is not redirected away from the targeted resource will eventually be impacted when the resource is deactivated and deallocated.
2. *Application fails to gracefully drain traffic/users from the targeted resource.* Preexisting traffic that is not gracefully transitioned or drained from the resource will be impacted when the resource is deactivated and deallocated.
3. *IaaS provider fails to deallocate the targeted resource.* If the deallocation request to the IaaS service provider fails, then the cloud consumer may continue to be charged for the resource even though it is presumably no longer productively engaged by the application.

7.7.6 Inadequate Slew Rate

Horizontal, vertical, and outgrowth will have different capacity growth slew rates. With a Slashdot effect or another dramatic event, the offered load could grow faster than the maximum growth slew rate. If that rapid traffic growth continues for long enough, then spare capacity will be consumed, and the service simply will not be able to keep up with the growth in demand, thereby producing an elasticity failure. Overload controls should assure that the service impact of inadequate slew rate is deliberately managed to minimize impact to priority users (e.g., serving active users with acceptable service quality and refusing new session requests until additional capacity is online) rather than compromising service quality, reliability, and availability for most or all users.

7.7.7 Tardy Capacity Management Decisions

Allocating, initializing, and engaging additional resources inherently take time to complete. Thus, capacity management decisions must anticipate trends and changes in load to request capacity changes before the capacity is actually needed and sufficient time is available to successfully grow service capacity. If capacity management decisions are not made fast enough, then there may be insufficient online capacity to serve the growing offered load. For example, after an extraordinary or Slashdot event occurs, the capacity management process should activate promptly so elastic growth actions can begin before the offered load completely overwhelms online capacity, possibly forcing newly allocated online capacity to immediately activate overload controls and shed load rather than normally serving the offered load.

7.7.8 Resource Stock Out Not Covered

Individual data centers and even particular IaaS service providers have finite physical resources. Resource allocation requests made after those resources are exhausted will fail, and applications should be prepared for those failures. Ideally, the applications will burst to another data center, possibly with another IaaS service provider. A less desirable strategy is to gracefully deny service requests according to business policies that cannot be served after IaaS capacity has been exceeded. Impacting service to existing users or crashing is unacceptable.

7.7.9 Cloud Burst Fails

An attempt to burst capacity to another data center, possibly a data center operated by a different IaaS service provider, could fail. Based on the cloud consumer's policy, applications might attempt to burst to an alternate data center or IaaS supplier, or they might gracefully deny service requests that cannot be served by the preexisting resource allocation. Impacting service to existing users or crashing is unacceptable.

7.7.10 Policy Constraints

Cloud consumers may impose constraints on maximum application capacity based on business policies. For example, software or intellectual property used in the application may have been licensed up to a maximum capacity, which is not permitting more than "X" users to simultaneously access some licensed content or software component. To avoid breaching these contractual terms, cloud consumers may cap elastic growth at a certain point and rely on overload controls to assure that the maximum online capacity is appropriately shared by priority users.

7.8 SECURITY AND SERVICE AVAILABILITY

One dimension of security attacks involves the impact it has on service availability. This section reviews the security risk impact on service availability, discusses DoS attacks, discusses estimating the service availability impact of security attacks, and concludes with recommendations.

7.8.1 Security Risk to Service Availability

The International Telecommunication Union's X.805 standard "Security architecture for systems providing end-to-end communications" recognizes that service availability

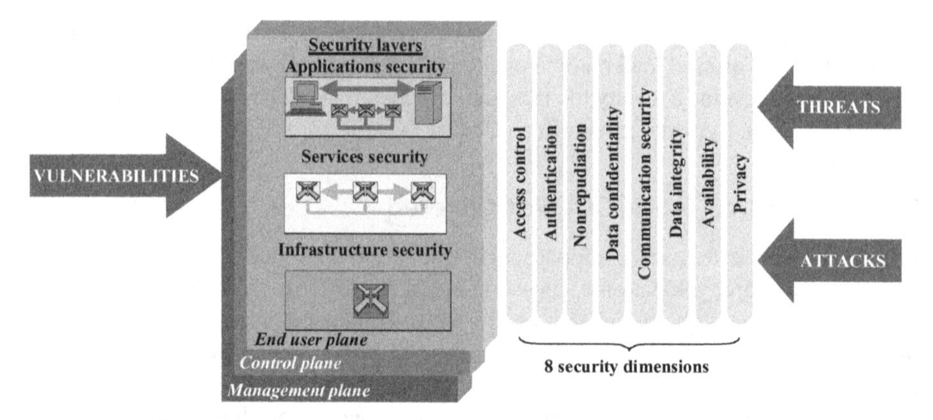

Figure 7.17. ITU X.805 Security Dimensions, Planes, and Layers.
Source: International Telecommunications Union [ITU-T G.114].

is one of the security dimensions that are vulnerable to attack. Figure 7.17 from [X805] visualizes security threats and attacks challenging eight security dimensions across end user data plane, control plane, and management plane of networked applications. Per [X805], "the availability security dimension ensures that there is no denial of authorized access to network elements, stored information, information flows, services and applications due to events impacting the network."

The availability security dimension is subject to two classes of threats:

1. *Destruction of Information or Other Resources.* Damage or loss of user or configuration information or other resource damage can prevent an application from delivering correct service to some or all users, thereby impacting service availability for affected users.
2. *Interruption of Services.* For example, DoS or distributed DoS (DDoS) attacks. A (distributed) DoS attack overwhelms the target system with service requests to drive the target into overload, and perhaps even collapse. When the system is overloaded with attack traffic, legitimate users are likely to experience increased service latency or be denied service, and as the attack increases, the system may deny service to even legitimate users to avoid total system collapse. For cloud-based services, DoS/DDoS attacks may ramp up traffic volumes faster than application capacity can be added, thus initially activating the elasticity failure described in Section 7.6.1, and eventually pushing traffic levels far above the maximum authorized application capacity (e.g., license limits of software components of the application).

The X.805 security model of Figure 7.17 recognizes three planes of concern for security, each with distinct security objectives:

- *Management Plane Security Objectives of Availability Dimension*, "Ensure that the ability to administer or manage the network-based application by authorized personnel and devices cannot be denied. This includes protection against active attacks such as denial of service (DoS) attacks as well as protection against passive attacks such as the modification or deletion of the network-based application's administrative authentication information (e.g., administrator identifications and passwords)" [X805].

- *Control Plane Security Objectives for Availability Dimension.* "Ensure that network devices participating in network-based applications are always available to receive control information from authorized sources. This includes protection against active attacks such as Denial of Service (DoS) attacks" [X805].
- *End User (Traffic) Plane Security Objective for Availability Dimension.* "Ensure that access to the network-based application by authorized end-users or devices cannot be denied. This includes protection against active attacks such as Denial of Service (DoS) attacks as well as protection against passive attacks such as the modification or deletion of the end-user authentication information (e.g., user identifications and passwords)" [X805].

7.8.2 Denial of Service Attacks

Denial of service and DDoS attacks seek to overwhelm a target application or network element with malicious service requests so that it is unable to service legitimate service requests and ends up crashing the target system. Interruption of service threats are currently a larger threat than destruction of information or other resources. The first five findings of the 2010 Worldwide Infrastructure Security Report [Arbor] are as follows:

- ***Network Operators Face Larger, More Frequent Attacks as Attackers Redouble Their Efforts****. . . . attackers have moved aggressively over* [2010] *to dramatically increase attack volumes—for the first time launching DDoS attacks breaking the 100 Gbps barrier.*
- ***Application-Layer DDoS Attacks Are Increasing in Sophistication and Operational Impact****. . . .*
- ***Mobile/Fixed Wireless Operators Are Facing Serious Challenges to Maintaining Availability in the Face of Attacks****. . . .*
- ***Firewalls and IPS*** [Intrusion Prevention System] ***Devices Are Falling Short on DDoS Protection****. . . .*
- ***DDoS Attacks Have Gone Mainstream****. The mainstream media has extensively reported numerous high-profile DDoS attacks motivated by political or ideological disputes*

One should begin by considering the theoretical upper limit of a DoS/DDoS attack. Applications ultimately execute on server hardware, and that server hardware is attached to a LAN in a data center via one or more network adapters. These network adapters often have nominal maximum speeds of 100 million bits per second (100 Mbps), 1 billion bits per second (1 Gbps), or even 10 billion bits per second (10 Gbps). While it is generally infeasible to achieve 100% of the nominal maximum speed of the Ethernet, achieving 80% of nominal maximum speed is often possible with appropriately configured IP infrastructure; this maximum practical throughput is often referred to as "wire speed" because communications is flowing at the maximum speed permitted by the "wire" (i.e., the network adapters, IP infrastructure, and physical media). While a 1 Gbps Ethernet adapter is now fairly inexpensive, the compute and storage resources necessary to support up to 800 Mbps of wire speed user traffic (e.g., hundreds of thousands of requests per second) is far more expensive. Traditionally, enterprises and service providers generally scale their compute and storage resources to meet the expected load, rather than scaling hardware to serve the sustained full wire speed of the network adapter(s). Hence, sustained wire speed request traffic can often drive an application far beyond engineered capacity so the system is forced to abandon legitimate traffic along with attacking traffic. Dedicated attackers can and do mount wire speed DoS/DDoS floods at target systems.

Beyond straightforward "brute force" wire speed flooding attacks on applications, attackers also mount syntax, semantic, and resource attacks.

- *Syntax attacks* deliberately send protocol request messages with syntactic errors, such as data overrun or under run, missing parameters, out of range parameters, and so on, with the expectation that these syntax errors will force the target system to parse the message, determine the syntax error, and construct an error response detailing the detected syntax error. All of that processing consumes CPU resources in an effort to crowd out service to legitimate user requests.
- *Semantic attacks* deliberately send protocol messages with invalid parameters, such as referencing transactions or web pages that don't exist. This forces the application to consume processing and disk resources searching for an object that won't be found, thus denying those processing and disk I/O resources to legitimate users.
- *Resource attacks* deliberately consume shared resources by means of an application installed to deplete all of the shared resources so that the "good" application cannot function.

7.8.3 Defending against DoS Attacks

Perimeter security elements like firewalls, deep packet inspection (DPI) engines, intrusion detection/prevention systems (IDPS), and other security appliances supported by a robust security policy are the primary defense against DoS/DDoS and many other security attacks.

Figure 7.18 expands on Figure 7.4 to illustrate how perimeter security and network infrastructure elements can be configured to rate limit the offered load to an application to assure that offered load does not exceed the maximum tested overload capacity. Ideally, perimeter security elements will block all attack traffic so that only legitimate user traffic flows to the application. If the perimeter security is evaded by attackers (e.g., via a new attack signature in a so-called "zero day" attack), then rate limits in network infrastructure and security elements themselves should still assure that traffic does not reach the maximum tested overload capacity.

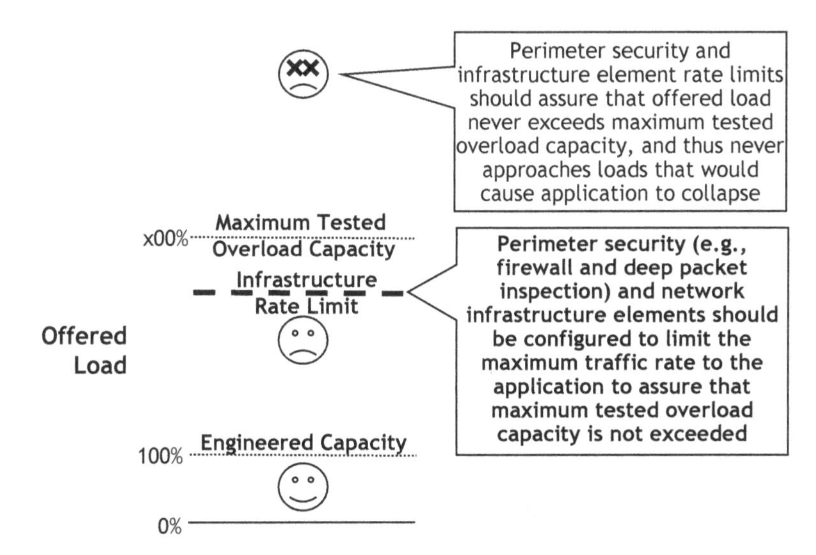

Figure 7.18. Leveraging Security and Network Infrastructure to Mitigate Overload Risk.

7.8.4 Quantifying Service Availability Impact of Security Attacks

In a successful DoS/DDoS attack, all traffic to the target application or host is likely to be impacted until the attack is successfully mitigated, thus producing a service outage. [Arbor] reports that most DDoS attacks are successfully mitigated in less than 30 minutes. The rate of DoS/DDoS attacks on deployed systems is driven by decisions of criminals and attackers based on economic and political considerations. Because the rate of security attacks cannot be generally estimated, it is inappropriate to make any quantitative estimates of the likely rate of DoS/DDoS attacks on a "typical" deployed system, nor of the likely service availability impact of the security attacks.

Accountability for service availability impairments due to security attack is a subtle topic. While the attacks are external attributable per [TL9000] because they represent "*...outages caused by third parties not associated with the customer* [enterprise or service provider] *or the organization* [supplier]," suppliers, service providers, and enterprises do have responsibility to minimize the security vulnerabilities exposed to attackers. Service providers and enterprises should protect networks and applications with appropriate firewalls and security appliances, establish appropriate password and authentication policies, promptly apply security patches, and follow other security best practices. There is always the risk of "day zero" and brute force DoS/DDoS attacks, but appropriate security diligence by suppliers, service providers, and enterprises can harden applications and thus make attackers efforts less effective.

Accountability for service availability impairments due to security attacks on cloud-based applications is even more complex than for traditional or virtualized application deployments for reasons including:

- *Rapid Elasticity Vulnerability.* Attackers can attempt to overload the rapid elasticity mechanisms of cloud-based applications by increasing attack traffic faster than additional service capacity can be engaged, and thereby presumably impact legitimate user traffic.
- *Attacks from within the Cloud.* Attackers can mount an attack from within the cloud hosting the target application, and thus potentially bypass at least some perimeter defenses.
- *Collateral Damage.* Attacks on applications sharing the same compute, storage, or other resources can indirectly impact service latency and service reliability of other applications because shared resources are consumed by the attack, and hence are unavailable to other applications.

7.8.5 Recommendations

The essential and common characteristics of cloud computing introduce some new security risks and expand some existing risks compared with traditional deployment models. Cloud consumers should work with their IaaS service provider to assure that adequate perimeter security defenses protect their applications and assure that best security practices and robust security policies are in place. Readers should refer to cloud security references like [ENISAa], [ENISAb], [NIST-D], [CSAa], [CSAb], [CSAc] and [CSAd] for further information.

7.9 ARCHITECTING FOR ELASTIC GROWTH AND DEGROWTH

Given both architectural limitations on elastic growth slew rate and business needs to control operating expenses, cloud consumers must decide how much spare capacity to keep online for variations in offered load and what high-water threshold conditions should trigger elastic growth events. Likewise, cloud consumers should also decide what low water event thresholds should trigger elastic degrowth events to release capacity. Thus, application architects should rethink their application architectures around the following assumptions:

- Application consumption of cloud resources (CPU, memory, and storage) should grow and shrink horizontally and/or vertically in reasonably quantized units (e.g., individual VM instances) across a reasonable range of offered loads.
- Applications should be architected so that independent application instances can be started in other data centers (i.e., cloud bursting) when offered load exceeds range of supported horizontal and vertical growth.
- Applications should trigger the cloud control software based on growth/degrowth policies. Alternatively, applications should make capacity monitoring information available to the cloud in order to allow the cloud to apply automatic elasticity policies.

- Applications should generate suitable management events (e.g., SNMP traps, log records) when triggering elastic growth or degrowth events.
- Cloud-based applications must implement robust overload controls to assure that if offered load exceeds the online capacity of any particular application instance, then the excess load is appropriately managed rather than simply allowing service latency and reliability to degrade for all users, and risking eventual service collapse and crash.

8

SERVICE ORCHESTRATION ANALYSIS

Service orchestration is a key component in the delivery of cloud services that meet customer and business requirements, particularly those requirements associated with reliability, availability, and latency. The chapter will begin with a definition of service orchestration. The chapter goes on to discuss how policy management and cloud management support service orchestration, the role service orchestration plays in mitigating some risks that could arise in the cloud computing environment, and ends with a summary.

8.1 SERVICE ORCHESTRATION DEFINITION

Service orchestration entails the linking together of architecture, tasks, and tools necessary to initiate and automatically manage a service. In the cloud environment, service orchestration includes linking together and automating tasks based on work flows, measured data, and policies, with the purpose of providing a service that meets the business needs based on the associated SLA if applicable. Based on U.S. National Institute of Standards and Technology standards (NIST) [NIST-C], service orchestration is responsible for the coordination and management of cloud components to provide

Reliability and Availability of Cloud Computing, First Edition. Eric Bauer and Randee Adams.

services that meet customer business requirements. That NIST definition of service orchestration implies a large scope that includes provisioning and managing services and assets, as well as scenarios such as rapid elasticity, cloud bursting, and disaster recovery. This service orchestration analysis will focus on the reliability, availability, and latency related aspects of service orchestration. Service orchestration encompasses the following functions:

- *On-Demand Self-Service.* On-demand self-service as explained in Section 1.1.1, "On-Demand Self-Service," is an essential cloud characteristic responsible for managing server and storage resources. Service orchestration provides the infrastructure, processes, and tools to support requests made to add or remove VM instances (or other software modules) or resources. In addition to the instantiation of a new application instance and its hardware resources, service orchestration supports the configuring of the network resources (i.e., bandwidth and virtual private network links), routing configuration, and the setting up of the firewall connections. Proper configuration of the application, its resources, and interfaces has a direct bearing on the reliability and availability of the application's service.
- *Resource Management.* Resource (or asset) management is responsible for the allocation and management of the resource pools discussed in Section 1.1.3, "Resource Pooling," for use by the software applications. Resource management includes assuring the application has sufficient resources, such as CPU, memory, and disk storage to meet its needs and responding to requests from on-demand self-service for the addition or removal of resources. Resource management is responsible for mapping virtual resources allocated by the applications onto the physical entities as discussed in Section 6.7, "Mitigating Hardware Failures via Virtualization." Resource management also include auditing for allocated resources that have gone unused for an extended period of time and that can be returned to the available resource pool. The allocation of sufficient resources, as well as the reporting of insufficient resources, has a direct impact on the availability and reliability of a service.
- *Service Monitoring.* Service monitoring is responsible for the collection and reporting of measurements of the key quality indicators (KQIs) and key performance indicators (KPIs). KQIs and KPIs can provide the basis for measured service as discussed in Section 1.1.5, "Measured Service," and serve as input for policies that are used to help manage capacity and measure service reliability. Although service monitoring does not directly impact service reliability and availability, it is an important component in measuring and reporting on system reliability to ensure conformance to SLAs. Measured service can be coupled with thresholds to trigger the growing of application instances or resources, which in turn will contribute to a positive impact on system availability (i.e., meeting higher capacity needs) and reliability (i.e., ensuring there are sufficient resources to successfully manage requests).
- *Service Distribution.* Service distribution, supported by policy management, is responsible for managing load distribution across the servers in the solution,

taking into account capacity, regulatory, latency, and security considerations. Distributing traffic in a manner that best meets customer requirements for service availability and reliability is an important responsibility for service orchestration. Service distribution includes ensuring the load is distributed to available application instances and rerouting if that instance is not able to manage its load due to failure or overload or due to the addition or deletion of application instances that can manage the load. In this way, it is a key factor in service availability, that is, making sure that all requests are successfully sent to a functioning application instance.

- *Service Provisioning.* Service provisioning is responsible for configuring subscribers and services to particular components in the cloud solution, taking into account locations that support low latency (e.g., locating the service near to the subscriber), high availability (e.g., provisioning a primary and secondary site), and disaster recovery (e.g., assigning primary and secondary sites that are sufficiently far apart that no single disaster will impact both sites). Incorrect service provisioning can lead to service availability and reliability issues, as well as potential latency problems.

Service orchestration can also facilitate the automation of support services, such as billing, but that is outside the scope of this analysis.

8.2 POLICY-BASED MANAGEMENT

Policy-based management is a key component in a service orchestration framework that provides a means to allocate resources based on defined policies. Policy-based management architecture and its uses have been specified in various IETF RFCs: [RFC3060] defines the policy information model, and [RFC3198] defines the policy management terminology and points to specific RFC references defining the various components and usages. Operational policies provide concrete specifications and input for operating, administering, and maintaining the cloud. This infrastructure that takes policies as an input and provides support for managing cloud services will be referred to as policy-based management.

Distributed Management Task Force (DMTF) policies and constraints are a useful way to define the cloud capabilities that a cloud service provider is offering. Per DMTF, the cloud consumer works with the cloud service provider (CSP) to customize a set of policies that will accompany the instantiation of the consumer's applications. The CSP provides a catalog of constraints, rules, and policies offered as part of the service. The cloud consumer can then request a customization of these constraints, rules, and policies in their instantiation request to meet their specific needs. The policies will then help govern the management of the cloud consumer's application. The CSP or the cloud consumer can also make changes once the service is running if the changes fit into the agreements made before instantiation of the service. The specifics are spelled out in [DSP0102].

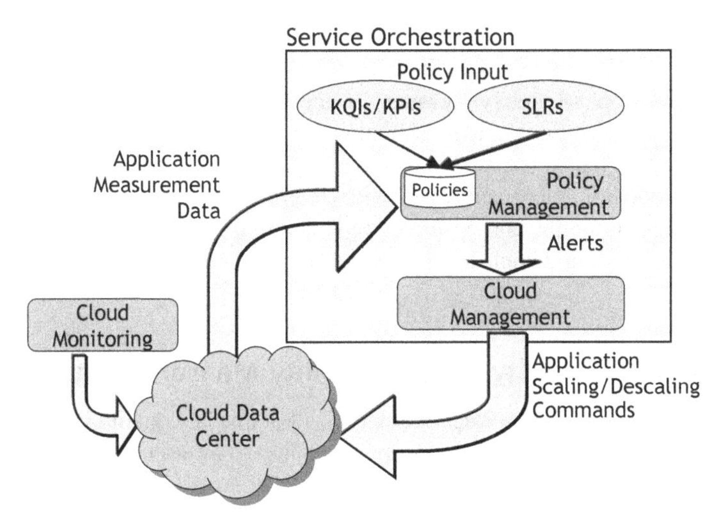

Figure 8.1. Service Orchestration.

Policy-based management takes input from Service Level Requirements (SLRs) (included in Service Level Agreements (SLAs)) and measurements (e.g., defined KQIs and KPIs) combined with associated thresholds on those measurements to create policies. Rules are constructed and input to the policy management infrastructure to describe the triggers and subsequent actions based on that policy. For example, if the SLR stipulates 100 successful transactions per second per VM and the associated measurements have been indicating 200 transactions per second per VM consistently over a period of time, then a policy defined to comply with the 100 successful transactions per second may automatically trigger the creation of a new VM instance to manage the additional traffic or issue a report (or alarm) to the customer to alert the customer to manually request more capacity through a web server interface. This manual request by the customer would then result in the creation of a new VM instance.

Figure 8.1 shows an example of a service orchestration workflow that manages the scaling or descaling of VMs based on usage data expressed by KPIs/KQIs. KPI/KQI data based on defined measurements and SLRs are inputs into policy management. Policy management uses policies to monitor data collected from the applications and issues alerts to cloud management as policies dictate (i.e., based on measurements exceeding a threshold dictated by an Service Level Requirement or Agreement). Cloud management (discussed in Section 8.3, "Cloud Management") determines whether the alerts require configuration changes, and if so, activates the appropriate mechanisms to add or remove VM instances from the cloud solution or to send a message to the cloud consumer to make a request to do so.

8.2.1 The Role of SLRs

Service-level requirements define the service expectations of a customer. SLRs are often included in SLAs, providing a means for the service provider and customer to agree upon and document the level of customer services to be supported, their quality goals, and the actions to be taken if the SLA terms are violated. Metrics, such as KQIs, are used to quantitatively measure service characteristics, such as transactions attempted, failed transactions, and so on that can then be monitored to validate product compliance to those expectations. Policies are built to report on or trigger events upon reaching or exceeding the customer service requirements based on the associated KQIs. This ability to define and ensure compliance to customer requirements is a key component in providing a high level of service to the customer in compliance with the SLAs.

8.2.2 Service Reliability and Availability Measurements

Sophisticated enterprises and service providers will define key quality indicator metrics that can be quantitatively measured for the most important aspects of service offered to users. Quantitative targets will be set for each of these KQI metrics, and bonus or incentive payments to enterprise or service provider staff may be tied to achieving those KQI performance targets. KQIs may be tied to SLA's with service providers, enterprises, or entities representing end users, and liquidated damages may be liable if KQI commitments are not met.

Service quality KQI's often include service reliability, service latency, or service availability metrics, but the KQI may be expressed in service specific language. For example, wireless telephony service providers often use *service accessibility* (probability that a user call attempt will succeed in an acceptable amount of time with acceptable voice quality) and *service retainability* (probability that a call will continue with acceptable voice quality until explicitly released/ended by one of the call participants). See Section 3.8.2, "Service Quality Metrics," for a discussion on these metrics.

Careful analysis of both the metric definition and the particulars of the tools and techniques that produce the quantitative results can often reveal the details necessary to precisely specify the service reliability and availability requirements that should be applied. For example, careful analysis of properly designed service KQI metrics should reveal precisely which protocol responses failed service requests and the portion of responses that could exceed the maximum acceptable service latency. These precise and quantitative definition details should be captured in reliability requirements for the virtualized application. See Section 13.7.1 for more details on measurements.

Since measurements represent key input for service orchestration, it is important that the measurements are well-defined and agreed upon by the cloud consumer. Policies can be built using these measurements and thresholds associated with them to trigger actions by cloud management.

8.3 CLOUD MANAGEMENT

Cloud management is another critical component used by service orchestration. As an interface to policy-based management, cloud management is responsible for growing and degrowing the configuration based on automatic (e.g., policy-based triggers) or manual input (e.g., web interface for adding/removing an application). Two of the mechanisms used for scaling and descaling are rapid elasticity and cloud bursting.

8.3.1 Role of Rapid Elasticity in Cloud Management

One of the roles of policy management is the triggering of defined actions to mitigate risks associated with exceeding engineered limits (e.g., service capacity or latency). Rapid elasticity offers the ability to add or remove cloud resources (VMs, storage devices, etc.) triggered by manual input (i.e., request for a new instance of a VM or application from a web-based GUI) or automatically based on a software trigger (e.g., reaction to a policy decision). Rapid elasticity may be used to mitigate the impact of service overload situations by rapidly adding and provisioning VM instances to manage increases in capacity through redistribution of the load to new VM instances. Conversely, rapid elasticity can scale back VM instances when there is much less traffic than supported by the current VM configuration. Note that policies may include information on where to locate the new VM instances based on regulatory issues, standards, or proximity to users to avoid increases in latency to ensure compliance to those requirements or standards.

8.3.2 Role of Cloud Bursting in Cloud Management

Cloud bursting (discussed in Section 7.4, "Cloud and Capacity") enables additional service capacity to be added outside the data center. An example of cloud bursting is when services running in a private cloud no longer have sufficient resources to meet their computing needs within the private cloud and must expand into a public cloud in order to obtain those resources. Service federation must be provided to include the necessary mechanisms to broker information on security-related identities, identity attributes, and authentication among the different security realms in the private and public clouds. Since there are security risks in expanding into another cloud, particularly a public cloud, this mechanism is often recommended for services that do not have to deal with sensitive information. There may also be risks of incompatibility in the public cloud infrastructure, making it more difficult for the service to run outside its private cloud. These disadvantages must be weighed against the promise of additional resources for the rare times when resources might be temporarily required. Figure 8.2 shows an example of how VMs in an Enterprise private data center are scaled into the public cloud for additional capacity.

8.4 SERVICE ORCHESTRATION'S ROLE IN RISK MITIGATION

Service orchestration provides a framework for managing compliance to customer expectations for reliability, latency, and security regulatory compliance through clear definition of customer requirements, careful service monitoring against those requirements, and mitigation actions when noncompliance issues arise. The following sections will discuss some of the risks and mitigations of those risks.

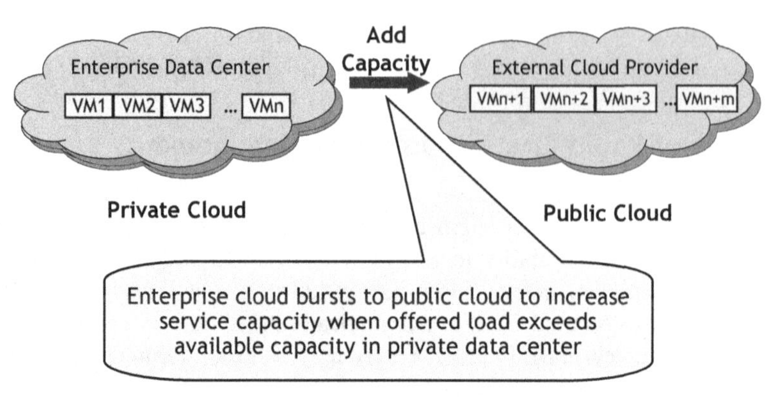

Figure 8.2. Example of Cloud Bursting.

8.4.1 Latency

There is the risk that traffic is directed to sites that are so heavily loaded that they cannot meet the service latency requirements. To identify latency issues, policies can be defined that use samples of transaction latency over a period of time to algorithmically identify trends that exceed the allowable rates defined in SLR's. If latency issues are identified, then these latency risks can be mitigated by:

- Using load balancing algorithms that attempt to route service to the server/data center closest to the user.
- Increasing bandwidth allocation to the servers with a heavy traffic load.
- Expanding service with rapid elasticity taking into account location of the primary users.
- Collocating primary data storage close to the accessing servers to minimize data access time.
- Collecting and reporting latency data so that any increases in latency can be monitored and managed.

Through careful monitoring and managing the latency, risks can be greatly reduced.

8.4.2 Reliability

The reliability of an application in the cloud environment can be compromised by risks associated with the sharing of resources, as well as the dynamic nature of its scaling and descaling. Policy management can help mitigate risks by actively monitoring key reliability indicators and providing escalation triggers and procedures when thresholds of key reliability indicators are exceeded. One of these key reliability indicators is defects per million (DPM). SLR's will generally indicate the maximum number of failures allowed over a period of time based on service expectations. A policy can then be defined, indicating that when a particular category of defects (i.e., failed transactions) reaches a certain limit over a period of time (e.g., 30 minutes), a critical alarm is generated. Service orchestration is responsible for taking this reliability policy as input, monitoring the system for defined defects, and issuing an alarm when the limit has been reached or exceeded. Reliability risks can be mitigated by moving traffic away from the failing component when it has neared the limit but before it has exceeded the limit. High availability mechanisms could execute a failover of the active VM instance to another VM in its failure group if the number of failures reaches a particular limit. Either way, policies could be used to trigger preventative actions based on service monitoring of events and data trends.

8.4.3 Regulatory

In the cloud environment, especially with automated mechanisms, such as rapid elasticity, there is a possibility that applications are scaled into areas that are outside regulatory boundaries. Rules and constraints need to be defined in policies to ensure compliance to regulatory requirements. The Open Data Center Alliance provides regulatory policy management guidance to help cloud consumers assess the regulatory requirements associated with their use of cloud services [ODCA]. Policy management can be used to mitigate this risk of regulatory noncompliance by defining policies that check for adherence to the rules and constraints established by the regulatory requirements and ensure compliance when selecting a site for VM instantiation. When architecting a system, it is important to understand the regulatory conditions around that service and countries it is being operated in to make sure that the system is properly configured and managed through the use of policies to meet those conditions. Rules for server location, data storage, and service expansion should be built into the policies to ensure that they do not fall outside of the bounds established and to trigger alerts or the inability to install VMs or data storage devices on servers that do not meet the regulatory requirements.

8.4.4 Security

Although outside the scope of this analysis it is important to note concerns around security. In the virtualization and cloud environments, there are additional security concerns beyond those of traditional systems due to multitenancy and rapid elasticity. In the case of multitenancy, it is important to manage access to the tenants and maintain their isolation from each other. In the case of rapid elasticity, it is important that new instances of the applications still meet the security requirements established between the cloud service provider and the customer. As with regulatory conditions, security requirements must be well understood when configuring a system, and security policies and constraints should be created to assure compliance to those security requirements.

DMTF provides the following examples of security policies in [DSP0102]:

- *Access Control.* Only specified cloud users have access to a particular service instance. Policies dictate which users can modify that service instance and where and when it can be deployed.
- *Network Security Policies.* These policies indicate how a service instance connects with other external service instances or resources (e.g., via firewall rules or packet inspection).
- *"Scope" of the Security Policies.* These policies specify in which regions or zones the instances are allowed to function.

These examples, as well as the supporting information in [DSP0102], provide architectural guidance in setting up a security infrastructure that can mitigate many of the risks found in the cloud environment. Service orchestration also includes service federation, particularly for cases in which cloud bursting is allowed and manages the implementation of the mechanisms needed to broker authentication across security realms as mentioned in Section 8.3.2.

8.5 SUMMARY

Service orchestration provides a framework for managing the complexity of the cloud environment through policies, data monitoring, automation, and cloud management to ensure compliance to customer requirements for availability, reliability, and latency. An effective service orchestration framework should consist of the following:

- *Mechanisms that collect and monitor measurement data against thresholds.* The thresholds may consist of multiple levels indicating how close the number of events within a time period is tracking against customer limits (e.g., percent failed transactions within a particular measurement period). Both the definition of the measurements as well as the thresholds must be agreed upon by the cloud provider and cloud consumer.
- *Policy management system* that includes a well-designed information model that supports the definition of rules, conditions, and actions to be taken. The system should be flexible enough to support complex rules involving multiple conditions and actions. Although template policies may be available specific service policies must also be agreed upon by the cloud provider, as well as the cloud consumer.
- *Cloud management system* that can perform, manage, and report on the actions dictated by the policy management system.
- *Automation* should be a key attribute of the service orchestration framework. Automation is key to minimizing operational complexity—and the associated procedural errors—as well as improving reliability and availability for those

cases in which policies resulted in actions that directed traffic away from VM's that were experiencing unacceptable service performance.

As part of the monitoring and management of the service, service orchestration is also able to mitigate some of the risks introduced by the highly dynamic cloud environment. Section 11.3.6, "Service Orchestration Considerations," will expound upon these mitigation techniques and provide recommendations for maximizing reliability, availability, and latency using service orchestration.

9

GEOGRAPHIC DISTRIBUTION, GEOREDUNDANCY, AND DISASTER RECOVERY

Traditional high availability arrangements deploy sufficient local excess capacity to promptly recover service following a single hardware or software failure, such as failure of a field replaceable hardware unit. Hardware must be operated in a physical location like a data center, and physical locations are inherently subject to catastrophic or force majeure events, like fires, earthquakes, floods, tornadoes, acts of war (including terrorism), and so on. These disastrous events can render some or all of the equipment deployed at the impacted site unavailable or inaccessible. The best practice to mitigate the service continuity risk of disaster events is to deploy redundant system configurations to a site that is geographically distant from the primary site to assure that no single disaster event will impact both sites. Geographically separated system redundancy is called geographic redundancy, or simply georedundancy.

The common cloud characteristic of geographic distribution is necessary but not sufficient to support georedundancy and disaster recovery. This chapter begins with an explanation of the differences between georedundancy and simple geographic distribution. Traditional disaster recovery principles are introduced, followed by a discussion of how virtualization and cloud computing offer improved options for georedundant disaster recovery. The chapter concludes with a discussion of potential service availability benefits of disaster recovery that are sometimes ascribed to georedundancy, and how these corollary benefits are impacted by virtualization and cloud computing.

Reliability and Availability of Cloud Computing, First Edition. Eric Bauer and Randee Adams.

9.1 GEOGRAPHIC DISTRIBUTION VERSUS GEOREDUNDANCY

Geographic distribution can be the basis for ad hoc or formal disaster recovery planning, but geographic distribution alone is insufficient to assure timely service recovery from disaster events that render a site unavailable or inaccessible. Georedundancy has two additional key requirements beyond simple geographic distribution of systems:

1. Plans and mechanisms are in place to rapidly migrate service away from a failed or impacted data center to an alternate data center in the event of catastrophic site failure. For example, application data must be securely stored at a geographically separated site and arrangements must be in place to assure that fresh backup data can be promptly restored to the recovery site.
2. Sufficient excess capacity is deployed so that the total engineered traffic load can be served indefinitely with acceptable service quality when any single data center is unavailable.

9.2 TRADITIONAL DISASTER RECOVERY

Traditional disaster recovery strategies were organized by the SHARE group [SHARE] into tiers offering better (i.e., shorter) recovery time objectives (RTO) and recovery point objectives (RPO) [TIPS0340] [Wikipedia]:

- *Tier 0: No Offsite Data.* Tier 0 enterprises have no disaster recovery plan and no saved data. Recovery time from disaster may takes weeks or longer and may ultimately be unsuccessful.
- *Tier 1: Data Backup with No Hot Site.* Tier 1 enterprises maintain data backups offsite but do not maintain a hot site. Backup data must typically be physically retrieved (so-called pickup truck access method, PTAM), and thus significant time is required to access backup media. Since Tier 1 enterprises may not maintain their own redundant servers to recover service onto, time may be required to locate and configure appropriate systems.
- *Tier 2: Data Backup with a Hot Site.* Tier 2 enterprises maintain data backups as well as a hot site, and thus recovery times are faster and more predictable than in Tier 1.
- *Tier 3: Electronic Vaulting.* Tier 3 enterprises maintain critical data in an electronic vault so that backup data is network accessible to the hot site rather than requiring backup media to be physically retrieved and transported to the hot site.
- *Tier 4: Point-in-Time Copies.* Tier 4 enterprises maintain more timely point-in-time backups of critical data so that more timely backup data is network accessible to the hot site, thus reducing the RPO.
- *Tier 5: Transaction Integrity.* Tier 5 solutions assure that transactions are consistent between production systems and recovery sites. Thus there should be little or no data loss from a disaster.

- *Tier 6: Zero or Little Data Loss.* Tier 6 solutions have little or no tolerance for data loss and thus must maintain the highest level of data consistency between production and recovery sites, including data not explicitly protected via transactions. Techniques like disk mirroring and synchronous I/O are generally deployed by Tier 6 enterprises to minimize RPO.
- *Tier 7: Highly Automated, Business-Integrated Solution.* Tier 7 solutions automate disaster recovery of Tier 6 solutions, thus shortening the RTO and with minimal RPO.

Each traditional tier successively will support increasingly better recovery point or recovery time objectives but at an additional cost for the business.

A geographically distant alternate site to recover service to following a disaster is obviously a critical ingredient of georedundancy. Traditional recovery site options are broadly classified as follows:

1. *Ad Hoc Site.* enterprise can simply plan to find a new facility after a disaster occurs, and have the replacement equipment delivered and installed at that new facility. This ad hoc strategy naturally yields the longest service disruption following a disaster.
2. *Cold Recovery Site.* ISO/IEC 24762:2008 [ISO24762] defines cold recovery site as a facility "with adequate space and associated infrastructure—power supply, telecommunications connections, environmental controls, etc—to support organization Information Communication Technology(ICT) systems, which will only be installed when disaster recovery(DR) services are activated."
3. *Warm Recovery Site.* ISO/IEC 24762:2008 [ISO24762] defines a warm recovery site as a facility "that is partially equipped with some of the equipment, computing hardware and software, and supporting personnel, with organizations installing additional equipment, computing hardware and software, and supporting personnel when disaster recovery services are activated."
4. *Reciprocal Backup Agreement.* Some governmental agencies and industries have mutual aid agreements to support each other in time of need (i.e., disaster recovery).
5. *Service Bureau.* Some companies offer processing capabilities for both ordinary and disaster recovery needs. Note that in the context of cloud computing, service bureaus might now be said to offer "Disaster-Recovery-as-a-Service" (DRaaS).
6. *Hot Site.* ISO/IEC 24762:2008 [ISO24762] defines hot recovery site as a facility *hot site*, that is fully equipped with the required equipment, computing hardware and software, and supporting personnel, and fully functional and manned on a 24 × 7 basis so that it is ready for organizations to operate their ICT systems when DR services are activated." Note that cloud computing data centers are generally hot sites.

Traditional geographic redundancy typically refers to an alternate data center that has sufficient equipment and facilities to promptly serve the critical traffic load with acceptable service quality and is separated far enough from the primary site to be unaffected by any single catastrophic event. Thus, georedundancy generally refers to a hot site somewhere with some prior arrangements to rapidly recover service in hours or minutes, rather than warm, cold, or ad hoc disaster recovery plans that may take weeks or months to recover service.

9.3 VIRTUALIZATION AND DISASTER RECOVERY

To assure rapid recovery time objectives with traditional, nonvirtualized deployments, it was generally necessary to deploy similar or identical hardware configurations—often with identical software releases installed and configured—to georedundant recovery sites. This inherently constrained the utility of the georedundant equipment because distinct hardware resources might be necessary for each application to be protected. For example, protecting application A1 in site 1 and B2 in site 2 on a third recovery site might require redundant instances of both systems (A3 and B3) on a third recovery site, because although no single disaster could impact both A1 and B2, both of these applications could not individually be rapidly and reliability recovered onto the same physical hardware.

Virtualization simplifies traditional disaster recovery by relaxing the compatibility requirements on hardware deployed to the recovery site, thus making it feasible for a virtualized pool of resources operated by a service provider to meet the disaster recovery needs of many enterprises. In traditional georedundancy, the hardware on the recovery site must be well matched to the primary site to assure that identical application and platform software can run with sufficient capacity to carry the entire traffic load served by the impacted site with acceptable service quality, reliability, and latency. With virtualization, the hypervisor masks minor differences in hardware configurations between the specific hardware configuration of both the primary and recovery system hardware. Thus, virtualization reduces the hardware compatibility requirements for recovery sites, which can make it easier to select and provision disaster recovery sites, even simplifying ad hoc recovery strategies.

Just as virtualization permits applications to be consolidated onto shared hardware resources, virtualization facilitates sharing of hardware for disaster recovery. Thus, although sufficient hardware capacity must still be engineered onto the georedundant site, virtualization enables those hardware resources to flexibly support disaster recovery of several critical applications from the same or perhaps from several different data centers, minimizing the need for nonshareable application-specific hardware. As a result, the capital expenses associated with configuring a georedundant site are reduced, along with the operating expenses for data center floor space, cooling, and so on.

While live migration is obviously an infeasible option for general disaster recovery because one does not typically have the luxury of advance warning before catastrophic site failure, recovery times might be shortened by activating paused or snapshot VM images on the recovery site rather than booting applications from scratch. Even if applications are booted from scratch for disaster recovery, the RTO onto virtualized platforms should be at least comparable with the RTO onto traditional cold standby configurations. If an active/standby or active/active georedundancy arrangement is used, then the RTO for native and virtualized deployments should be essentially the same.

9.4 CLOUD COMPUTING AND DISASTER RECOVERY

Some public clouds are the modern version of "service bureaus" discussed in Section 9.2, "Traditional Disaster Recovery"; private clouds can be a modern instantiation of "hot sites" for disaster recovery, and community clouds may offer similar features as "reciprocal backup agreements." By prearranging for disaster recovery as a service, formal plans can be put in place to assure that resources necessary to recover service can be allocated in the same facility that enterprise data is vaulted in so that wide area network (WAN) bandwidth bottlenecks between the data site and the new service site do not prolong RTO. In addition, clouds offer storage services, including electronic vaulting, so enterprises can eliminate the burden of moving and managing physical media containing backup data.

Alternately, enterprises can consciously plan to rely on the rapid elasticity and geographic distribution offered by cloud service providers to recover users impacted by a site disaster by redirecting their traffic to one or more geographically separated data centers. While this is often technically feasible, successful disaster recovery plans must address the following requirements:

- *All data from primary site must be vaulted or replicated to a remote site.* If the recovery site is not the same as the vault/replication site, then recovery data must traverse a WAN that could slow disaster recovery if the data set is large.
- *Service capacity on recovery site(s) must be able to grow fast enough to meet RTO objective*, including time to import necessary data from electronic vault or other repository. For session-oriented services, it may take significant processing effort to authenticate and authorize each impacted user individually, as well as effort to (re)build session context to recover service. While under normal circumstances, users may log on to the application across a broad window of time—thereby keeping the logon/session setup load modest—disaster recovery is likely to prompt a very large number of users to attempt to log on/recover to the recovery site essentially simultaneously. As a result, it may be necessary to engineer peak capacity to support the unusually high rate of user logons and session setups in disaster recovery scenarios. As described in Section 4.9, "Expectations of IaaS Data Centers," the Open Data Center Alliance defines four classifications of infrastructure as a service (IaaS) providers, and the recoverability expectations from [ODCA-SUoM] are given in Table 9.1.
- *Mechanisms must be in place to redirect user traffic to recovery site.* These mechanisms should be transparent to users and require no user changes or reconfiguration of user programs.

TABLE 9.1. ODCA IaaS Recoverability Objectives [ODCA-SUoM]

SLA Level	Description
Bronze	Reasonable efforts to recover the IaaS service (e.g., access to boot volumes and ability to reboot the cloud subscriber's virtual environment again) with up to 24 hours of data loss (e.g., loss of boot disk updates due to no intraday backup), and up to 24 hours of recovery time. No site disaster recovery (DR). Note that the focus is on recoverability of the underlying service, after which cloud subscriber still has their own recovery to complete.
Silver	Provisions made to recover within 4 hours, with up to 24 hours of data loss. (No DR for full site disaster.)
Gold	Enhanced recovery capability to recover within 2 hours for hardware failure, 24 hours for site failure, and no more than 4 hours of data loss.
Platinum	Highest recovery focus to provide as close to continuous nonstop availability as possible, aiming for <1-hour recovery and <15-minute data loss even in the event of full site failure.

Source: Open Data Center Alliance.

- *Nonservice-impacting migration back to recovered primary site shall be possible.* While there may be a service impact due to the disaster event itself and the subsequent georedundant recovery, there should be little or no service impact for the planned and graceful service migration back to the repaired primary site.
- *Periodic (e.g., annual) disaster drills shall be possible to verify that disaster recovery mechanisms and plans meet RTO and RPO expectations, and that graceful service migration back to the primary site has minimal impact on user service.*

Since rapid elasticity, on-demand self service, and geographic distribution are characteristics of cloud computing, some enterprises can plan to rely on ad hoc recovery from disasters. In addition to dramatically increasing the risk that service itself and enterprise data will ultimately not be successfully recovered, the recovery times themselves will inevitably be significantly longer than if formal disaster recovery plans had been put in place and tested prior to the disaster event. Fundamentally, since ad hoc recovery cannot be tested and debugged effectively before a real disaster event, enterprises that opt for ad hoc disaster recovery inevitably rely simply on the best efforts of their staff in the very stressful and chaotic postdisaster period to salvage enterprise service and data—and often the enterprise itself—following a disaster event.

The cloud computing ecosystem assures that a wide variety of cloud computing service providers enable enterprises to select data storage options that meet the enterprise's RPO requirements and prearrange for emergency resource availability to meet RTO requirements. Cloud-based disaster recovery strategies do not eliminate the need for enterprises to execute periodic (e.g., annual) disaster drills to assure that disaster recovery works as planned and that RTO and RPO requirements are met.

9.5 GEOREDUNDANCY RECOVERY MODELS

Georedundancy is activated by detecting a failure at a primary site and redirecting impacted traffic to a redundant site that is configured and ready to recover service for the impacted traffic. There are fundamentally three georedundancy recovery strategies:

- *Manually Controlled Recovery.* This is the traditional disaster recovery strategy: a business leader formally declares a disaster and a well-defined disaster recovery plan is executed to manually transition operations to a recovery site.
- *Server-Driven Recovery.* Redundant servers or other systems monitor the health of servers and upon detecting failure of a system serving user traffic a server automatically takes actions to recover service to a redundant server instance to mitigate service impact without requiring manual action by maintenance engineers.
- *Client-Initiated Recovery.* The client application, device, or user detects the failure and explicitly initiates a recovery action. A simple example of client initiated recovery is a human user detecting a stuck or nonresponsive web server, then clicking "cancel" on their browser followed by "reload" to recover from a web server failure which hopefully retries the request to web server that is available. Readers can easily imagine client application architectures that automate the failure detection and automatic service recovery to an alternate application instance (i.e., a different IP address), which is located in a different data center.

9.6 CLOUD AND TRADITIONAL COLLATERAL BENEFITS OF GEOREDUNDANCY

Some enterprises assume that traditional georedundancy offers several benefits beyond disaster recovery: reducing planned service downtime; mitigating catastrophic element failure; and mitigating uncovered (i.e., failures that are not detected and recovered) and duplex element failures. While these traditional benefits are feasible with cloud deployments, they are no longer necessarily tied to georedundancy. Let us consider each of the assumed collateral benefits of georedundancy:

9.6.1 Reduced Planned Downtime

Major activities, such as growing or degrowing hardware configurations of systems or physically moving equipment, have less risk of impacting service if traffic is gracefully drained from the systems prior to beginning the activity. Orderly migration of traffic away from the target site to the georedundant site is a traditional way to quiesce a site so it can be taken offline prior to executing planned activities. Obviously, there is far less risk that a successful or failed activity will disrupt user service if there is no traffic flowing through the element when the activity is performed. Virtualization and cloud computing offer options for minimizing planned downtime without requiring georedundancy; this is discussed in detail in Chapter 5, "Reliability Analysis of Virtualization."

9.6.2 Mitigate Catastrophic Network Element Failures

Occasionally a system will experience a catastrophic or duplex failure (i.e., both redundant units are simultaneously unavailable) that defeats high availability mechanisms and will require hours to repair. For example, water—even rodent urine—in a rack of equipment might physically damage multiple hardware units in a single system, thus overwhelming the high availability design of the system and requiring a service impacting hardware repair before service can be restored. Traditionally, if service can be recovered to a georedundant instance of the impacted network element with significantly less overall impact than repairing the damaged network element, then georedundant recovery may be a good option. Note that traffic is sometimes migrated back to the primary system after it is repaired, so the potential service impact of eventual traffic migration after repair should be considered when deciding whether or not to engage georedundancy to mitigate an element failure. Cloud computing dramatically changes the economics of hardware redundancy arrangements by making it possible to distribute individual instances of redundancy to different hypervisors on physical servers that are separated within a data center far enough that nothing short of a catastrophic site failure would impact all redundant module instances of a single system. For example, one can imagine a roof leak compromising a single chassis or rack of equipment, thus rendering a traditional system unavailable; but one cannot imagine a single roof leak simultaneously, compromising both redundant application instances running on hypervisors on physical hardware at opposite sides of the same data center. For example, Amazon Web Services uses the concept of "availability zones" to mitigate the risk of catastrophic failures within a data center by assuring that each zone is physically distinct with independent networking, power, and cooling infrastructure [AWS08].

9.6.3 Mitigate Extended Uncovered and Duplex Failure Outages

Some types of properly configured solutions can even be engineered to leverage georedundancy to mitigate more common uncovered and duplex failure downtime, thereby boosting overall service availability seen by users. Specifically, automatic failure detection and recovery by client applications can be engineered to mitigate uncovered and duplex failure downtime by having clients automatically switch to georedundant system instances if requests are not properly served by the primary system instance with acceptable service latency, quality, and reliability. This topic is considered in detail in "Beyond Redundancy: How Geographic Redundancy Can Improve Service Availability and Reliability of Computer-Based Systems" [Bauer11]. Traditionally, an active redundant system instance would be used to mitigate uncovered or duplex failure outages via client-initiated recovery because client-initiated recovery often offers a more practical way to rapidly detect and recover from these failures. Virtualization and cloud computing changes the economics of (virtual) hardware, so alternate solution architectures can be considered to mitigate the risk of uncovered and duplex failure events that enable clients to recover service to redundant system instances collocated with the failed instance, rather than requiring them to switch to a more distant data center.

9.7 DISCUSSION

Fundamentally, cloud computing providers offer distributed hot sites for disaster recovery, and virtualization coupled with measured service and rapid elasticity means that georedundant solutions can be deployed via cloud computing at lower expense than traditional georedundancy. Ultimately, the feasible service availability benefits of cloud-based georedundancy are essentially the same as the feasible benefits of comparable traditional georedundancy deployments. Virtualization does offer a slight incremental benefit over traditional georedundancy in its ability to reduce downtime via live migration and to reduce the negligible risk of catastrophic physical failure of a traditional network element by physically distributing hardware resources supporting the application across a data center, but these are small benefits that are not usually even quantified. The economics of cloud computing permits at least some enterprises to shift from the traditional "active" primary site plus "standby" disaster recovery site model to an all-sites active model, which offers availability benefits that are discussed in Chapter 11, "Recommendations for Architecting a Reliable System."

RECOMMENDATIONS

10

APPLICATIONS, SOLUTIONS, AND ACCOUNTABILITY

Cloud computing is inherently a more complicated arrangement than tradition computing; instead of suppliers offering equipment and applications directly to enterprises that will operate the equipment and applications, cloud computing separates the cloud consumer enterprise that rents computing resources from the cloud service provider who owns and operates the computing resources. In addition to suppliers, cloud service providers, and cloud consumers, there are likely to be several communications service providers hauling IP traffic between end users and cloud data centers. All of these players are accountable for some service impairments that can impact the quality of experience for end users. This chapter offers canonical service downtime budgets and models to help understand how accountability changes as traditional applications migrate to the cloud. This chapter also frames the broader challenge of end-to-end service availability via several standard service measurement points.

10.1 APPLICATION CONFIGURATION SCENARIOS

Virtualization enables deployment flexibility beyond the options of traditional application deployment. In rough order of increasing complexity, these virtualization scenarios are as follows:

Reliability and Availability of Cloud Computing, First Edition. Eric Bauer and Randee Adams.

- *Traditional or Native Deployment (i.e., No Virtualization Is Used).* A software application is installed and integrated with an operating system running directly on nonvirtualized physical hardware.
- *Hardware Independence Usage Scenario.* virtualization reduces or eliminates an application's dependence on the specifics of the underlying physical hardware in the hardware independence usage model. While the application may still require the same machine instruction set (e.g., Intel), virtualization can decouple the physical memory, networking, storage, and other hardware-centric details from the application software so the application can be moved onto modern hardware rather than being tied to legacy hardware platforms.
- *Server Consolidation Usage Scenario.* In the server consolidation usage scenario, virtualization is used to increase resource utilization by having multiple applications share hardware resources. In some cases, this provides the ability to take advantage of otherwise underutilized hardware resources. Moore's law assures that the processing power of servers grows steadily over time, yet the processing needs of individual application instances does not necessarily grow as rapidly. Thus, in many cases, the growth in available processing power may not be effectively used by a single application running on the server hardware. In these cases, applications may nominally oversubscribe hardware capacity and the hypervisor relies on statistical usage patterns to make resource sharing work well.
- *Multitenant Usage Scenario.* A multitenant deployment permits multiple independent instances of a single application to be consolidated onto a single virtualized platform. For example, different application instances can be used for different user communities, such as for different enterprise customers; web service and electronic mail are examples of common multi-tenant applications as multiple independent instances of the same application may be running on a virtualized server platform to simultaneously serve different web sites or users from different enterprises. While some applications are explicitly written to be multitenant, other applications were written with the design assumption that a single application instances on a single hardware platform supports a single user community. Virtualization can facilitate making these single system-per-user community applications support multitenancy configurations in which several distinct user communities peacefully coexist on a shared, virtualized hardware platform.
- *Virtual Appliance Usage Scenario.* The virtual appliance notion of the Distributed Management Task Force [DSP2017] represents one ultimate vision of virtualization. In the appliance vision, applications are delivered as turnkey software prepackaged with operating systems, protocol stacks and supporting software. The supplier benefits by being able to thoroughly test the production configuration of all system software, and the customer benefits from simpler installation and maintenance, and should enjoy the higher quality enabled by having their field deployment software configuration be 100% identical to the reference configuration that was validated by the appliance supplier.

- *Cloud Deployment Usage Scenario.* The cloud deployment usage scenario provides the most flexible configuration, which is able to grow and degrow automatically along with changing workloads. With the flexibility of cloud deployment comes increased complexity, which is mitigated by service orchestration and elasticity, which provide automation guided by policies and usage data. Cloud deployment risks and mitigations are primarily considered in Chapter 13, "Design for Reliability of Cloud Solutions."

Undoubtedly not all usage scenarios will apply to all applications. As a practical matter, some organizations will integrate virtualization into their existing applications over several releases by supporting different usage scenarios in different releases. For example, an application might be engineered and tested to support server consolidation in one release; engineered and tested for multitenant and cloud deployment in another release, and eventually offered as a virtualized appliance in a later release.

10.2 APPLICATION DEPLOYMENT SCENARIO

Neither traditional nor virtualized systems are useful in isolation; to deliver useful service to users, some physical hardware must be installed in a suitable physical environment and supplied with both power and IP connectivity. Operationally, this is generally achieved by deploying applications into a data center (see Section 1.3.1, "What Is a Data Center?"). Organizations do not generally deploy applications by simply connecting a traditional or virtualized server hosting an application into a data center to the public internet. Instead, there is usually a security appliance like a firewall or deep packet inspection server to enforce a security perimeter to protect the application from external attack. Within the security perimeter is often a load balancer to distribute the offered load across the application's front-end servers. Many applications are architected with multiple tiers to simplify scalability, such as supporting user interface and client interaction in a tier of front-end servers, implementing application logic and business rules in a middle tier, and maintaining application data in a third tier of database servers. As critical applications are generally designed to remain operational even during routine maintenance and repair, these elements are often deployed across redundant instances. All of this physical hardware is installed in a data center that provides power, a suitably controlled environment, and network connectivity to all of the elements, including the routers that connect the data center to the public Internet. This canonical application deployment architecture is illustrated in Figure 10.1. Note that although the diagram shows pictures of server hardware elements, software on routers, perimeter security, load balancers, application front-end, application back-end, and database servers is implicitly assumed to be included in this deployment diagram.

Thus, the service availability seen by a user outside of the data center implicitly integrates the downtime of the data center's routers, perimeter security, load balancers, power, environment and IP interconnection infrastructure, as well as the target application, and all this equipment and infrastructure is inevitably subject to failures, just as the target application is. The service availability seen from the public Internet for one

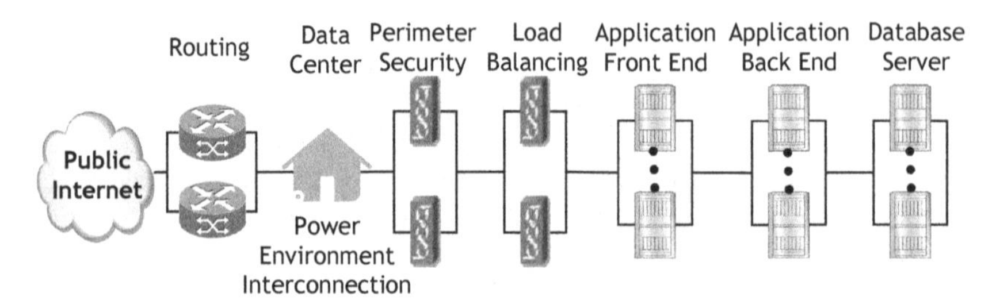

Figure 10.1. Canonical Single Data Center Application Deployment Architecture.

application instance in one data center is inevitably lower than that of the product-attributable service availability of standalone applications.

10.3 SYSTEM DOWNTIME BUDGETS

System architects and reliability engineers use downtime budgets to manage service availability. For example, a system with "five 9's" service availability is budgeted to have annual service downtime of 5.26 minutes per year; 5.26 minutes per year is the multiplicative product of the number of minutes per year (i.e., 365.25 days per average year times 24 hours per day times 60 minutes per hour) multiplied by 0.001% (99.999% uptime means 0.001% downtime). As with any budget:

- the expected downtime "expenses" are categorized;
- each category is assigned a reasonable allocation of the overall downtime budget;
- category allocations are adjusted to reach an acceptable and optimal total "cost";
- architecture, design, and test plans are managed to achieve the individual downtime allocations; and
- if the downtime budget is missed in one measurement period (e.g., release), then it can be altered, and/or additional effort can be invested in the next period to meet the downtime budget.

Thus, the question of whether or not a virtualized system instance can achieve the same service availability as a native configuration comes down to the question of whether it is feasible and likely that a virtualized deployment can achieve a long-term average downtime budget that is equivalent to the downtime budget of a native system. We consider this question in three steps:

1. review the product-attributable downtime budget of a sample traditional high availability system;

2. alter the traditional product-attributable downtime budget for a hardware-independence virtualized deployment scenario and assess implications for an infrastructure as a service (IaaS) supplier; and
3. revise the hardware independence budget for cloud deployment scenario.

10.3.1 Traditional System Downtime Budget

Traditionally, system downtime expectations and predictions offered by suppliers covered only product-attributable service downtime, which is largely due to software and hardware failures (see Section 3.3.6, "Outage Attributability"). Downtime caused by factors not attributable to the system supplier or the product itself (e.g., power failures, network failures, and human mistakes by the customer's maintenance staff) are generally excluded from product-attributable system availability measurements and predictions because they are allocated to other categories (e.g., customer-attributable downtime). Likewise, the measurement typically covered only agreed service time, so scheduled or planned downtime periods were excluded (see Section 3.3.1, "Service Availability Metric," and Section 3.3.7, "Planned or Scheduled Downtime").

Traditional system downtime budgets allocate unplanned product-attributed service downtime across three broad categories: hardware, software, and planned/procedural (sometimes called "human"). A traditional "five 9's" system budget generally allocates 10% of the budgeted 5.26 prorated minutes (315 prorated seconds) of annual service downtime to hardware, meaning that hardware attributed causes typically gets about 30 seconds of prorated annual downtime. The vast majority of the remaining downtime will be allocated to unplanned software failures, but some of the remaining 4 minutes and 45 seconds might be budgeted to unsuccessful planned and procedural activities like failed software upgrades. The software downtime may be further budgeted either by architectural layer (e.g., application software vs. platform software) or by functional module (e.g., front-end software processes vs. back-end software processes), or may be factored in some other way. Table 10.1 gives a sample "five 9's" product-attributable downtime budget for an application.

Note that while five 9's technically means 5.26 annualized down minutes, downtime budgets typically round this to 5.25 down-minutes for simplicity. This tiny rounding error is likely to be far smaller than the uncertainty in the estimates of individual downtime categories, so it does not materially affect the utility of the budget.

10.3.2 Virtualized Application Downtime Budget

The hardware independence and server consolidation usage scenarios insert a hypervisor, and perhaps a host OS instance, between the guest OS supporting the target application and the underlying hardware. While hardware-attributed downtime doesn't simply vanish in these scenarios, accountability for hardware-attributed downtime may be different. In particular, since the virtualized platform and underlying physical hardware may be supplied separately from the application software, the virtualized application budget explicitly considers software and virtualized hardware downtime separately.

TABLE 10.1. Sample Traditional Five 9's Downtime Budget

Product-Attributable Downtime Category	Annualized Target for 99.999% Seconds	Annualized Target for 99.999% Minutes	%
Hardware sttributable—target: 30 seconds = 0 minute 30 seconds			
Hardware failure downtime—service downtime triggered by hardware failures.	30	0.5	10
Software attributable—target: 225 seconds = 3 minutes 45 seconds			
Software failure downtime—service downtime due to software failures of platform and/or application software.	225	3.75	71
Planned and procedural attributable—target: 60 seconds = 1 minute 0 second			
Successful scheduled activities—service downtime "by design" for successful upgrade, update, retrofit, hardware growth, and other scheduled or planned maintenance activities.	0	0	0
Unsuccessful procedural activities—service downtime attributed to unsuccessful or botched maintenance activities such as upgrade, update, retrofit, hardware growth, and provisioning.	60	1	19
Total	315	5.25	
Availability	99.999%		

Table 10.2 modifies the sample traditional five 9's budget of Table 10.1 for hardware independence or server consolidation usage scenarios as follows:

- "hardware failure downtime" becomes product-attributable "virtualized hardware platform downtime";
- "software failure downtime" is unchanged; and
- planned and procedural is refactored from outcome-based attribution (i.e., successful scheduled activities vs. unsuccessful scheduled activities) to downtime attribution based on application versus virtualized platform.

Thus, the virtualized application retains the entire software failure downtime attribution, as well as a portion of the planned and procedural attribution, while the hardware failure downtime and hardware-related planned and procedural downtimes are explicitly separated. This enables one to explicitly address the feasibility and likelihood of a virtualized application achieving its product-attributable availability target by independently considering the feasibility and likelihood of the virtualized software achieving its downtime budget over the long term and the feasibility and likelihood of the virtualized hardware platform achieving its product-attributable downtime budget over the long term as well.

TABLE 10.2. Sample Basic Virtualized Five 9's Downtime Budget

Product-Attributable Downtime Category	Annualized Target for 99.999% Seconds	Annualized Target for 99.999% Minutes	%
Hardware attributable—target: 30 seconds = 0 minute 30 seconds			
Virtualized hardware platform downtime—service downtime attributed to virtualized hardware resources (e.g., virtual CPU, memory, disk, and networking).	30	0.5	10
Software attributable—target: 225 seconds = 3 minutes 45 seconds			
Software failure downtime—service downtime due to software failures of platform and/or application software.	225	3.75	71
Planned and procedural attributable—target: 60 seconds = 1 minute 0 second			
Application software-related planned and procedural downtime—product-attributable service downtime attributed to successful and unsuccessful planned and procedural activities associated with application	45	0.75	14
Virtualized platform-related planned and procedural downtime—product-attributable service downtime attributed to successful and unsuccessful planned and procedural activities associated with the virtualized hardware platform.	15	0.25	5
Total	315	5.25	
Availability	99.999%		

10.3.3 IaaS Hardware Downtime Expectations

For a virtualized application with the canonical five 9's budget of Section 10.3.2 to achieve five 9's product-attributable service downtime on an IaaS platform, the IaaS platform should offer comparable product-attributed hardware downtime to assure comparable service availability. Architecturally, the IaaS-attributable downtime budget challenge comes down to this: *can the product-attributable downtime of the IaaS provider's infrastructure achieve comparable service downtime to the system's traditional high availability hardware configuration?* Note that for consistency with traditional system downtime budgets and predictions, this allocation considers only hardware and software downtime causes directly associated with emulating the traditional system hardware; this means that the other categories of impairments generally attributed to IaaS service providers (e.g., power, environment, and human) are not included in this product-centric budget. Downtime due to rapid elasticity and other aspects of cloud computing are considered in the next section ("Cloud Based Application Downtime Budget").

Figure 10.2 shows a reliability block diagram (RBD) of sample blade-based high availability system architecture, and Figure 10.3 shows a RBD of equivalent sample high availability IaaS infrastructure. Logically, the Ethernet switch blades of the blade-based system of Figure 10.2 are replaced by pairs of top-of-rack (TOR) and end-of-row

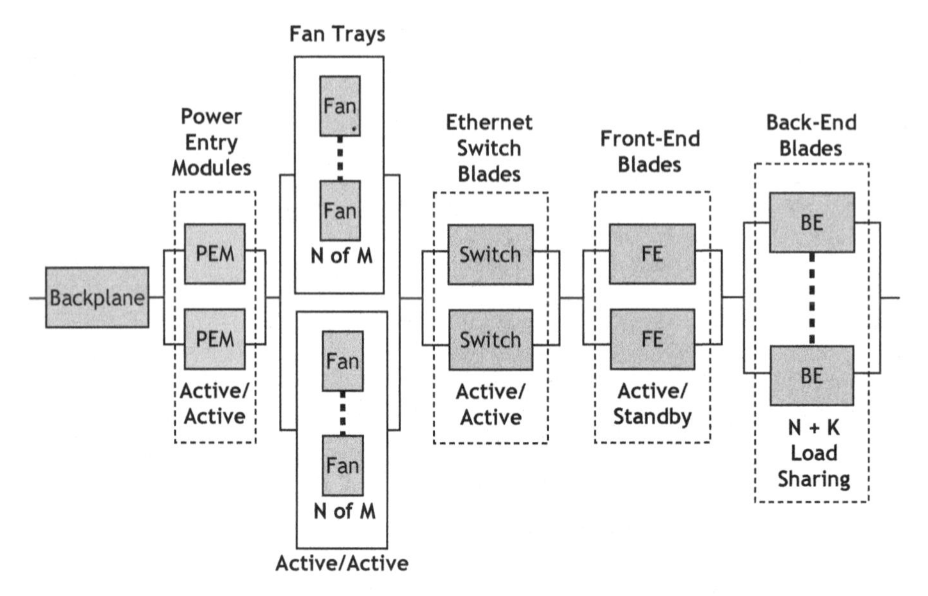

Figure 10.2. RBD of Sample Application on Blade-Based Server Hardware.

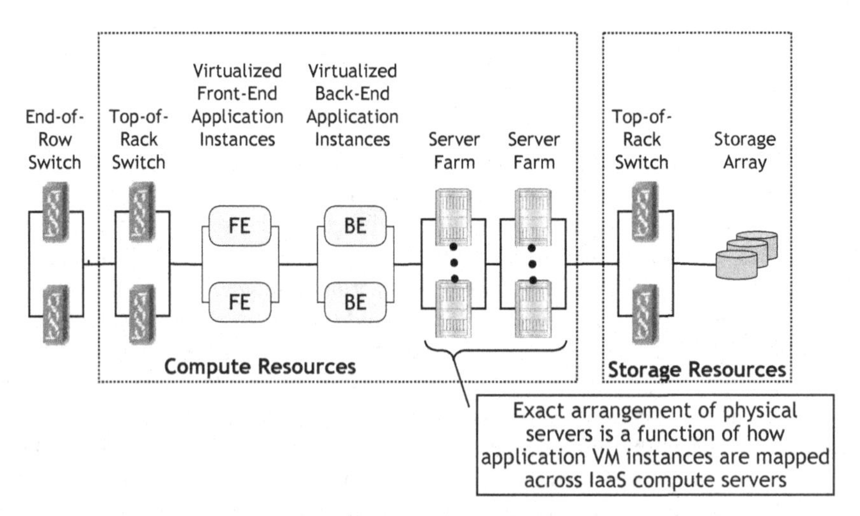

Figure 10.3. RBD of Sample Application on IaaS Platform.

(EOR) Ethernet switches; the compute blades are replaced by portions of the IaaS provider's server farm and disk arrays; and the power entry modules and fan trays are integrated with each of IaaS provider's components. The backplane element in Figure 10.2 logically separates into the internal chassis or backplane arrangements within each of these IaaS components (which are considered in availability modeling of each component) and the Ethernet cables that interconnect the EOR switch, TOR switch, server resources, and storage array; Ethernet cables are not considered in availability modeling. The exact arrangement of compute servers in the RBD is a function of how the application's virtual machines are mapped onto physical hardware. Typically redundant VM instances (e.g., active and standby instances) will explicitly be mapped onto different physical servers so that no physical server is a single point of failure. Note that as the storage array may be in a different rack (or row) than the server resources, a separate pairs of TOR switches (or additional IP infrastructure) may be required to connect the compute and storage resources. Traditional high availability systems are developed to achieve a long-term average of 30 seconds of annualized downtime across Figure 10.3; the question becomes whether IaaS providers can architect their high availability infrastructure (e.g., Figure 10.3) to achieve comparable product-attributable downtime. This question is considered in detail in Section 11.5, "Minimizing Hardware-Attributed Downtime." Fundamentally, while there is likely to be more hardware—and hence a higher hardware failure rate—in the more flexible IaaS deployment configuration than with an optimized traditional native hardware deployment, more effective failure detection, redundancy, and recovery mechanisms can at least partially compensate for this slightly higher underlying hardware failure rate.

10.3.4 Cloud-Based Application Downtime Budget

Cloud deployment scenarios are fundamentally different from the hardware independence usage scenario because the cloud service provider's virtualized hardware platform is offered separately from the application software. While the application supplier is not accountable for the root cause of any hardware failures, the application supplier is responsible for promptly and automatically recovering service following typical hardware failures when the application is deployed in a high availability configuration. For example, virtualized servers hosting application VMs will occasionally fail, thus impacting application users being served by the affected VM instances. Application suppliers are expected to configure their software to automatically detect and recover from these inevitable hardware failures with minimal impact on user service. As at least some of these hardware events are likely to cause brief user service impact, application suppliers can only be expected to achieve their service availability expectations when the underlying hardware offered by the cloud service provider is acceptably reliable and the cloud infrastructure behaves robustly. For example, if the cloud service provider's RAID storage system fails to properly mitigate a hard disk failure and renders application data unavailable for a period, then that downtime should be attributed to the cloud service provider. Likewise, if the cloud service provider's hardware experiences epidemic hardware failures well beyond the prescribed failure rate (e.g., as indicated in the SLA), then the application supplier should not be held accountable for the excess downtime resulting from the application's high availability mechanisms being forced to activate frequently, thus accruing more than the budgeted amount of hardware-attributed service downtime. Alternately, if the cloud service provider's virtualized platform management software fails to maintain proper control of their platform infrastructure and meet the needs of the application VM instances, then it will be very difficult for the application supplier's software to recover service in the required amount of time (e.g., in seconds). Thus, application suppliers should retain a modest cloud platform downtime budget to cover detecting and recovering from ordinary cloud provider platform failures. However, since extraordinary cloud service provider failures are beyond the reasonable ability of application software to address rapidly and automatically, that downtime should be assigned to the cloud service provider rather than the application software. The canonical cloud-based application budget should include a modest allocation for application recovery from ordinary cloud service platform failures, but no application-attributable downtime need be budgeted for planned and procedural downtime of the cloud platform because that is entirely the responsibility of the cloud service provider unless there are application specific mechanisms built in to the planned operations, such as volatile data synchronization.

Cloud deployment introduces expectations for the application to support rapid online elasticity, service orchestration, and perhaps live migration of VM instances to enable the cloud service supplier to better manage their physical resources. By definition, live migration and online capacity changes are executed while the system is online and servicing users, so any failure that impacts service availability for those users should be counted as product-attributed downtime. Although the root cause of product-attributable failures of elastic growth, elastic degrowth, live migration, and other IT service management (ITSM) activities is ultimately likely to be software, the authors recommend creating a new downtime category called "cloud service management." As this new category is an evolution of the traditional application software related planned and procedural category, that downtime allocation is carried forward into "application-attributable cloud maintenance activities." In addition, as the application software supplier is not generally accountable for planned and procedural downtime of the cloud service platform, the downtime that was used for product-attributable planned and procedural downtime of the virtualized platform in Section 10.3.2 can be reallocated to application-attributable cloud maintenance budget to help cover the additional downtime due to elastic growth and degrowth and other cloud-related maintenance actions. The 60 seconds budgeted for product-attributed cloud service management downtime will likely be subdivided by activity such as elastic growth and degrowth versus software release management activities and so on.

A canonical application-attributable downtime budget for a "five 9s" cloud-based application is given in Table 10.3.

Note that while the risk of failure—and expected recovery time—for any particular ITSM operation (e.g., elastic capacity growth) should be fairly constant, increasing the frequency of ITSM actions naturally increases the likelihood of service downtime. For example, if the application supplier estimates one elastic capacity change per week will produce a long-term average of 10 seconds of prorated product-attributable service downtime per application instance per year, then executing an average of two elastic

TABLE 10.3. Canonical Application-Attributable Cloud-Based Five 9's Downtime Budget

	Annualized Target for 99.999%		
Application-Attributable Downtime Category	Seconds	Minutes	%
Cloud platform attributable—target: 30 seconds = 0 minute 30 seconds			
Application downtime recovering from ordinary XaaS failures—service downtime for application to detect and recover from ordinary XaaS platform failures.	30	0.50	10
Software attributable—target: 225 seconds = 3 minutes 45 seconds			
Application software failures—service downtime due to software failures of platform and/or application software.	225	3.75	71
Cloud maintenance attributable—target: 60 seconds = 1 minute 0 second			
Product-attributable cloud maintenance activities—chargeable service downtime for: • elastic capacity growth and degrowth; • software upgrade, update, retrofit, and patching • live migration; and • other IT service management activities.	60	1.00	19
Total	315	5.25	
Availability	99.999%		

capacity changes per day is likely to accrue roughly an order of magnitude more service downtime. Thus, suppliers should make reasonable assumptions for the rate of ITSM actions and accept accountability for meeting downtime expectations based on those assumptions. If the cloud consumer or cloud service provider performs more ITSM actions than were reasonably assumed, then the consumer or service provider should be accountable for the excess downtime.

10.3.5 Summary

Table 10.4 summarizes the evolution of nominal downtime budgets proposed by the authors from a traditional five 9's budget to virtualized deployment (e.g., hardware independence or server consolidation usage scenario) to cloud deployment. The key insight of this table is that downtime allocations stay fairly consistent:

- Hardware related failures still occur and accrue about 10% of overall service downtime.
- (Unplanned) software failures still occur and should accrue about the same amount of overall (prorated) service downtime.
- Product/application-attributed service management (e.g., procedural and maintenance) activities still carry some downtime risk.

TABLE 10.4. Evolution of Sample Downtime Budgets

Sample "Five 9's" Product or Application-Attributable Downtime Budgets			Annual Down Minutes	%
Traditional Deployment	Virtualized Deployment	Cloud Deployment		
Hardware related—target: 30 seconds = 0 minute 30 seconds				
Hardware failure downtime—service downtime triggered by hardware failures.	*Virtualized hardware platform downtime*—service downtime attributed to virtualized hardware resources (e.g., virtual CPU, memory, disk and networking).	*Application downtime recovering from ordinary XaaS failures*—service downtime for application to detect and recover from ordinary XaaS platform failures.	0.50	10
Software attributable—target: 225 seconds = 3 minutes 45 seconds				
Application software failures—service downtime due to software failures of platform and/or application software.			3.75	71
Procedural and maintenance attributable—target: 60 seconds = 1 minute 0 second				
Successful scheduled activities—service downtime "by design" for successful upgrade, update, retrofit, hardware growth, and other scheduled or planned maintenance activities. *Unsuccessful procedural activities*—service downtime attributed to unsuccessful or botched maintenance activities, such as upgrade, update, retrofit, hardware growth, and provisioning.	*Application software-related planned and procedural downtime*—product-attributable service downtime attributed to successful and unsuccessful planned and procedural activities associated with application software. *Virtualized platform-related planned and procedural downtime*—product-attributable service downtime attributed to successful and unsuccessful planned and procedural activities associated with the virtualized hardware platform.	*Product-attributable cloud maintenance activities*—chargeable service downtime for: • elastic capacity growth and degrowth; • software upgrade, update, retrofit, and patching • live migration; and • other IT service management activities.	1.00	19
		Total	5.25	
		Availability	99.999%	

10.4 END-TO-END SOLUTIONS CONSIDERATIONS

In engineering, a *solution* is a design that fulfills (solves) the requirements (constraints) of a business need (problem). In this context, a solution refers to an arrangement of products, facilities, policies, and services that fulfills an information related need. In the context of the eight-ingredient model presented in Section 3.2, a solution consists of hardware, software, networking (both applications protocols and IP), power, environment, humans, and policies. As a practical matter, solutions largely integrate existing and new products (hardware plus software and application protocols), which are typically installed in existing data centers (environments with power and IP networking) that communicate over largely existing IP networks and often leveraging standard application payload syntaxes and semantics. The value add of the solution is exactly how these ingredients are integrated, as well as the policies that govern how these ingredients will function and be operated, administered, maintained, and provisioned by human staff. Solution design for reliability (discussed in Chapter 13) assures that the integration of these eight ingredients to meet a business need for information or communications meets the expectations for service reliability and service availability.

10.4.1 What is an End-to-End Solution?

An end-to-end solution includes the equipment and facilities that connect an end user to an application instance hosted in a data center. Figure 10.4 illustrates a sample end-to-end solution for a user accessing an application instance via a smartphone. For the user to successfully access the application, all of the following components and facilities must be available:

- *End User Device* (smartphone, in this case). Must be fully operational, meaning that hardware and software must be up, battery must be adequately charged, business arrangements (i.e., a service contract) must be in place to assure access to a wireless network, and so on.

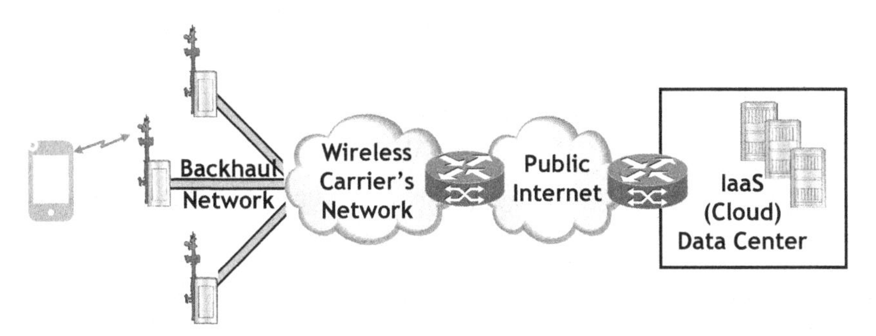

Figure 10.4. Sample End-to-End Solution.

- *Wireless Base Station.* At least one of the carrier's wireless base stations with sufficient capacity and acceptable wireless coverage to the user's device must be available.
- *Wireless Backhaul.* Facilities and equipment must be available to backhaul IP traffic from the carrier's base station to their core network.
- *Wireless Carrier's Core Network.* Must be available to pass traffic to and from the public Internet.
- *Public Internet.* The Internet is not a single monolithic entity, but rather an internetworking arrangement between many network operators. In this context, "public Internet" is shorthand for the one or more service providers that carry IP traffic between the wireless carrier's core network and the data center hosting the application instance serving the user. Note that high reliability data centers will be engineered with connections to several Internet service providers, and wireless carriers will also have network connections to multiple internet service providers. Thus, there are likely to be several redundant IP paths to connect the end user's wireless carrier's core network with the data center hosting the application instance serving the end user.
- *Data Center Infrastructure and Facilities.* The data center hosting the user's application instance, as well as routers, security appliances, load balancers, compute and storage servers, IP infrastructure, and so on, must be available.
- *Target Application.* Obviously, the application itself must be available to serve requests from the end user.

Thus, the service reliability, availability, and latency experienced by the end user can be impacted by far more than merely the application instance itself and the data center hosting the application instance. In fact, equipment and facilities closest to the end user tend to have less redundancy and lower reliability than equipment and facilities in the core of the network and in cloud data centers. For example, while there is often full redundancy in IP networking equipment and facilities in high reliability data centers and carrier's core networks, an end user often has a single (nonredundant) device to access a service with, sometimes has patchy wireless coverage to one, or perhaps several base stations, and each base station may backhaul traffic to the wireless carrier's core network over infrastructure that could be simplex (i.e., nonredundant).

10.4.2 Consumer-Specific Architectures

Cloud computing facilitates more diverse solution architectures than were traditionally deployed. For example, one can imagine an enterprise opting to use cost-effective compute resources offered by a regional cloud service provider but choosing to maintain enterprise data in their private data center. Applying this requirement to the canonical deployment model of Figure 10.1 yields Figure 10.5. While the assumed customer-specific requirements need not change the solutions elements (e.g., the same

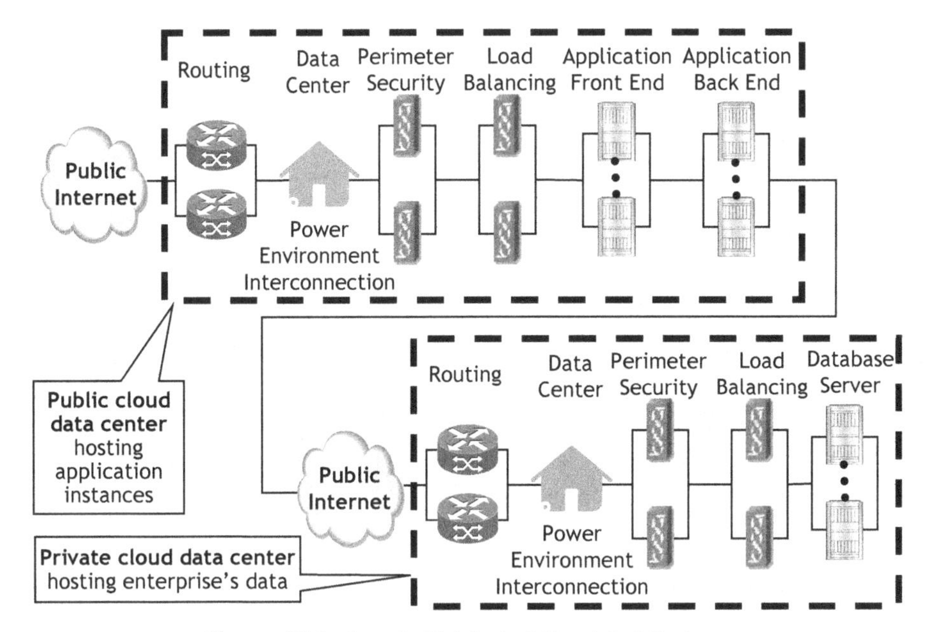

Figure 10.5. Sample Distributed Cloud Architecture.

database server may be used), the customer specific requirements do add the following elements to the critical service delivery path:

- wide area network (WAN) connectivity from public data center hosting application instances to private data center hosting enterprise data;
- second pair of routers in the private data center hosting enterprise data;
- private data center power, environment, and interconnection facilitie;s;
- second pair of security appliances in the private data center; and
- second pair of load balancers in the private data center.

Adding more equipment and facilities to the critical service delivery path naturally increases the risk of service unavailability due to failure or unavailability of those additional elements.

10.4.3 Data Center Redundancy

Cloud computing makes it easier to deploy an application to multiple data centers because the cloud consumer avoids the huge capital expense of building a second data center and merely pays for the resources actually used in each data center. Because data centers are inherently subject to external and force majeure risks, as well as ordinary failures, the best practice is to deploy redundant instances of critical applications to a geographically separated data center to assure that user service can be promptly recovered following a disaster or catastrophic failure. When properly configured, geographic redundancy can mitigate at least some of the service downtime accrued by applications deployed in individual data centers, as discussed in Chapter 9, "Geographic Distribution, Georedundancy, and Disaster Recovery." Section 11.2 discusses how to maximize service availability across multiple data centers.

Redundant data centers also permit a variety of service recovery strategies to mitigate critical failures of applications or equipment in a data center. Consider Figure 10.6, in which the canonical data center application of Figure 10.1 is deployed in both cloud data center "A" and cloud data center "B." If the application back-end servers in site "A" become unavailable due to duplex/multiple failure (or other reason), then the application front-end servers in site "A" can redirect their traffic to back-end servers in site "B," which will be supported by database servers in site "B," thereby offering service continuity for active users. High availability solutions will be architected to rapidly locate and use available resources to mitigate service impact of failures

Solutions will also distribute applications across multiple data centers (e.g., in multiple regions) to improve the quality of experience for globally distributed end users. With appropriate engineering, these globally distributed data centers can provide both business continuity via disaster recovery and boost service availability by mitigating application or facility failures.

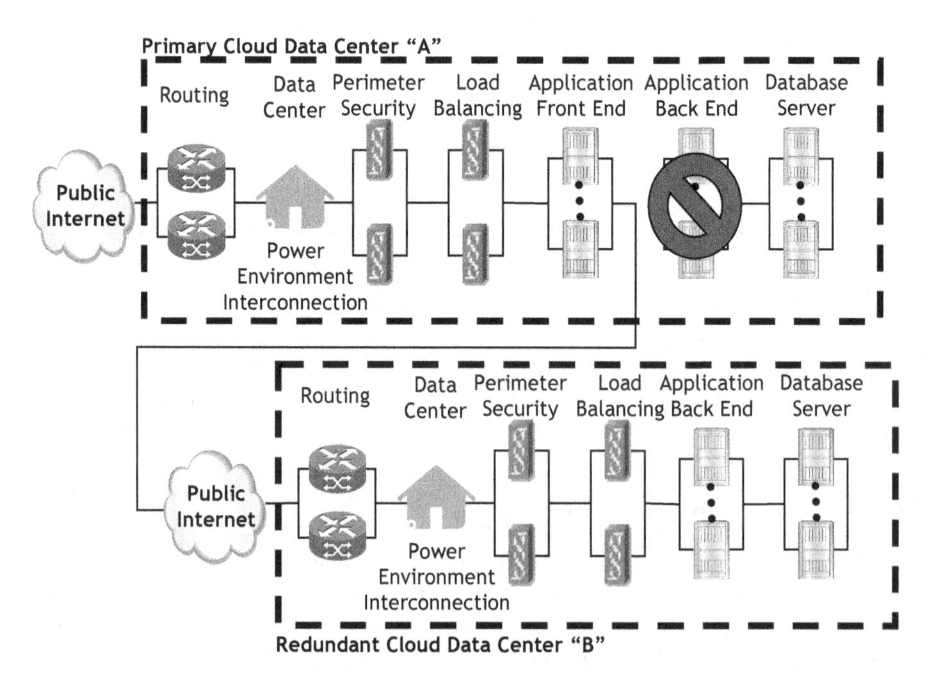

Figure 10.6. Sample Recovery Scenario in Distributed Cloud Architecture.

10.5 ATTRIBUTABILITY FOR SERVICE IMPAIRMENTS

As explained in Section 3.3.6, "Outage Attributability," the telecommunications industry has traditionally classified service outages into three buckets:

- *Product-or supplier-attributable events*, such as software and hardware failures;
- *Customer or service provider-attributable events*, such as human errors by customer's or service provider's staff when provisioning services or performing maintenance actions;
- *External-attributable events*, such as lightning, accidents, or deliberate acts by third parties, and natural disasters that damage network facilities and infrastructure.

The cloud computing service model is inherently more complicated than traditional telecom networks because:

1. accountability previously held by customer or service provider is often split in cloud context between the cloud consumer and the XaaS cloud service provider; and
2. many more service providers may be in the service delivery path.

Thus, the simple three bucket attribution model is no longer sufficient. The authors offer the following more comprehensive attributability model for cloud-based services. While it may be appropriate to add additional buckets or consolidate some buckets based on the deployment model or other details of a particular cloud based service, these categories are a useful starting point.

- *End User Attributable.* Some service impairments will be attributable to operation, configuration or failure of the end user's device. For example, service delivery will be impacted if a user operates their wireless device to the point of battery exhaustion. This category also includes user equipment that is in front of the commercial access network serving the end user. For example, if the end user's microwave oven impacts the Wi-Fi link between their wireless device and their Wi-Fi access point, then that impairment would be end user attributable.
- *Access Network Attributable.* The end user's IP traffic is carried by a service provider over LTE, 3G, DSL, GPON, cable, satellite, or some other wireless or wireline IP access technology to the IP WAN that connects to the IaaS provider's data center. Access and backhaul networks are inherently subject to facility and infrastructure failures due to equipment failure, accidents, lightning and so on. While the specific access network outage events might be product attributable, service provider attributable, or external attributable, that detail is unimportant to the end user; regardless of the root cause, their service was impacted.

- *WAN Attributable.* Typically, one or more Internet providers will carry the end user's IP traffic from their ISP's access network to the IaaS data center hosting the application. Note that while IP packets may flow across several carriers' networks, the IaaS data center may be served by multiple carriers, and each of those carrier's core IP networks is likely to include redundant facilities and equipment. Thus, despite the technical and commercial complexity of WAN networking, this category is likely to contribute little service impairment for applications hosted by top-tier IaaS service providers.
- *Cloud Service Provider Attributable.* Individual data centers and IaaS/platform as a service (PaaS) infrastructure are inherently at risk of catastrophic failures that impact availability and reliability of applications hosted in those data centers.
- *Application Software (or Product) Attributable.* Defects in the application software, including software provided by software as a service (SaaS) service providers and software suppliers, can cause reliability and availability impairments.
- *Cloud Consumer Attributable.* Errors by the cloud consumer to properly provision, configure, or operate their application can directly impact service offered by end users. For example, if user account data are misconfigured by cloud consumer's provisioning data entry staff, then users might be erroneously denied access to services that they are entitled to.

Note that in some cases, a single organization may be accountable for more than one attributable category. For example, a telecom service provider may be accountable for both the access network and the WAN, and an application service provider may be accountable for both the cloud consumer and application software categories.

While this attribution framework is not perfect for every type of cloud-based application with every service and deployment model, it is a reasonable starting point. For example, specific applications may add additional attribution categories (e.g., external attributable for force majeure and other events that are beyond the reasonable control of any of the explicit parties in the basic attribution framework).

Specific outage responsibilities will vary based on both the cloud service model (IaaS, PaaS, and SaaS) and contractual terms between the cloud consumer and cloud service provider. Figure 10.7 visualizes hypothetical outage (and hence downtime or availability) accountability for key elements in the service delivery path of a cloud-based application offered by an IaaS or PaaS provider from a single data center. Actual accountabilities will vary based on contractual agreements between cloud consumers and their cloud service providers. Consider accountability on an element-by-element basis:

- *Routing.* Cloud service provider is typically responsible for IP routing and connectivity throughout the data center to the demarcation point with one or more Internet access providers.
- *Data Center.* Cloud service provider is responsible for all aspects of data center operation.

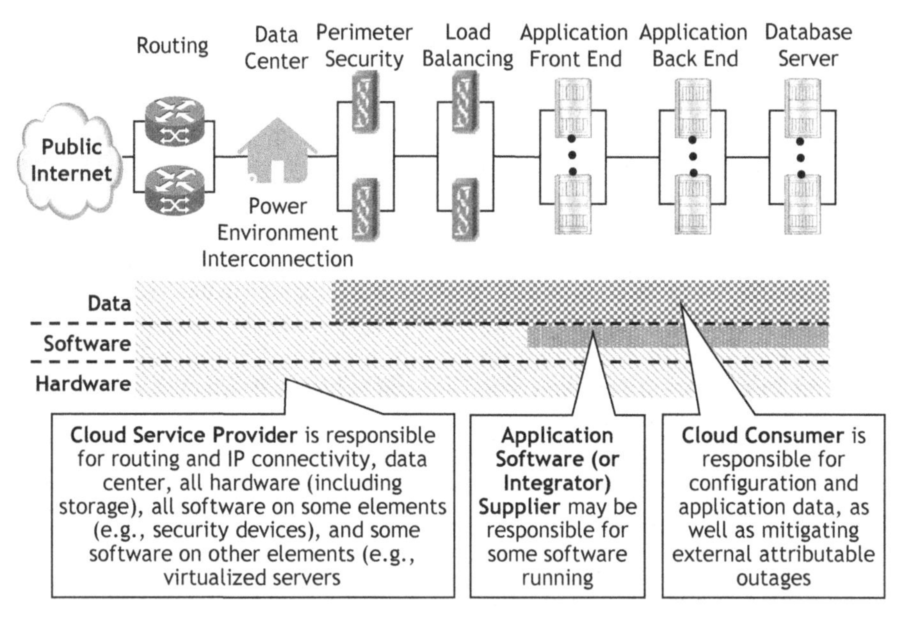

Figure 10.7. Simplified Responsibilities for a Canonical Cloud Application.

- *Perimeter Security.* While the cloud service provider is typically responsible for the perimeter security element hardware and software, cloud consumers have at least partial responsibility for configuring the perimeter security, such as configuring what traffic is allowed to pass through the security perimeter and opening firewall pinholes.
- *Load Balancing.* Cloud service providers often take responsibility for load balancing hardware and software, but expect the cloud consumer to appropriately configure the load balancers so that traffic is directed to operational application front-end server instances.
- *Application Front-End, Back-End and Database Server.* Cloud service provider is responsible for all aspects of server hardware and virtualization platform. More or less application software will be supported by the cloud service provider depending on the particulars of the IaaS, PaaS, or SaaS arrangement. There may be more or less cloud-consumer-specific application software running above the cloud service provider's software depending on the specifics of the consumer's application; responsibility for that software may be retained by the system integrator or distributed across several software suppliers and/or retained by the cloud consumer themselves. Responsibility for provisioning and correctness of application data is retained by the cloud consumer, while the cloud service provider has responsibility for assuring that data written to virtualized storage are continuously available.

10.6 SOLUTION SERVICE MEASUREMENT

Characterizing reliability and availability of network-based user services is inherently complex because end users are separated from the running application instances via IaaS data center infrastructure, WAN facilities and infrastructure, backhaul networking, wireless or wireline access, and some user equipment, as illustrated in Figure 10.4. Different users are likely to access the service via different devices (e.g., smartphones, tablets, laptops, and set-top boxes) over different wireless or wireline access networks operated by different service providers. Popular and highly available applications are likely to be offered from several geographically distributed data centers, so individual users will be served from different application instances running in different data centers. As these different data centers will be geographically distributed, various WAN facilities and infrastructure will be engaged to pass IP traffic between the end users' ISP and the IaaS data center serving them.

A key service measurement challenge is deciding where in the service delivery path to collect data because this drives how much of the end-to-end impairments are covered by the measurement. For the most expansive end-to-end measurement, service measurement data can be gathered from end users' devices (e.g., smartphones and set-top boxes) to integrate the impact of all service impairments across access, backhaul, WAN, IaaS data center, and application instances. Unfortunately, it is inherently difficult to determine and attribute the root cause of any service impairments based on end device data because the impact of so many elements and facilities is implicitly convolved together. At the other extreme, one can query or probe specific application instances from within the data center hosting the application's VM instances to determine their reliability and availability as seen from within the data center hosting the specific application instance. While this type of focused local measurement makes it very easy to characterize the reliability and availability of specific application instances, it does not reliably estimate the true end-to-end service quality, reliability, and availability experienced by end users.

A common engineering compromise is to query or probe all nominally operational application instances from a single fixed site (e.g., the cloud consumer's enterprise network or a specific cloud data center) to characterize the service likely to be experienced by end users with the highest quality IP networking service. End users with wireless access will undoubtedly experience somewhat poorer service quality due to networking impairments associated with wireless networking technology, service quality and coverage issues with their wireless service provider, and other issues (e.g., battery exhaustion of their wireless device). Nevertheless, data from a fixed site outside of the target data center offer a good reference point to characterize service reliability.

10.6.1 Service Availability Measurement Points

As shown in Figure 10.8, there are three natural points to consider the measurement of service availability of a cloud-hosted application:

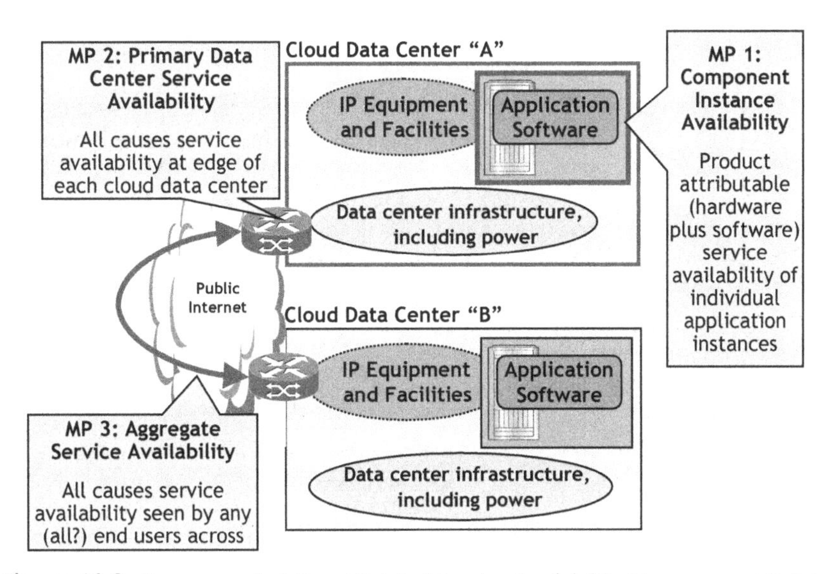

Figure 10.8. Recommended Cloud-Related Service Availability Measurement Points.

- *Measurement Point 1 (MP 1): Component Instance Availability.* Product-attributable service availability for each component or system critical to application service delivery in the cloud data center. An application's measurement should be taken with minimal IP routing, switching, and facility infrastructure between the server hosting the application and the measurement point to eliminate all impairments not associated with the application and directly supporting hardware. MP 1 ratings for routers, security appliances, load balancers, and other infrastructure configurations can be considered separately. MP 1 does not consider the service availability benefit of georedundancy.
- *Measurement Point 2 (MP 2): Primary Data Center Service Availability.* All causes application service availability per data center seen by the public internet (i.e., on the carrier side of the cloud data center's router). MP 2 characterizes the performance of individual application (or solution) instances along with the performance of the hosting data center, but MP 2 does not consider the service availability benefit of georedundancy.
- *Measurement Point 3 (MP 3): Aggregate Service Availability.* Service availability across multiple data centers to mitigate any impairment of individual application instances, IP equipment and facilities, and data center infrastructure (including power and environmental factors) that may impact any single data center. MP 3 incorporates the service availability benefit of georedundant application instances deployed across multiple cloud data centers.

MPs 1 and 2 can be easily overlaid onto the canonical data center deployment of Figure 10.1 to create Figure 10.9. For simplicity, we consider the performance of

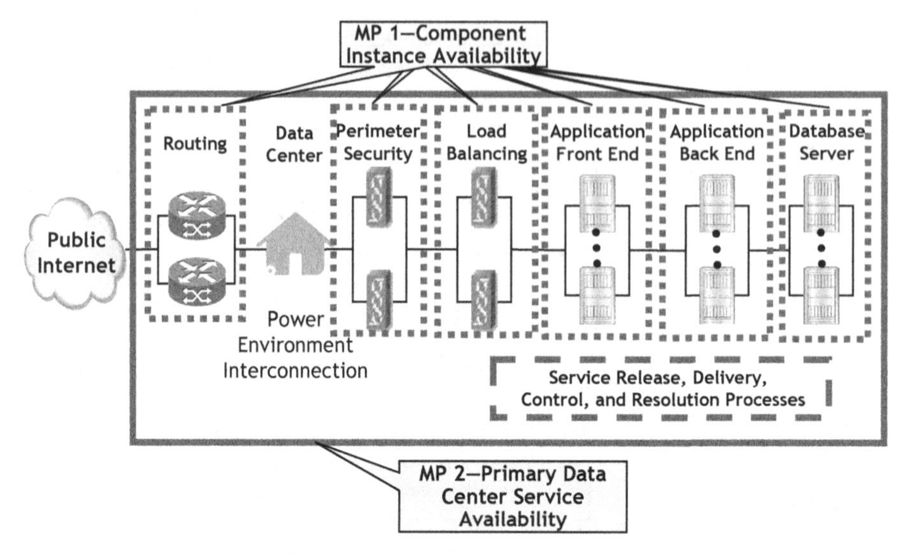

Figure 10.9. Canonical Example of MP 1 and MP 2.

application front-end servers, back-end servers, and database servers separately, as well as routers, load balancers, and security appliances. One can thus see in Figure 10.9 how MP 2 is logically the sum of MP 1 values of all elements in the service delivery path, as well as additional downtime associated with the data center itself and human operations and provisioning of all equipment, facilities, and applications.

MP 3 is inherently more complex and subtle to model and understand because it integrates the benefits across data center redundancy, as well as sophisticated service load distribution strategies and client-initiated recovery mechanisms. In some cases, georedundant recovery is performed by the client, such as when the client times out (or explicitly cancels) transactions to one data center and retries their service request to a different IP address for a georedundant data center. In other cases, like the example of Figure 10.6, more sophisticated recovery strategies are employed. Thus, MP 3 is impacted not only by the fundamental rate of critical MP 2 failures but also by:

- the speed and accuracy of detecting MP 2 impairments;
- the georedundant recovery latency and effectiveness; and
- any potential user service disruption when user service is migrated back to the recovered primary data center.

Predicting MP 3 is a subtle and complex subject that is considered in [Bauer11].

In addition to these general measurement points, there is the overall end-to-end service availability. Overlaying MP 1, MP 2, and MP 3 from Figure 10.8 onto a generalized version of Figure 10.4 produces Figure 10.10, which highlights the scope of the end-to-end service availability, which the authors will call measurement point 4

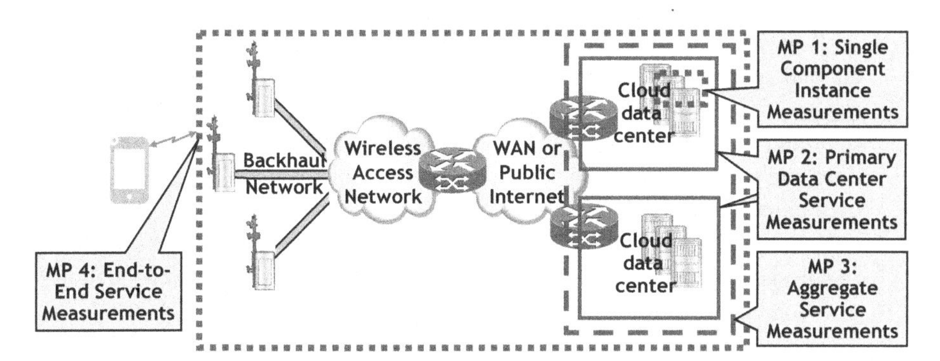

Figure 10.10. End-to-End Service Availability Key Quality Indicators.

(MP 4) for *end-to-end service availability*. Operationally, MP 4 integrates the MP 3 aggregate service availability across georedundant data center metric with the end-to-end access, backhaul, and wide area networking between the user's equipment and the cloud data center serving the user. MP 4 is obviously highly dependent on the particulars of the access network serving the end user, so different end users, even in the same geographic area, might experience very different MP 4 performance.

MP 4 performance is naturally impacted by the physical location of each individual end user and the networking equipment and facilities between them and the serving cloud data center. For example, the quality of service experienced by a user via their smartphone on the streets of London for a cloud application served from a data center in the United Kingdom may be very different from the quality of service experienced via the same smartphone in, say, East Africa. While two cloud data centers is the minimum number necessary to enable good business continuity and disaster recovery for critical services, far more (smaller) cloud data centers can be distributed logically and physically closer to end users, thereby boosting users' quality of experience by shortening transport service latency. In addition to shortening transport latency, the shorter end-to-end service delivery path to the closest (small) distributed data center instance should boost service availability be eliminating removing WAN facilities and infrastructure, which are inevitably subject to failure and contribute downtime.

10.7 MANAGING RELIABILITY AND SERVICE OF CLOUD COMPUTING

The remainder of the book is organized as follows:

- Chapter 11, "Recommendations for Architecting a Reliable System," presents architectural recommendations to maximize the service reliability and availability of virtualized applications and cloud-based solutions.
- Chapter 12, "Design for Reliability of Virtualized Applications," explains how design for reliability diligence should be altered to assure that virtualized

applications can meet, and perhaps exceed, the availability expectations of traditional application deployments.

- Chapter 13, "Design for Reliability of Cloud Solutions," explains how traditional solution design for reliability diligence should be tailored so cloud based solutions can meet or exceed the service reliability and availability expectations of traditional solution deployments.
- Chapter 14, "Summary," reviews the key insights of Part I, "Basics," and Part II, "Analysis," and summarizes the key take aways of Part III, "Recommendations."

11

RECOMMENDATIONS FOR ARCHITECTING A RELIABLE SYSTEM

The earlier chapters provided reliability and availability analyses for virtualization and cloud, as well as a lower level analysis of key areas, such as software, hardware, capacity, and service orchestration. The purpose of this chapter is to draw from the information in the earlier chapters to provide recommendations for architecting a highly reliable, highly available application and solution architectures for virtualized or cloud environments. The recommendations will also provide input for specifying a design for reliability framework that aligns with expectations for applications in the virtualized and cloud environments. The chapter begins with some key architectural decisions that need to be made, such as how to map software into virtual machines (VMs), optimize service load distribution, and choose the optimal data management mechanism. The chapter goes on to discuss some other key topics, such as hardware downtime, rapid elasticity, service transition activity management, and disaster recovery. The chapter concludes with a discussion of optimal reliability and availability of cloud-based applications.

11.1 ARCHITECTING FOR VIRTUALIZATION AND CLOUD

This section discusses the factors and tradeoffs to consider when architecting for virtualization and the cloud such as designing for high availability, multitenancy, and coresidency.

Reliability and Availability of Cloud Computing, First Edition. Eric Bauer and Randee Adams.

11.1.1 Mapping Software into VMs

Virtualization provides a means of packaging software so that it may be easily installed and scaled. One of the decisions in architecting an application with virtualization is determining how to best map the VMs for the appropriate user model to maximize ease in setting up redundancy, scaling, and distribution and meeting latency requirements. Some of the factors to take into consideration are:

- *Growth within the Failure Group.* If one of the components is active–standby and the other is $N + K$, then the growth patterns are different for those two components and suggest separate VMs so that another "N" instance can be created without also needing to add the active-standby component as well.
- *Ease of Growth.* If growth of one component always necessitates the growth of the other, then a single VM may be more appropriate to facilitate application scaling.
- *Architecture for the Application.* If the application is made up of separate component types, such as a front end and a back end that don't share any resources, then that might suggest a clear separation of VMs; however, if the components do share resources or have tight latency requirements, then a single VM may work better.
- *Affinity and Antiaffinity Rules.* Affinity and anti-affinity rules guide which VMs can be coresident on a single hypervisor and which VMs should be deployed on different hypervisor instances on physically separate compute hardware. VMs that frequently communicate with each other may coreside on the same server to decrease message latency. Some VMs will be deployed to different hypervisors to assure that no virtualized server becomes a single point of failure for a critical service; see Section 12.4.1, "SPOF Analysis for Virtualized Applications" for more details. VMs such as the high availability manager and the applications may choose to reside on different servers to reduce recovery time due to the implications of a simultaneous failure of the VMs or the failure of the server they are coresiding on. See Section 11.1.4, "Software Redundancy and High Availability Mechanisms."
- *Hardware Agnosticism.* Although VMs should be hardware agnostic, the types or configurations of the hardware nodes may need to be taken into account to ensure the VM can perform well on that node.

11.1.2 Service Load Distribution

One of the key advantages of cloud computing is the ability to seamlessly distribute service load across multiple servers and across multiple locations and even across multiple cloud providers with the assistance of load balancers and policies. Load distribution in the cloud environment can be highly complex as it needs to take into account factors, such as subscriber affinity, redundancy, latency, availability, security, regulatory issues, and capacity.

From a customers' point of view, service must meet their specific requirements, such as latency, capacity, availability, security, as well as recovery time objective (RTO) and recovery point object (RPO) in the case of failure. Determination of an appropriate load distribution architecture that meets those requirements will consider:

- number of application instances;
- redundancy of applications and data; and
- proximity of application instances to end users.

Policies must be clearly defined to manage service distribution in accordance with latency, regulatory and security requirements.

When architecting the server configuration to maximize load distribution, the distance between data centers should be considered, particularly if there is frequent data exchange between data centers (e.g., data synchronization across databases). When multiple cloud data centers offer disaster recovery protection, one must also assure that the data centers are far enough apart that no single force majeure event will impact more than one of the data centers.

11.1.3 Data Management

Data—both static (e.g., subscriber data) as well as dynamic (e.g., state or transaction level data)—are a critical part of a reliable service. Regardless of the type, all data must be redundantly stored and managed to survive failure of a component. Static data should be automatically provisioned to avoid errors, and any changes to the data should be logged or journalled. Data management is very complex in the cloud environment, since transactions can span multiple application instances and be stored in multiple locations. There are two types of mechanisms generally used to keep data synchronized: ACID (*atomicity, consistency, isolation, and durability*) and BASE (*B*asically *A*vailable, *S*oft state, and *E*ventual consistency).

Mechanisms that possess *ACID* properties ensure transactional reliability. Many relational database systems provide ACID capabilities. These mechanisms should be used when transactional reliability and immediate consistency are essential to meet customer needs, as these mechanisms can be very resource intensive and may introduce latency into transactions.

When data consistency is required but can be performed over a longer period of time, mechanisms that support *BASE* may be used to provide a simpler, less resource-intensive solution. Many web services can take advantage of the less complex BASE properties. As an example, email services do not have to be instantly up to date, while many banking services do need immediate consistency when managing their transactions. Nonrelational databases, such as *NoSQL* (not only SQL [Structured Query Language]), are recommended for use in cloud configurations to provide better performance and scalability. NoSQL, described in [NoSQL], is a distributed database management system that is nonrelational and does not use SQL as its query language. It has a simple infrastructure (i.e., based on a key-value model), has BASE properties, and thus lends itself to lower latency and increased scalability. Facebook developed a NoSQL-type distributed storage system called Cassandra to manage its great amounts of data across many servers with high availability, high reliability, and scalability [CASS]. Cassandra supports a configurable level of replication and a failure detection mechanism shared by the nodes in the system, which offers a high availability solution. Google developed its own distributed storage system called Bigtable [Bigtable] to manage large amounts of data for many of its applications, such as Google Earth.

Georedundancy can also enhance the availability of the data by providing data access at multiple locations thereby mitigating the impact of a failure of one of the locations. Data sharding or partitioning of the data into multiple smaller databases rather than one large database is a way to provide added robustness as well as improved performance in accessing data due to the smaller table size. An example of sharding is to create instances of the database based on subscribers' last names (e.g., [A–G], [H–N], [O–T], and [U–Z]). If the server that is storing the subscriber data for [A–G] fails, then its redundant copy can serve the associated users; even if there is only one copy of the [A–G] subscriber information or a dual failure occurs associated with the data for that set of subscribers, then subscribers with the last names starting with [H–Z] should not be impacted at all by the failure(s).

Another performance improvement that is gaining popularity in the industry is the use of in-memory databases. With virtualization, it is feasible that an application can request sufficient RAM so that it can store its data in-memory and realize improved access latency and reliability.

11.1.4 Software Redundancy and High Availability Mechanisms

Services must be architected with redundancy (even georedundancy) at the software and hardware levels, and with high availability mechanisms at their foundation. Robust application platforms generally provide some type of failure detection, reporting, and recovery mechanism as discussed in Section 3.6, "Redundancy and High Availability." Virtualization software assisted by the hypervisor often also provides capabilities that detect hardware failures and recover the application on a different server as discussed in Section 5.4.2, "Virtualized Recovery Options." The hypervisor can also detect VM failures and restart the failed VMs. In order to meet customer reliability and availability requirements, an assessment must be performed on the virtualized application and its internal high availability mechanisms to determine whether it can meet customer requirements with its internal mechanisms alone or whether the virtualization high availability mechanisms should be added to handle those failures not being covered by the internal high availability mechanisms. If the application does not meet availability requirements with internal mechanisms alone, then the high availability and fault tolerance capabilities of the virtualization software should be evaluated to see whether those capabilities would better ensure that requirements could be met.

Another architectural consideration for high availability management is ensuring that the high availability manager is itself redundant, preferably active–active. In the case of an active–standby high availability manager, if both the high availability manager as well as the application instance were to fail at the same time due to software or hardware failures, then the impact might be greater and the recovery time might be longer since the high availability manager would need to recover before executing any necessary service recovery actions.

In addition to making sure that sufficient high availability mechanisms are in place, the notion of "design for failure" is important. Design for failure puts the emphasis on recovering from failures rather than simply trying to prevent failures. Particularly in the case of very large cloud-based systems, failure of some kind (software or hardware) is very likely so it is important to be able to quickly and automatically detect and recover from those failures. Netflix uses the colorful term "Rambo architecture" to describe this aggressive design-for-failure vision that "*each system has to be able to succeed, no matter what, even all on its own*" [Netflix10]. Failure mode effects analysis (FMEA) of all likely failure scenarios is a best practice. It is also critical to provide manual means of recovery in the event that the automated mechanisms are not performing properly or that the failure was uncovered, that is, not automatically detected and recovered. Since manual intervention is often error prone (especially when executed in the stressful context of an emergency recovery action), it is essential that procedures are well documented and well tested by the responsible maintenance staff.

The recovery-oriented computing (ROC) described in [ROC] also discusses the notion of recovering from failures rather than avoiding failures. ROC emphasizes several key areas for building robust software:

- redundancy and isolation—isolating a failed component, letting its redundant mate function while it is being recovered;
- support for undo—recognizing that humans will make mistakes and providing a way to back out of an erroneous operation;
- diagnostic support—in order to identify the failure and its cause quickly to facilitate recovery;
- verification of recovery mechanisms—to ensure that the recovery mechanisms are solid and robust; and
- model availability/dependability of the product—as a way to gauge how well it is performing.

In addition to the above, preventative actions should be taken such as:

- Provide frequent backups of software images and data across sites.
- Automate and regularly perform disaster recovery testing so that the tools and staff are prepared in the event disaster recovery is needed.
- Create health checks that verify the health of the system components.

- Perform data integrity check and correction audits.
- Verify automatic failure detection and recovery mechanisms by periodically introducing faults or failures into the system to ensure that proper recovery takes place. These failures may include less graceful scenarios, such as failing an instance without first draining its work load.

Even in the best designed system, failures will occur in the hardware, software, and network. Architecture, design, implementation, and testing must thus consider failures at every level.

11.1.5 Rapid Elasticity

Rapid elasticity is a powerful cloud mechanism that can provide automatic scaling and descaling of hardware resources resulting in more efficient usage of resources, as well as mitigation of the risk of overload conditions. As indicated in Section 7.4, elasticity can result in a VM increasing its resources (vertical growth) or an increase in the number of VMs (horizontal growth within the data center or outgrowth into a different data center).

To maximize the efficiency of rapid elasticity, resource monitoring, metrics, and thresholds must be put in place, as well as hysteresis to detect when resources are reaching their capacity so that the needed growth mechanism (as dictated by policy) is triggered. For example, if an application has reached its threshold of CPU usage, then an elasticity mechanism should be triggered to allocate additional resources.

Although there are risks associated with rapid elasticity as explained in Section 7.7, there are ways to mitigate those risks. Rapid elasticity requires applications to be designed:

- to manage scaling and descaling;
- to provide accurate monitoring and recording of resources and performance; and
- to support well-defined policies and robust trigger mechanisms to automate and reliably accomplish the growth and degrowth of the application.

System testing must be performed with varying levels of offered load, generating traffic peaks to trigger application growth and low traffic points to trigger degrowth scenarios to ensure that the applications and service orchestration infrastructure can manage the changes in configuration reliably.

11.1.6 Overload Control

In the dynamic scaling environment of cloud, it is possible to reduce the amount of time a system is in overload by taking advantage of rapid elasticity to meet spikes in offered load. Overload control mechanisms in traditional systems issue alarms based on meeting or exceeding defined capacity thresholds resulting in the rejection or shedding of traffic according to the severity of the alarm. Service orchestration can take these same alarms and trigger cloud management mechanisms, such as rapid elasticity, to instantiate a new instance of the application and reloadshare traffic to include the new instance. In this way, the overload condition could be mitigated and perhaps even eliminated. Native overload control mechanisms should still be in place for situations in which elastically grown application capacity does not come online fast enough and as a backup when the scaling attempt does not succeed.

11.1.7 Coresidency

The server consolidation model provides a means for multiple applications to reside on the same physical server. Server consolidation has the following benefits:

- Operational expenditure (OPEX) savings based on reduced ongoing support of the equipment (e.g., reduced hardware, power, cooling, and space costs, as well as staff to monitor and manage the equipment).
- Capital expenditure (CAPEX) savings in the acquisition of the equipment (non-cloud deployment, only).

Challenges include:

- Increased impact of server failure due to increased number of applications supported on a particular server.
- Potential vulnerability to "noisy neighbor" applications that impact access to resources for the target application.
- Hypervisor becomes a single point of failure that impacts all VMs under its control.

The challenges should be mitigated through the selection of a high availability architecture (see Section 3.6, "Redundancy and High Availability") that supports server or site failure recovery and rapid elasticity to manage changes in capacity. Thorough robustness testing of the configuration should be performed, including testing of the workflows with failure and very high capacity loads to verify the robustness of the high availability and elasticity mechanisms.

11.1.8 Multitenancy

Multitenancy entails the sharing of hardware resources by independent applications with different user populations. Similar to the coresidency case, it is based on the server consolidation model. Multitenancy has the same cost benefits as coresidency, as well as the same challenges.

The challenges should again be mitigated through the selection of a high availability architecture (see Section 3.6, "Redundancy and High Availability") that supports server or site failure recovery and rapid elasticity to manage changes in capacity. Thorough robustness testing of a multitenant configuration should be performed to insure tenant isolation, including testing of the workflows under various failure scenarios (of an individual tenant, multiple tenants, and the entire server) and with high capacity loads (of an individual and multiple tenants) to verify the robustness of the high availability and elasticity mechanisms.

11.1.9 Isochronal Applications

Isochronal applications have special needs since it is essential that they do not suffer latencies that disrupt the quality of their service. As discussed in Section 7.6.3, "Service Latency Risk," resource contention, real-time notification latency, and virtualization overhead can contribute latency for virtualized configurations. While virtualization overhead is likely to be fairly consistent on a particular virtualized platform, latency due to resource contention and real-time notification can vary based on the behaviors of the other applications that are sharing processing, storage, networking, and memory resources with the target application. Latency variations over time naturally make it harder to maintain the strict real-time requirements of isochronous services.

Thus, architects for isochronous services should:

1. Explicitly characterize the real-time isochronal expectations for the virtualized platform, such as maximum notification latency characterizing how "late" a real-time notification interrupt can be and notification jitter characterizing the maximum acceptable variation in notification latency.
2. Determine if it is technically feasible for the target platform or IaaS service to meet these requirements.
3. Determine what architecture and configuration is recommended for optimal isochronous performance on the target platform or IaaS service.
4. Prototype and test the isochronous application service to validate the technical feasibility of meeting the application's service requirements on a virtualized platform.
5. If the quality of service offered by the prototype is acceptable across a range of test scenarios, then move forward with the architecture. If the quality of service is not consistently acceptable, then reconsider the application architecture and/or the decision to virtualize the isochronous/real-time service(s).

11.2 DISASTER RECOVERY

As indicated in Section 9.3, "Virtualization and Disaster Recovery," virtualization simplifies traditional disaster recovery by relaxing the compatibility requirements on hardware employed on the recovery site and allowing applications to share servers and server resources with other applications. Cloud offers additional mechanisms, such as elasticity and disaster recovery as a service, as explained in Section 9.4, "Cloud Computing and Disaster Recovery." To maximize the capabilities of virtualization and cloud and minimize RTO and RPO times, the following recommendations should be considered:

- Ensure that target sites are located a distance away from the failed site and meet any legal and regulatory requirements assumed for the service in accordance with policies.
- Choose target servers (i.e., server where recovered application will be installed) that meet the computing, network, and storage requirements for the applications (not usually necessary for cloud deployments).
- Ensure that copies of the application software and data are easily available (e.g., vaulted) for prompt recovery on the target server at the disaster recovery site.
- Provide disaster recovery tools and procedures that can meet the RTO and RPO requirements for that application.
- Ensure that disaster recovery plans are well defined, documented, and tested.
- If rapid elasticity at the georedundant cloud data center is being used for the service to recover in a disaster scenario, then the requirements listed in Section 9.4 must be met.

11.3 IT SERVICE MANAGEMENT CONSIDERATIONS

Cloud-based solutions share the same types of service management activities as traditional systems. Just as with virtualized systems, cloud-based mechanisms can be used to mitigate service impact during these activities by manually migrating active VM instances to other servers while maintenance activities, such as hardware or software upgrades, are being carried out. If service must be returned to the original server, then the VM can be migrated back to that server once the maintenance activity has completed. In order to leverage virtualization and cloud mechanisms, care should be taken to ensure that there are available resources to support an active system while parts of the system are undergoing maintenance (e.g., not putting the standby instance on the same server as the active instance during an activity that requires the entire server to be taken out of service).

11.3.1 Software Upgrade and Patch

Virtualization provides the ability to manage VMs on different software versions, even on the same hardware server. This capability greatly facilitates software upgrade and patch activities through the installation of VM(s) on the new software version alongside the VM(s) running on the old software version. The new software version VM(s) are activated (and synched if needed with the old version), and a portion of the traffic is directed (e.g., via routing table updates) toward the new version instances. Once the new version has been verified to work sufficiently well, current traffic is allowed to complete on the old version instances and remaining traffic is routed to the new instances. Service impact should be minimal with this mechanism.

11.3.2 Service Transition Activity Effect Analysis

A service transition activity effect analysis is usually organized as a table similar to a FMEA (see Section 5.1.3, "Failure Mode Effects Analysis") table, with service transition activities as rows and service impact in the columns. As with a FMEA table, cells showing unacceptable service impact should be highlighted, and system architects should consider options to mitigate the unacceptable service impact. Table 11.1 gives a sample service transition activity effects analysis for a virtualized application in a cloud environment that takes advantage of the cloud mechanisms of live migration and on-demand self service to minimize service impact. Since virtualization provides containment for the VM instances, maintenance activities directed to a particular VM instance should only affect that VM instance, while activities directed toward the server, hypervisor, and host components may affect all hosted VMs. With mechanisms such as live migration, most service transition activities are capable of being executed

TABLE 11.1. Example Service Transition Activity Failure Mode Effect Analysis

Event	VM or Cloud Mechanism Used	Impact on Affected VM Active Transactions
HW upgrades	Live migration of all VMs	Minimal impact (migration time)
HW growth/degrowth	Live migration of all VMs	Minimal impact (migration time)
Firmware patches and upgrades	Live migration of all VMs	Minimal impact (migration time)
Hypervisor software upgrades and patches	Live migration of all VMs	Minimal impact (migration time)
Host OS security patches	Live migration of all VMs	Minimal impact (migration time)
Application software upgrades and patches	Creation of new release VM instance on server	Minimal impact
VM instance growth/ degrowth	On-demand self-service	No impact

with little or no service impact (e.g., service impact from time suspending active VM to time traffic is rerouted to newly activated VM). Note that some service transition activities are nonservice impacting with traditional techniques; however, the table will indicate situations in which live migration and on-demand self service are used. Hardware growth and degrowth activities will likely require load balancing and rerouting of traffic. If degrowth takes place traffic served by the impacted component will be drained and redirected to remaining components. Many of the service transition activities entail upgrades (major changes) or patches (minor changes) to a particular VM (or its components such as the application or guest OS) or shared components (such as the hardware, firmware, hypervisor, and host OS).

11.3.3 Mitigating Service Transition Activity Effects via VM Migration

Service transition activities for traditional deployments, such as hardware or network upgrades may entail extensive reconfiguration and often produce service impact and accrue service downtime that is not acceptable to the end user. Virtualization can mitigate downtime and reduce or eliminate any user service disruption through the live migration of VM instances to another compatible host computer just prior to the time the maintenance operation is to be performed. This frees up the server so that the administrator can complete the needed maintenance when no production VM instances are executing on the target system. The maintenance activities themselves will be less complex since they do not have to worry about disrupting service while the activity is being performed since service is being provided on other computers. Since the activities are planned, the resources can be set up in advance, and a graceful migration of traffic can be put in place to further minimize service disruption of existing transactions. Using live migration the hypervisor moves the VM instance upon request to a different host computer. Live migration supports dynamic load balancing of virtualized resources and dynamic failover support to ensure little or no service impact during the migration. Live migration is depicted in Figure 11.1. VMs are suspended on the source server and

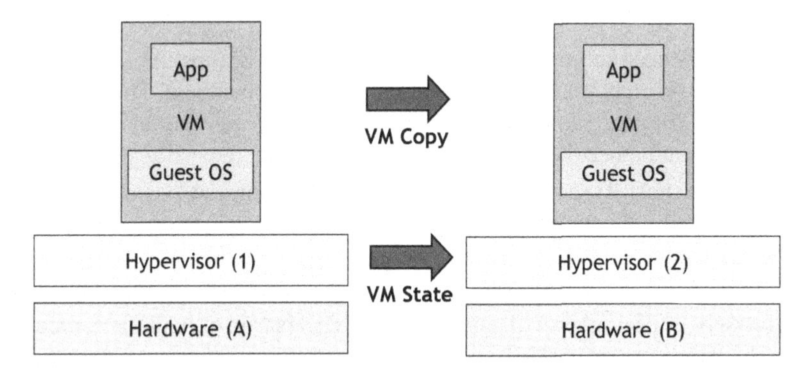

Figure 11.1. Virtual Machine Live Migration.

resumed on the target server. Hypervisor and VM state information is copied as well to provide a seamless transition for users involved in active transactions during the migration.

The basic steps for performing a live migration are:

1. A live migration of virtual machine hosting "App" between two hosts "A" (active) and "B" (alternate) within the same failover cluster is requested.
2. An alternate VM instance is created on host "B."
3. The initial memory state is copied from the VM instance on host "A" to the VM instance on Host "B" over the live migration network.
4. Any memory pages that were changed during the copy process (dirty pages) are marked, and the pages are copied over iteratively.
5. The VM instance is suspended or paused on host "A," and the state of the VM instance is copied to host "B." The VM instance is activated on host "B," an address resolution protocol is issued to update routing tables, and the VM instance on the "A" is removed. Since the VM instance on node "A" is not paused until the VM instance on node "B" has been installed and synched with the source, users should experience no service disruption.

Key benefits of using live migration to minimize service transition downtime include:

- Minimizing service impact on users. Service transition activities (e.g., extensive hypervisor software, hardware or IP configuration changes) that are traditionally service impacting and result in long periods of downtime can be performed without service impact, thereby meeting a common customer requirement for continuous service availability.
- Freeing up the server resources allocated to the VM so that it can be serviced without trying to maintain some level of service for that VM. This should make the procedures much less complex.

The challenges of the live migration approach are:

- Sometimes need to migrate back to the source server (host A) once the maintenance activity has completed (e.g., due to latency concerns). This usually needs to occur during the same maintenance interval; however, it is an optional activity and may not be necessary in most cloud solutions.
- The target (host B) server must have sufficient resources to successfully manage service during the maintenance activity.
- The live migration must not disrupt the other VMs that reside on the server. Robustness testing must verify that live migration of a single VM does not adversely impact other VMs on the server. This is true of both the source as well as the target servers.

11.3.4 Testing Service Transition Activities

Standard robustness testing includes routinely testing basic reliability mechanisms (e.g., failovers, process restarts, component reboots) and random injection of faults at various levels (e.g., process, VM, server, network, and data) to ensure that the system can recover properly and maintain some level of service during the recovery and that the monitoring systems correctly report failures and status. In cloud environments, configurations are more fluid and apt to change once put into service. As a result, some applications would benefit from exercising these reliability mechanisms in active systems to make sure they are still properly functioning in the changing configurations under varying traffic patterns. Note that this may not be recommended for highly critical services, but does provide a means of verifying that high availability mechanisms are working well so that weaknesses are identified and resolved before they result in service outages. As an example, Netflix deploys their so-called "chaos monkey" [Netflix11] to occasionally kill running VM instances in their production system, and thus assure that high availability mechanisms are running at top performance. Per [Edberg], Netflix deploys a "simian army" of agents, including a latency monkey alongside the chaos monkey, to impair aspects of their solution to assure that automatic mechanisms perform optimally and give operations staff more practical experience with the high availability behavior of the Netflix solution. [Hamilton] goes even further with the following blunt advice: "[I]f testing in production is too risky, [then] the script isn't ready or safe for use in an emergency."

11.3.5 Minimizing Procedural Errors

Procedural errors arise due to one or more of the following:

- Documented or undocumented procedures executed by human was wrong, ambiguous, or misleading.
- User interface was ambiguous, misleading, or wrong.
- Human erroneously entered wrong input.
- Fatigue especially during the night shift or panic during emergency recovery.
- Human executed wrong action, neglected to execute correct action, or executed actions out of sequence.
- System failed to check input or system state prior to executing requested operation.

Best practice for designing highly reliable procedures is to focus on three broad principles:

1. Minimize human interactions.
2. Help the humans to do the right thing.
3. Minimize the impact of human error.

Service transition tools and procedures should have the following requirements:

- Automation (e.g., service orchestration) shall be used to replace manual procedures and make use of mechanisms, such as live migration when appropriate.
- The reliability requirements associated with the service transition activity (e.g., *x* seconds of service downtime) must be fulfilled.
- Use of the Open Virtualization Format (OVF) or similar format should be used to provide configuration information so that it can be clearly defined and validated by the tools. The OVF provides a file format and detailed specification that makes it easy to configure the virtual infrastructure, including CPU, memory, networking, and storage for each VM in a standards-based way so that it will work under any hypervisor that is OVF compliant. By using the OVF format or another similar format, each VM instance or infrastructure component specified can also be validated during installation. In this way, many of the errors that occur in configuring the system can be eliminated, lessening the probability of procedural errors arising from errors in configuration.
- It must be possible to create and configure an updated instance of the application while running the old version, and to seamlessly stop the old version and activate the new once it is ready. This may be performed on the same server or on a different server dependent upon the type of maintenance activity and the availability of resources.
- Clear, accurate documentation and training must be provided for those managing the service transition activity.
- Thorough testing of the maintenance procedures must be performed to ensure the procedures meet the reliability requirements for service transition activities. Virtualization can be used to perform these procedures on a test application instance.

Cloud computing provides mechanisms that automate many of the procedures, such as VM creation and installation, thereby minimizing human interactions. The mechanisms try to help the humans do the right thing by providing easy-to-use graphical user interfaces to guide the users through the operations and audits to verify their correctness. Activities, such as software upgrade, may be performed on separate VM instances on different servers so as not to disrupt the active instances that are still providing service. This has the added benefit of minimizing the impact of human error since the maintenance is being performed on nonactive instances and can be validated before activation.

As with all procedures, it is very important to provide clear, accurate documentation and training to further mitigate the risk of procedural errors. Virtualization can be used to practice procedures on dummy VM instances so humans can verify their understanding of the procedure and establish a baseline understanding of successful procedural execution.

Virtualization adds a layer of complex software between the application, platform and (guest) OS software, and the physical hardware resources that support that VM's software. There is an inherent risk of failure when configuring this complex software and its associated virtual devices. Human errors introduced while executing these procedures can result in service outages. However, the potential increase in procedural failure rate due to inevitable operations, administration, maintenance, and provisioning activities supporting the virtualization platform itself may be somewhat offset by available tools that use a graphical user interface to help set up the configurations, templates for the VMs, audits to validate the configurations, and remediation capabilities in the event of failures. Thus, it is difficult to make general statements about the overall procedural failure rate implications of virtualization compared with native deployment.

In addition to procedural failures, there may be lost or incomplete transactions during activities, such as live migration, when traffic is diverted from an active VM to another instance during the maintenance activity. In the case of a successful migration, the number of failed transactions will be minimal, but if the migration fails, there could potentially be a large number of transactional failures.

11.3.6 Service Orchestration Considerations

Service orchestration can be a powerful tool for managing work flow efficiently and reliably and should be used by applications that are using rapid elasticity or anticipate the need to create many instances of an application. Service orchestration can be used to instantiate and provision new VM or application instances and allocate their hardware resources. Conversely, it can also degrow an application instance and release the resources. Service orchestration must robustly support the following functions:

- *Automation.* Service orchestration must be able to automate tasks based on input, such as SLAs, KQIs, and policies, and coordinate the tasks needed to instantiate and provision a VM with its needed computing, storage, and network resources in accordance with that input. This is true whether it is creating a single instance of the VM or many instances of the same VM.
- *Managing Complexity.* Service orchestration must ensure that resources are correctly configured for the VMs and do not conflict among the VMs.
- *Manual Request.* A front end must be provided to input the service requirements, configurations, policies, and service requests to support initial requests for service as well as to make manual requests to grow or degrow the service.

Particularly because service orchestration is capable of automatically creating instances of the application, it is important that it works flawlessly and must be robustly tested at high loads through normal, as well as failure scenarios verifying that policies are followed, and activities such as growth and degrowth can be managed with no service impact. Clear error reports must be provided if there are failures during the orchestration, and changes must be able to be reversed if there is a critical problem.

11.4 MANY DISTRIBUTED CLOUDS VERSUS FEWER HUGE CLOUDS

One of the essential characteristics of cloud computing is resource pooling (see Section 1.1.3), in which a pool of servers, storage devices, and other data center resources are deployed and shared across many users to reduce costs to the service provider and cost to the enterprise that pays for cloud services. Fewer huge clouds may be the configuration that can best take advantage of resource pooling's ability to more efficiently allocate resources and save installation and hardware costs, as well as the service provider's ongoing operational costs for floor space in a data center, electricity, cooling, and hardware maintenance. However, users who are physically farther away from a data center are likely to experience higher service latency due to:

- *Transmission Latency.* Light takes more than five microseconds to travel a mile, so transmission latency accumulates for each mile of fiber, coaxial cable, twisted pair, or air that data must travel between the user's device and the cloud data center.
- *Equipment Latency.* Every router, switch, repeater, firewall, security appliance, and other interworking element in the transmission path between the user's device and the application adds packet latency.

Cloud data centers that are physically closer to the users they serve generally have less transmission latency and fewer interworking elements that add packet latency.

While it is undoubtedly simpler and probably cheaper for a cloud consumer to host a single instance of their application in only one data center, this means that users who are distant from that one data center are likely to experience higher service latency than users who are physically close to the data center. The alternative is to deploy the application to multiple data centers that are physically closer to end users. When deciding how many data centers to use and geographically where those data centers should be located, cloud consumers should consider:

- *Service Latency.* Distributing applications to data centers physically close to end users can reduce service latency, jitter, and packet loss and improve network throughput by minimizing the networking equipment and facilities between the serving data center and the end user. Having many distributed cloud data centers provides the ability to place more servers closer to their users, while fewer huge clouds suggest that it is more likely that many users will be further from the sites. In addition, fewer huge clouds are likely to suffer heavier network throughput on those sites and have a higher likelihood of overload conditions.
- *Service Reliability and Availability.* Along with the improvement to service latency indicated for many local clouds, minimizing the networking equipment and facilities between the serving data center and the end user can also have a positive impact on service reliability and availability by reducing the number of components that can potentially fail while providing service to a particular user. From a total solution point of view, fewer data centers means there are fewer

servers to add into the availability calculation; however, the loss of any one of those will have a bigger impact on the solution availability and greater impact to the user population, as more of the users will have to be rerouted to another cloud. Conversely, more local clouds means adding more servers into the availability calculation; however, the loss of any one will have less impact on the total user population and will be less disruptive to the entire user population during recovery. Failure scenarios, such as a site disaster, will have a greater impact in the fewer huge cloud data centers configuration since it is more likely that a larger set of users will be impacted and will have to recover to other clouds, possibly resulting in a reregistration storm or overload on the other site(s). The more local clouds configuration provides a more geographically diverse environment, so the likelihood that multiple local clouds are impacted by the disaster is much less.

There are pros and cons to each approach. When architecting for the solution, reliability, availability, latency, and cost need to be taken into account and prioritized, as well as location of the user community. For critical services where reliability, availability, and latency are of the highest priority and the user community is spread over many locations, many local cloud data centers may be the right configuration. If cost is paramount or users are clustered in a few areas, fewer huge cloud data centers may be a better fit.

11.5 MINIMIZING HARDWARE-ATTRIBUTED DOWNTIME

Five 9's system availability expectations means that prorated product-attributable service availability (i.e., hardware plus software and application software) will have a long-term average of less than 5.26 minutes per system per year. Hardware-attributed downtime is generally allocated one-tenth of the downtime budget for high availability systems, which gives 30 seconds per system per year to downtime attributed to hardware. This implies that a 99.999% system is generally built on a 99.9999% hardware platform. Note that five 9's availability expectations for systems—or six 9's for hardware platforms—applies only to product-attributable causes, rather than to any of the impairments to power, networking, physical environment, human, policy, and other factors that may contribute downtime attributed to the data center, the enterprise, or external factors (e.g., force majeure). Thus, one must not confuse product-attributable service availability with "all causes" service availability, which aggregates downtime for product-attributable, service provider-attributable, and external-attributable causes.

Typically, traditional systems are built with a fairly optimal hardware configuration from which superfluous components and assemblies have been omitted to minimize the capital expense for hardware. In addition to reducing cost, this also helps to reduce the number of hardware components that can fail (i.e., lowering the FITs or increasing the hardware MTBF). In contrast, cloud computing—and to a lesser extent virtualization—puts more hardware into the service delivery path to maximize operational flexibility of the hardware resources. For example, traditional systems may rely on hard disks that are internal to the server blade or rack-mounted hardware so there is minimal (failable) hardware between the processor and the nonvolatile data storage. In contrast, since cloud computing architectures are built with resource sharing and elasticity as primary goals, nonvolatile data storage is inherently decoupled—often to storage area networks (SANs) or network attached storage (NAS)—rather than relying on nonvolatile data storage within the computer servers. Decoupling storage from compute resources inserts IP networking infrastructure and storage control hardware between the compute resource and the physical device offering nonvolatile data storage; obviously, that IP networking infrastructure and storage control hardware is vulnerable to failure in the cloud computing configuration. Thus, a key architectural question becomes: is it feasible and likely that an application executing on virtualized hardware resources can have comparable hardware-attributed downtime to (often simpler) traditional high availability system hardware configurations?

We address this question in two steps:

1. What contributes to predicted hardware-attributed downtime in traditional system architectures?
2. How can virtualization in the context of cloud computing minimize each of the predicted categories of hardware-attributed downtime categories?

The answers to these two questions drives recommendations for minimizing hardware-attributed downtime in cloud computing.

11.5.1 Hardware Downtime in Traditional High Availability Configurations

Let us consider the canonical example of a simple high availability system architecture built from an active–standby pair of servers. We will use the active–standby Markov model of Figure 11.2 and make the following canonical modeling assumptions:

- Each server has 100,000 hour MTBF ($1/\lambda$).
- Ninety percent of hardware failures on both active (C_A) and standby (C_S) units will be detected and recovered in 15 seconds ($1/\mu_{FO}$) with a success probability of 99% (F_A).
- Uncovered hardware failures of active hardware will be detected in 30 minutes ($1/\mu_{SFDTA}$).
- Uncovered hardware failures of standby hardware will be detected in 24 hours ($1/\mu_{SFDTS}$).
- Manual hardware repair takes 30 minutes ($1/\mu_{REPAIR}$) with a success probability of 99% (F_M).
- Repairing duplex hardware failures takes 4 hours ($1/\mu_{DUPLEX}$).

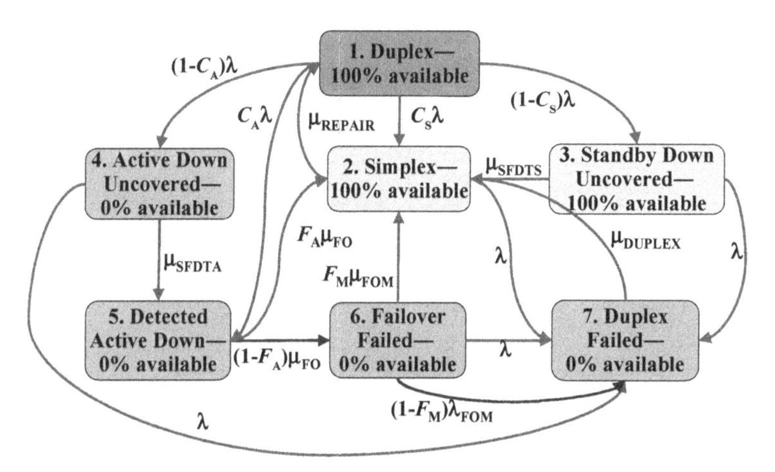

Figure 11.2. Active–Standby Markov Model.

TABLE 11.2. Canonical Hardware Downtime Prediction

State	Time (Minutes)	Percentage
4—Active down uncovered	0.263	84
5—Detected active down	0.022	7
6—Duplex failed	0.004	1
7—Failed failover	0.026	8
Overall downtime	0.315	100

Solving the Markov model yields the downtime predictions in Table 11.2; the downtime prediction is shown as a pie chart in Figure 11.3.

Now consider what each of these downtime predictions in Table 11.2 means, in order of descending downtime contribution:

- *State 4: Active Downtime Uncovered—Nominally 84% of Predicted Hardware Downtime.* This state captures the time service is unavailable because the hardware has failed but the system has not yet detected the failure and thus no recovery actions have been taken. An uncovered failure is sometimes called a "silent" or "sleeping" failure, for obvious reasons. Uncovered downtime also includes the more challenging "dreaming" failure situations in which the hardware incorrectly reports that it is fully operational (i.e., dreaming that it is healthy) when in fact it has failed. Note that even when it becomes apparent that service is unavailable to maintenance engineers, the fact that the hardware failure

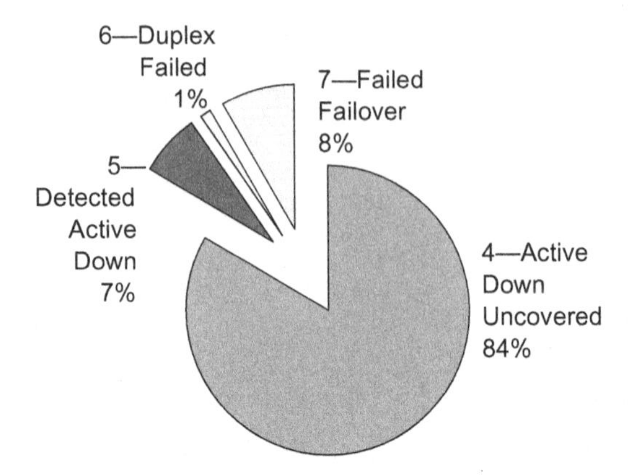

Figure 11.3. Pie Chart of Canonical Hardware Downtime Prediction.

is not immediately visible via alarms and other standard fault management tools means that additional time must be spent to manually troubleshoot the problem, isolate the failure to identify the proper recovery action to execute, and at least a portion of this excess troubleshooting time is captured in this state.

- *State 7: Failed Failover—Nominally 8% of Predicted Hardware Downtime.* this state captures the downtime when automatic switchover to standby hardware has failed and manual recovery actions are necessary to successfully recover service onto the standby unit.
- *State 5: Detected Active Down—Nominally 7% of Predicted Hardware Downtime.* This state captures the downtime when the high availability mechanism functions properly: the hardware failure is promptly detected, the proper recovery activation is executed, and service is recovered automatically in $1/\mu_{FO}$ (i.e., 15 seconds).
- *State 6: Duplex Failed—Nominally 1% of Predicted Hardware Downtime.* This state captures the rare state in which both active and standby hardware units are down simultaneously. Enterprises with good operational policies will repair failed primary units fast enough that the window of simplex exposure (i.e., operation with a single, unprotected hardware unit) is short enough that this sequential failure scenario is very rare. The more likely scenario is that the standby unit experiences an uncovered hardware failure that does not promptly present an alarm, so the system is actually simplex exposed when the active hardware unit fails; the first indication the enterprise has that the standby unit had failed is when service fails to recover following failure of the active unit. The standard mitigation for this risk is to periodically execute routine switchovers to verify that the standby unit is fully capable of serving users. The more frequently the routine switchovers are executed, the lower the risk of unknowingly being simplex exposed.

Now consider how each of these canonical downtime predictions is affected in a virtualized cloud computing environment:

- *State 4: Active Downtime Uncovered—Nominally 84% of Predicted Hardware Downtime.* Virtualization adds an additional layer of system software between the underlying hardware and the application software, and it is possible—although perhaps not likely—that the virtualization layer will identify some hardware failures that the guest OS and platform plus application software fail to detect. In some cases, mature native platforms or applications will include sophisticated auditing mechanisms that continually probe the hardware to detect failures or degraded performance before a service outage occurs; these mechanisms boost the hardware failure coverage of platforms and applications that support them. If the virtualized platform does not support the same level of hardware auditing that was supported on the native platform and the application does not appropriately adapt and deploy the auditing on the virtualized platform, then the effective hardware coverage for that application could be lower (worse) than for native configurations. When multiple applications are consolidated onto a virtualized hardware platform, such as in server consolidation or cloud computing usage scenarios, then the effective hardware coverage should ideally be the product of the hardware coverage factors of the applications sharing the virtualized hardware. Hopefully, the first application to encounter failed hardware on a virtualized platform (e.g., by attempting to access a failed hardware device or resource) should raise an alarm, and thus cause the hypervisor and infrastructure software to proactively trigger appropriate hardware recovery actions and thereby minimize time spent by applications in the "active downtime uncovered" state due to hardware failures. Theoretically, a more diverse suite of applications sharing hardware resources can effectively boost the hardware coverage factor enjoyed by all of the applications. Virtualized and cloud configurations in which only a single application executes on a hardware resource (e.g., a specific compute blade or rack mounted server) are likely to experience essentially the same hardware coverage as the application would have with native deployment on identical hardware.
- *State 7: Failed Failover—Nominally 8% of Predicted Hardware Downtime.* Although the hypervisor adds another layer of monitoring and control in the service recovery path, it is likely to be difficult for the hypervisor to differentiate a failed failover from a successful failover. Thus, failed failover downtime is unlikely to be impacted by virtualization or cloud deployment.
- *State 5: Detected Active Down—Nominally 7% of Predicted Hardware Downtime.* The downtime accrued by successful automatic failure detection and recovery will be nominally the same for both traditional and virtualized deployments. Note that it is theoretically possible that this time might be slightly reduced if the hypervisor detected a hardware failure faster than the traditional deployment scenario might, such as if another application running on the hardware platform possessed a faster and/or more effective hardware coverage than the target application. In that case, the target application might enjoy an

availability boost because the neighbor applications are more vigilant than the target application.

- *State 6: Duplex Failed—Nominally 1% of Predicted Hardware Downtime.* Cloud computing should reduce this category because a physical hardware repair (e.g., 4-hour MTTR action) should never be the only way to make required hardware resources available to an application; there should always be some spare online or nearline capacity that can be rapidly engaged to address traffic spikes or hardware failures. In addition, hardware resource sharing by multiple applications should assure that even if the standby instance of the target application is not very effective at detecting failures of the underlying hardware, one of the active neighboring applications should promptly detect any hardware failure and trigger the hypervisor and cloud infrastructure software to migrate applications to operational hardware so the target application spends minimal time simplex exposed in state 2 "standby down uncovered."

In addition, virtualization enables higher availability hardware configurations like dual network interface cards (NICs) and redundant array of inexpensive (or independent) disk (RAID) to be used with unmodified applications. Cloud computing eliminates that capital expense that often discouraged enterprises from deploying RAIDs and redundant networking infrastructure for traditional application deployments. More importantly, the higher resource utilization enabled by cloud computing makes it easier to justify deploying higher availability hardware to support more applications. For example, with traditional application deployments, an enterprise must explicitly invest in hardware redundancy (e.g., RAIDs and dual NICs) for the hardware resources supporting each individual application. The resource-sharing aspect inherent in cloud computing means that redundant hardware like dual NICs and RAIDs can be amortized across more applications, thereby reducing the cost for each application to benefit from higher availability hardware configurations.

Therefore, although virtualization and cloud computing often exposes application instances to more fallible hardware, the overall downtime contribution from that larger aggregate hardware failure rate is often lower than the downtime contribution of traditional deployments because of the following factors:

1. *Higher availability hardware configurations like dual NICs and RAID may become more cost effective in cloud computing than in traditional deployments*, so a system deployed with application level file replication in traditional configurations may be deployed with RAID in the cloud, thereby reducing the number of failures that must be recovered via the application level file replication mechanism.
2. *Higher effective hardware coverage may be experienced for both active and standby application instances because neighboring applications can also detect hardware failures*, thus boosting the effective hardware coverage enjoyed by individual applications

TABLE 11.3. Summary of Hardware Downtime Mitigation Techniques for Cloud Computing

Predicted Downtime State	Predicted Traditional Hardware Downtime Contribution (%)	Cloud Computing Mitigation Strategies
4—Active down uncovered	84	Running diverse applications on virtualized platform instances to maximize likelihood of prompt hardware failure detection and having hypervisor emulating detected hardware failures for VM instances that may not have already detected hardware failure.
7—Failed failover	8	Essentially unchanged.
5—Detected active down	7	Essentially unchanged, unless hypervisor can detect and emulate hardware failures faster than guest OS and application platform plus software mechanisms can.
6—Duplex failed	1	Maintaining sufficient spare online or nearline capacity so that manual hardware actions are never required to engage hardware resources to mitigate a hardware failure Configuring diverse application instances—including both active and standby instances—on each virtualized hardware platform to minimize risk of "silent" hardware failure impacting standby application instances

3. *Higher hardware coverage will be realized from purpose-built highly reliable storage chosen for cloud-based solutions.*

Strategies to mitigate downtime by predicted state are summarized in Table 11.3.

11.6 ARCHITECTURAL OPTIMIZATIONS

Virtualization and cloud technologies enable new architectures and business models for cloud-based applications to provide services that are more powerful and flexible than with traditional methods. Since applications are no longer tied to specific hardware resources, those resources can be added or removed automatically based on changing workloads, service capacity, latency, and availability trends across the network. By leveraging this flexibility, optimal architectures can be designed to meet the needs of the application based on its data (both volatile and nonvolatile), network, CPU, and storage usage, as well as its estimated capacity, latency, availability, and reliability requirements. Note that service providers and cloud consumers have somewhat different views of optimization. Cloud consumers want to maximize the service reliability and availability performance for their applications despite inevitable failures at reasonable cost; cloud providers want to maximize their yield/revenue at minimum cost while delivering acceptable (but not perfect) service quality, reliability, and availability. This section considers optimization from the cloud consumer's perspective; yield management of IaaS is not considered. A set of reliability and availability criteria are offered, followed by a discussion of how various aspects of application service architecture and operation can optimize those criteria. A case study for an enterprise web server is then provided that applies many of these mechanisms and strategies to maximize service reliability and service availability to help clarify their usage.

11.6.1 Reliability and Availability Criteria

There is no one size fits all architecture. Various categories of applications have fundamentally different requirements that permit different architectural options to tolerate and mitigate inevitable failures. For instance: financial applications require very high data consistency while web search has weaker data consistency requirements. Likewise, anonymous services like web search that require no identification, authentication, and authorization can recover user service to redundant servers or alternate data centers faster and easier than services that require users to identify and authenticate themselves before being authorized to access application- and user-specific data.

From a reliability and availability point of view, the following service reliability and availability criteria are considered for optimization:

- *Accessibility.* The service should be continuously accessible to users, which means that no routine or emergency maintenance activity like software upgrade, hardware replacement, network reconfiguration, capacity growth, or degrowth should cause service unavailability or downtime for users. In other words, the service should be continuously available 24 hours a day, 7 days a week, every week of the year, forever; this is sometimes called 24 by 7 by forever.
- *Retainability.* User sessions should be perfectly retained with no perceptible disruptions or loss of functionality from initial service access (e.g., logon) to orderly session completion (e.g., logoff), particularly after single-element failures or failover to another element within the data center.
- *Quality and Reliability.* Correct application transaction results, streaming, and data content, and so on, shall be delivered to the user with excellent service quality, including low service latency. Note that data consistency can impact the reliability of transaction results; an application that delivers a properly structured result to a user based on inconsistent (e.g., old and no longer valid) data may not be a correct result.

- *High Availability.* Any single failure can be automatically detected, isolated, and recovered with less than the maximum acceptable service disruption to users with negligible if any loss of volatile context information.
- *Disaster Recovery.* Application service shall be recoverable following disaster events within the application's RTO to an alternate data center, and nonvolatile data will be recoverable within the application's RPO. Unlike recoveries within the data center, active user sessions and volatile user data will likely be lost during disaster recovery due to the complexity and bandwidth requirements for maintaining such data across data centers.
- *Moderate Operating Expense.* Operating expense should be minimized through frugal use of cloud resources, including timely release of cloud resources that are no longer required, and automation of maintenance activities and recovery actions.

These criteria should be met with an efficient solution that maximizes resource usage and minimizes operating expense.

11.6.2 Optimizing Accessibility

Continuously accessible systems are designed with redundancy and no single point of failure so that no planned maintenance action (e.g., software upgrade, hardware growth) or unplanned failure forces service to become unavailable or inaccessible for an unacceptable period of time. In addition to component redundancy, services are built to detect, isolate, and recover from all types of failure, and often leverage the following techniques:

- *Load balancers intelligently direct traffic* to an application instance based on criteria such as application availability, workload, and proximity to user, and dynamically update the routing information based on automatic growth and degrowth realized by rapid elasticity.
- *All component instances are accepting some traffic* to both eliminate recovery latency associated with startup and activation of redundant components and to maximize the likelihood that failures are promptly detected (rather than waiting until a "standby" application instance is activated to discover a previously undetected failure).
- *Aggressive protocol recovery is implemented* through maximum retry counts to mitigate lost, damaged, and late IP packets.
- *Resilience is built into the service* on the client side or on the application side using one or more of the following techniques:
    - Managing resilience on the *client* side: Messages are buffered on the client side and resent to the same or alternate application instance upon detection of a

failure on the targeted instance or a response time out. This is managed with protocol support for retry (e.g., SIP and HTTP), short guard timers, limited number of retries (improved with heuristics on retry behavior), and client initiated recovery. [Bauer11] provides details on how to maximize recovery time using client initiated recovery.

  - Data needed to maintain service is cached on the *client* or on the *application* to mitigate reliance on data access during times of service or communication disruption.
  - *Applications* are architected for failure as discussed in Section 11.1.4, "Software Redundancy and High Availability Mechanisms." Since any instance can manage the workload of another, failures are quickly detected by the client, and traffic is seamlessly directed to another instance for handling. The failures are also detected and recovered internally so that the instance is once again available for service.

- *Nonvolatile data are replicated and available to all application instances.* Data are loosely consistent—with BASE properties rather than the more restrictive ACID properties used when the data are relatively static or the service does not require the most recent version to operate effectively. Services that do maintain and rely on transaction-level volatile data for reliable service will likely require databases with ACID properties to ensure data consistency.
- *Rapid elasticity is leveraged* so that instances of the application and its resources (including volatile and nonvolatile data) are easily and quickly added or removed automatically based on the customer traffic needs. This is driven through service orchestration leveraging real-time data on traffic loads, and operates based on thresholds and policies. Instances and resources can be added vertically within the server, horizontally within the data center, and through outgrowth into additional data centers as described in Section 7.4.2, "Elasticity Expectations." Hardware resources, including CPU, memory, and disk storage, are added or removed automatically based on changing application needs. This is supported by virtual resources and managed by the hypervisor.
- *Software and hardware upgrades can be performed seamlessly.* VM containment supporting the ability to instantiate a new version of the application while the old version is still operating and live migration providing the ability to copy memory to another server that has an instantiated instance of the new version of the application to support more extensive upgrades can support this capability. Graceful termination of a VM instance is available so that transactions are allowed to continue and terminate normally and are not lost during the migration.
- *Procedural tasks that are complex and error prone are automated* and made robust through the addition of health checks on the hardware resources to ensure they are ready for the procedure and audits on the data to ensure it is not corrupted before or after the procedure.

Advanced architectural techniques that optimize accessibility:

- *Clients send a single request to multiple application instances in parallel*, and use some intelligent algorithm to select the response (e.g., first response, first confirmed response, plurality, or majority vote). This technique greatly simplifies routing of requests but adds complexity in selecting and managing the responses and reduces capacity (or increases load, depending on your perspective) because each transaction is processed by two (or more) application instances in parallel.
- *Client applications may even establish and maintain authenticated sessions with alternate servers* to eliminate reauthentication latency if the primary server fails and the client must redirect traffic to the alternate server instance.
- *"Rambo" architecture.* Systems are built to expect and tolerate some failures that represent a degradation of some services but allow other services to perform. This technique provides increased availability for some services but does add the complexity of detecting, isolating the failure, deferring recovery, and determining that service can continue despite the failure. Designing with degradation allowed also entails understanding what the customer will tolerate for degraded services and which services are absolutely critical and not subject to degradation. For example, live streaming of video may be a critical service, while sorting through supporting movie reviews is a nonessential service.

11.6.3 Optimizing High Availability, Retainability, Reliability, and Quality

High availability mechanisms exist to detect, isolate ,and recover from failures. Service architecture should be assessed using techniques such as FMEA, as discussed in Section 5.1.3, "Failure Mode Effects Analysis," to flesh out all possible failure scenarios involving one or more of the components and confirm that failures can be rapidly detected and properly handled, including preserving state and context data when applicable. Service retainability and reliability are managed by making state or other transaction-related information available to other application instances from a common storage area or by pushing state out of server instances into client and/or shared registries. Fault tolerance mechanisms help ensure that any data (volatile as well as nonvolatile) critical to service retainability is maintained and kept consistent.

Overload detection and control mechanisms are put in place to manage short-term spikes in traffic when rapid elasticity cannot instantly accommodate changes in traffic. Overload mechanisms should shed or turn away some traffic during peaks to maintain the quality of most (or at least some) traffic rather than degrading all traffic or causing a component collapse or service failure.

11.6.4 Optimizing Disaster Recovery

Application instances are deployed to geographically distant redundant data center(s) to mitigate the risk of a force majeure event that renders a data center destroyed, inaccessible, or otherwise unavailable. Data centers are located far enough apart so that no single disaster or force majeure event (e.g., earthquake, hurricane, and tsunami) will impact both the primary and recovery sites. All nonvolatile application and user data are replicated to a georedundant site frequently enough to meet the application's RPO requirement, and sufficient resources are available in the recovery data center (e.g., bandwidth to replicated data, and rapid availability of compute capacity and network bandwidth) to meet the application's RTO requirement. Since incidents requiring recovery to another data center are infrequent, they are generally designed to allow slower synchronization of nonvolatile data and loss of volatile state information during the recovery than in the intra-data center recovery cases to minimize wide area network (WAN) traffic between the data centers. Service architectures which include many small data centers provide additional redundancy that can support a single data center failure, as well as the ability to optimize service latency by routing to the data center closest to the user. See Section 11.4 for details.

11.6.5 Operational Considerations

In addition to architecting, developing, testing, and deploying a robust solution, it is essential that high-quality IT service management policies and practices be used to manage operation of the solution. Beyond standard IT service management best practices (e.g., ITIL), the following operational topics warrant special consideration:

- *Proactive Management of Spare Online and Elastic Resource Capacity.* Sufficient spare online application capacity should be configured to simultaneously recover from typical failure scenarios (e.g., crashed VM instance) and to accommodate traffic spikes and growth. Note that the size of the spare online capacity is influenced by the expected and committed elastic growth slew rate supported by the IaaS provider and the application itself, as well as the application's elastic growth trigger points. Sufficient spare online capacity should always be available to assure that all users are served with acceptable service quality and latency, even when traffic grows and elastic growth is initiated.
- *Routine Testing of Robustness and Recovery Mechanisms.* Just as periodic disaster drills are a best practice to assure that disaster recovery mechanisms, plans, and procedures are correct, and give staff training and practice in executing those procedures, periodic validation of high availability mechanisms verifies that automatic mechanisms operate correctly and that staff can recognize successful automatic recoveries compared with unsuccessful recoveries that require manual corrective actions. Netflix's chaos monkey and simian army [Netflix11] are examples of the best practice for routinely testing robustness and recovery mechanisms.

11.6.6 Case Study

Service reliability and availability optimization concepts are best understood via an example. Although not all of the optimizations apply to all applications, this example

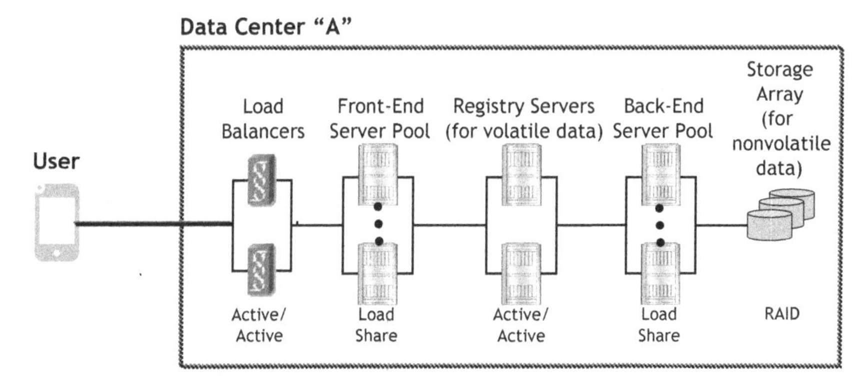

Figure 11.4. RBD for the Hypothetical Web Server Application.

offers some specificity and clarity to the concepts discussed previously in the case of a particular type of application. This section will show how these key concepts might apply to a hypothetical enterprise data centric web server application as depicted in the reliability block diagram (RBD) in Figure 11.4. The hypothetical applications consists of a front-end/back-end arrangement in which front-end servers interact with client applications (e.g., web browsers) and back-end servers mediate transactions with nonvolatile application data. A particular user of the hypothetical application instance is served from data center "A." User traffic passes through a pair of load balancers, which distribute requests across a pool of front-end server instances, which handle user identification, authentication, authorization, session and transaction context, user interface, and so on. Front-end server instances push copies of volatile user data into a pair of registry servers so that if one front-end server instance fails, then any other front-end server instance can rapidly retrieve the user's volatile information from one of the registry servers. A pool of back-end server instances operates on the application data that is maintained in a storage array. Front-end servers pass request messages to back-end servers that are protected with short guard timers, and a front-end can retry a failed, lost, or late operation to another back-end server instance. Nonvolatile data storage is maintained in a RAID storage array.

Rapid elasticity is leveraged to improve accessibility through growth and degrowth of resources that are in line with the changing service needs. Figure 11.5 illustrates how the hypothetical application instance can grow horizontally. Front-end server instances can be created or destroyed, and the load balancers will intelligently distribute the workload across available front-end application instances. Note that even if the front-end server instance that was serving a particular user is destroyed to reduce front-end capacity, another front-end server instance can seamlessly recover the user's volatile application state from one of the pair of registry servers. Likewise, back-end server instances can be created or destroyed, and front-end server instances will intelligently distribute their requests across the pool of available back-end servers; front-end servers will retry requests that may have been lost if a back-end server instance was destroyed to shrink back-end capacity.

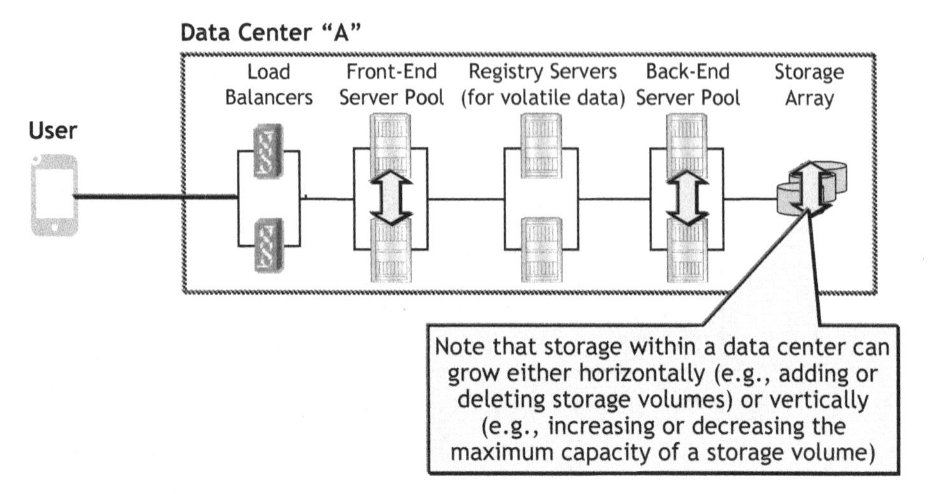

Figure 11.5. Horizontal Growth of Hypothetical Application.

If the hypothetical application needs to grow beyond data center "A" based on its exceeding the capacity of data center "A" or to provide georedundancy for the service, then an additional set of load balancers, front-end servers, registry servers, back-end servers, and storage array are set up in data center "X" to support service outgrowth as depicted in Figure 11.6. The application users' clients must receive an indication of the availability of the new data center via DNS, a traffic redirection indication from the data centers, client reconfiguration, or through other mechanisms. For example, if we assume that each application instance is limited to a pair of load balancers and a pair of registry servers, then when the workload approaches the limits of either of those components the application should activate outgrowth and balance the workload to the outgrown application instance.

Components within each application instance rely on aggressive guard timers and retry strategies to assure that client requests are correctly served in less than the maximum acceptable service latency. Figure 11.7 illustrates this aggressive failure detection and recovery by a front-end server instance "F1" for a failure of back-end server instance "B1." Front-end server instance "F1" receives a client request that requires a back-end server instance to respond to a request. "F1" sends a request to back-end server instance "B1." Assume that "B1" fails prior to replying to a request from "F1." "F1" awaits a response from "B1" until a brief guard timer expires, and then retries the request to "B1." Since "B1" has failed and has probably not recovered service yet, the guard timer for the retry to "B1" will also expire. Rather than making a second retry to "B1" (which is unlikely to have recovered service in another few milliseconds if it failed to respond to the two previous attempts), "F1" sends the request to an alternate back-end server "B2." Assuming "B2" is up, "B2" should respond properly to the request before the guard timer expires. Note that since the guard timers

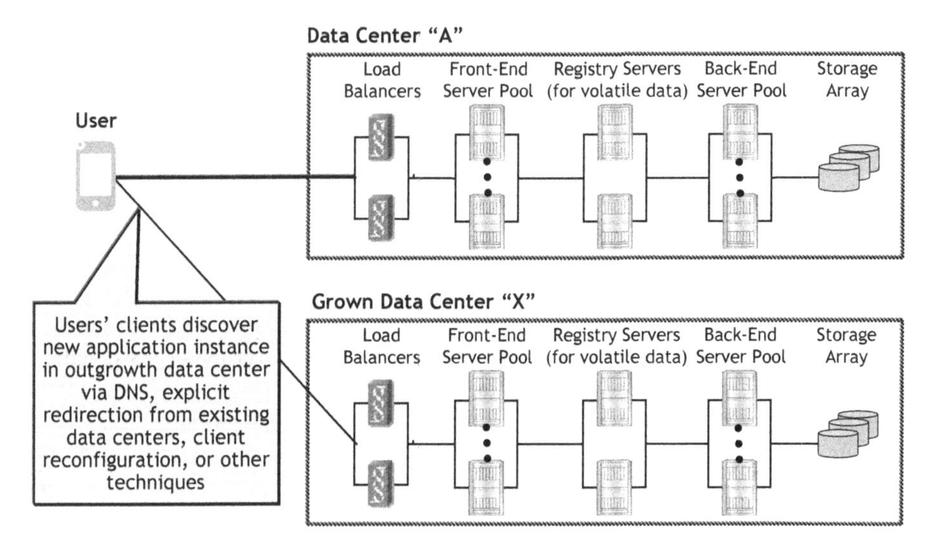

Figure 11.6. Outgrowth of Hypothetical Application.

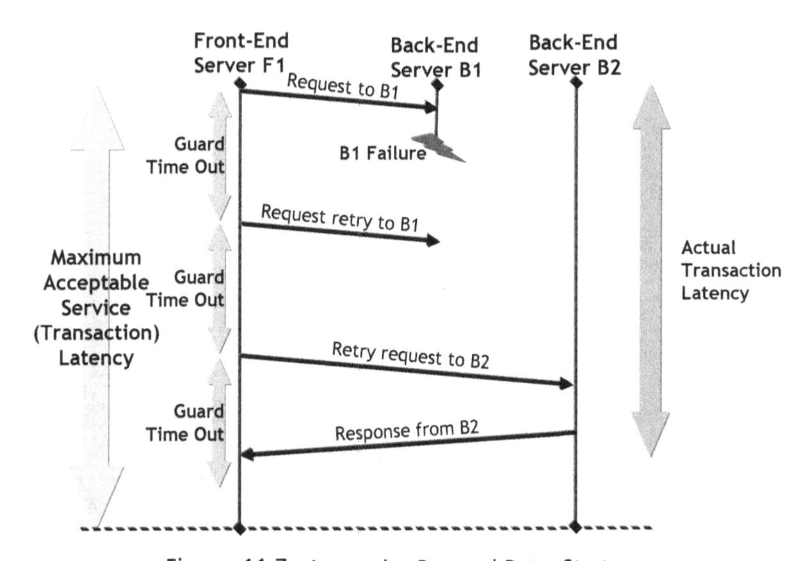

Figure 11.7. Aggressive Protocol Retry Strategy.

were set to be much shorter than the maximum acceptable service latency, "F1" had sufficient time to retry once to the original back-end server "B1" and make one request to an alternate back-end server "B2." There are a wide variety of retry strategies that can be considered; the point is that the guard timers and retry strategy should be constructed so that failures can be detected and service recovered without exceeding the maximum acceptable service latency so the failure is masked from the user.

Note that volatile data (e.g., user session state) are replicated to a pair of registry servers so that if a front-end server instance fails and the load balancer redirects a user's request to an alternate front-end server instance, then the alternate server instance can retrieve the user's volatile data from the registry server to recover the user's context to mask the effects of the front-end server failure.

Disaster recovery is supported via georedundancy and replication of non-volatile application data across multiple active application instances. Accessibility is further enhanced through replication of nonvolatile data to other data centers, allowing clients to send requests to those active georedundant application instances. Timely replication of nonvolatile data also factors into the RPO associated with disaster recovery. Volatile application data is generally not replicated between data centers due to WAN traffic concerns. Figure 11.8 shows data replication between data centers.

With georedundancy and data replication between sites in place, if a disaster were to occur, disaster recovery plans and procedures can be activated automatically or manually to recover service in line with RTO and RPO requirements. Figure 11.9 illustrates the redirection of client service to data center "B" when data center "A" fails. As indicated, volatile data are lost since they are not replicated across sites, but nonvolatile data are preserved up to the time of the last data synchronization between

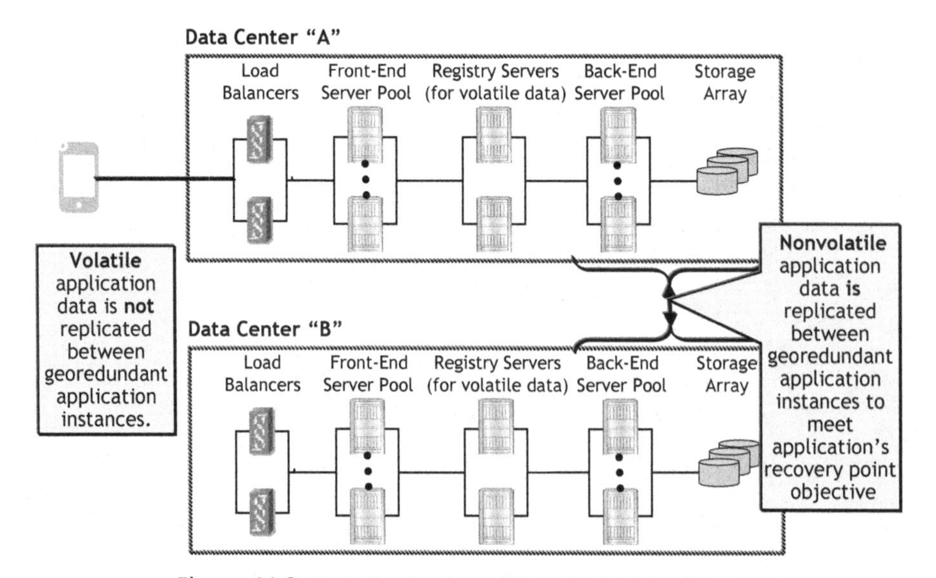

Figure 11.8. Data Replication of Hypothetical Application.

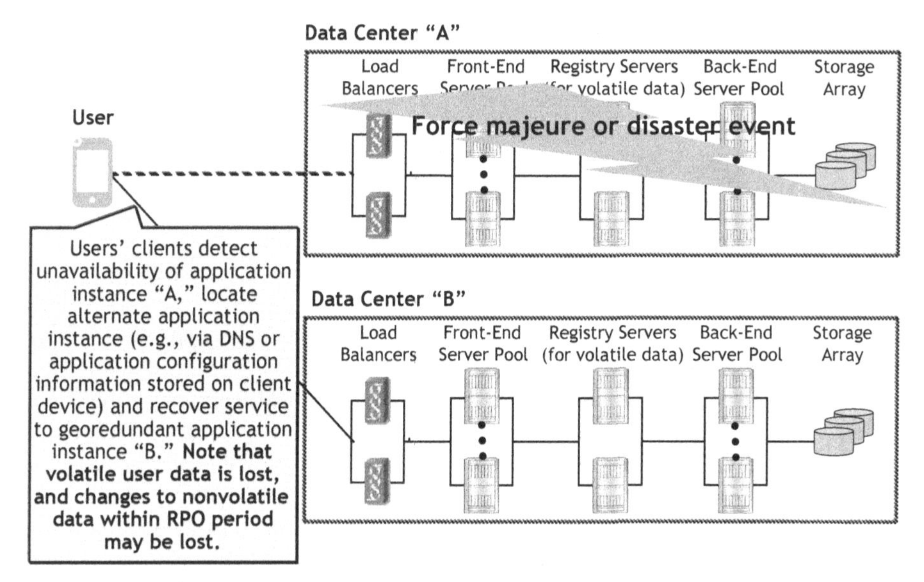

Figure 11.9. Disaster Recovery of Hypothetical Application.

sites. Depending upon the cloud customers' application requirements, the interval of the nonvolatile data synchronization can be reduced so that minimal nonvolatile data are lost in these situations.

11.6.7 Theoretically Optimal Application Architecture

The theoretical limit of service availability across a pair of elements is derived in [Bauer11] and elsewhere as Equation 11.1.

$$\text{Availability}_{\text{Pair}} = 2 \times \text{Availability}_{\text{Element}} - \text{Availability}_{\text{Element}}^{2}.$$

Equation 11.1. Maximum Theoretical Availability Across Redundant Elements

Essentially, this theoretical maximum assumes that a client has perfect knowledge of element availability and thus is always able to use whichever application instance is available at any time. The simplistic model underlying Equation 11.1 is based on the following technical assumptions:

1. Instantaneous detection of all failures with 100% accuracy.
2. Instantaneous and flawless identification of the failed element so the proper recovery action can be initiated.
3. Instantaneous and flawless service recovery onto the redundant element.

Unfortunately, real-world applications cannot meet all of these assumptions. Mathematically, this means that aggregate service availability across a pair of data centers (measurement point 3 or MP 3) must be less than—and typically far, far less than—the theoretical availability based on primary data center availability (measurement point 2 or MP 2). This is explicitly captured in Equation 11.2.

$$\text{MP } 3 < (2 \times \text{MP } 2) - (\text{MP } 2)^2.$$

Equation 11.2. Maximum Theoretical Service Availability

Thus, the question becomes *what service architectures most closely approach the theoretical limit of Equation 11.2?*

Figure 11.10 illustrates a configuration of the hypothetical web server application with an advanced client that can approach the theoretical limit of Equation 11.2. Essentially, the client maintains simultaneous authenticated sessions with a pair of georedundant application instances (i.e., in data center "A" and in data center "B"), to either send requests to both application instances in parallel (presumably using the first correct response received) or to send requests to one application instance with short guard timers and aggressively retry the request to the georedundant application instance if the initial application instance failed to respond correctly before the guard timer expired. Obviously, many traditional application protocols and client implementations will not support failure detection and recovery strategies that are this aggressive, but one can

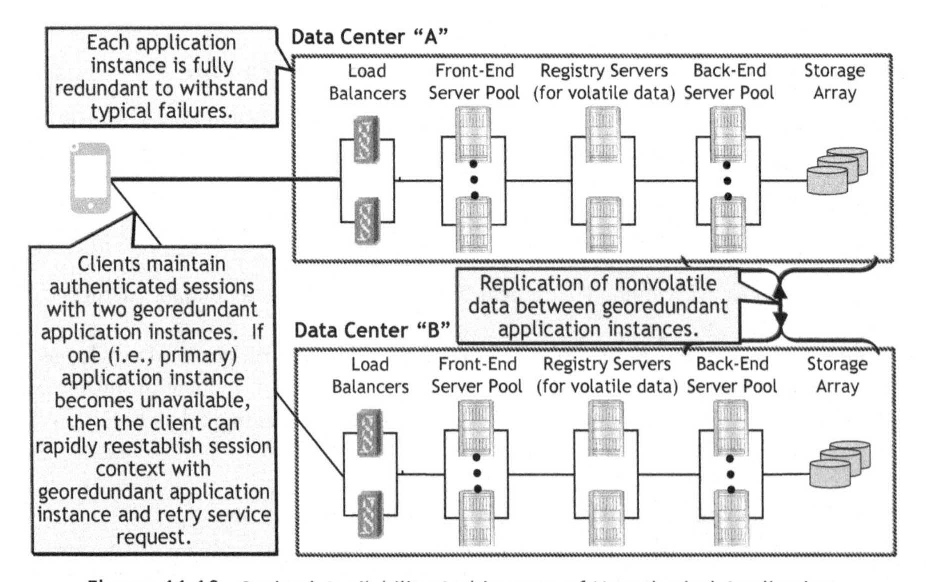

Figure 11.10. Optimal Availability Architecture of Hypothetical Application.

imagine how over time, application protocols and client implementations will evolve to support more aggressive strategies when optimal service reliability and availability is required. This architecture also assumes full internal redundancy within each data center and enough reserve capacity for one site to seamlessly manage all traffic for both data centers if one of the data centers fails meeting near perfect availability at the expense of additional resources and complexity to ensure no loss of service or data.

12

DESIGN FOR RELIABILITY OF VIRTUALIZED APPLICATIONS

While traditional applications can often be installed and run on a virtual machine (VM) with little or no modification, some additional design for reliability (DfR) diligence is appropriate to assure that it is both feasible and likely for the virtualized deployment to have reliability and availability as good as with traditional application deployment. More comprehensive design for reliability diligence—especially architectural work, analysis, development, and testing diligence—may even make it feasible and likely for the reliability and availability of a virtualized deployment to exceed that of traditional deployment. This chapter considers how design for reliability diligence changes for a virtualized application compared with a traditional native deployment of that same application.

12.1 DESIGN FOR RELIABILITY

Design for reliability of traditional information and computer-based systems is well understood and documented in [Bauer10] and visualized in Figure 12.1. The activities of traditional design for reliability are:

Reliability and Availability of Cloud Computing, First Edition. Eric Bauer and Randee Adams.

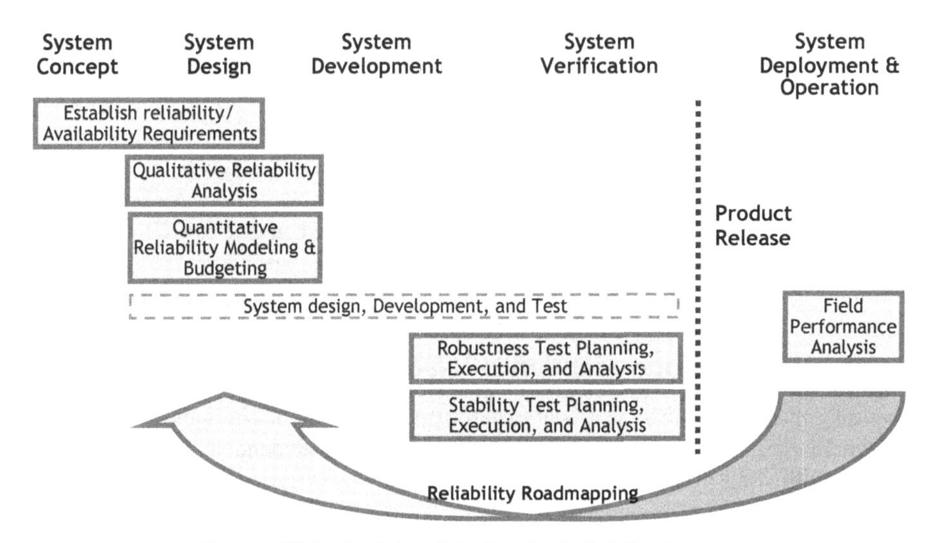

Figure 12.1. Traditional Design for Reliability Process.

- *Establish reliability and availability requirements* because all designs should be defined by clear and verifiable specifications.
- *Qualitative reliability analysis* assures that the high-level design is consistent with general, qualitative reliability and availability requirements.
- *Quantitative reliability modeling and budgeting* assures that it is technically feasible for the design to achieve the quantitative reliability and availability requirements over the long term. Budgeting quantitative targets for various system characteristics makes it more likely that overall system level targets will be achieved.
- *Robustness testing* verifies the effectiveness of failure containment and high availability mechanisms by confronting the system with likely failure scenarios to assure that the system automatically detects, contains, and recovers the failure without unacceptable service impact.
- *Stability testing* verifies that the system is stable and meets reliability expectations under a sustained heavy and mixed traffic load.
- *Field performance analysis* compares actual field performance with requirements and modeled results. Gaps with requirements and expectations drive reliability improvement roadmaps; gaps with model results drive refinement of mathematical models and parameters estimates.
- *Reliability roadmapping* captures planned and proposed reliability/availability improving features, testing and other actions to mitigate gaps between customers' expectations and actual or predicted system performance.

In addition, hardware reliability diligence was traditionally necessary to assure that the hardware platform supporting the application and platform software was likely to maintain an acceptably low hardware failure rate throughout the designed service lifetime of the system.

12.2 TAILORING DfR FOR VIRTUALIZED APPLICATIONS

The virtualization application usage scenarios enumerated in Section 10.1 introduce new risks to service reliability and service availability that are addressed by tailoring traditional system DfR. Let us consider each of these usage scenarios individually:

12.2.1 Hardware Independence Usage Scenario

Assuming the hypervisor is very stable and offers a faithful and reliable virtualization of the hardware platform being virtualized, the hardware independence usage scenarios offer the following reliability risks beyond traditional application deployment:

- *Not Making Virtualized Hardware a Single Point of Failure (SPOF).* As discussed in Section 5.1.2, "Single Point of Failure Analysis," high availability systems do not include any single points of failure. While traditional design for reliability diligence assures that there are no SPOFs in native deployment, one must be careful to assure that software modules that were deliberately tied to separate hardware in native deployment to prevent SPOFs are not mapped to VMs that could be configured on the same virtualized hardware platform (e.g., via antiaffinity rules), thereby making that hardware platform an SPOF. DfR treatment of this risk is covered in Section 12.4.1, "SPOF Analysis for Virtualized Applications."
- *Meeting Hardware Downtime Budget.* Reliability diligence must verify that the expected "hardware independence" deployment configuration does not accrue significantly more hardware downtime than the native deployment. This risk was discussed in Chapter 6, "Hardware Reliability, Virtualization, and Service Availability"; DfR treatment of this risk is covered in Section 10.3.1, "Traditional System Downtime Budget"; and design considerations were covered in Section 11.5, "Minimizing Hardware-Attributed Downtime."
- *The Application's High Availability Architecture Is Less Effective in Virtualized Configuration.* Just as careful design and testing is necessary to verify performance and reliability of native application's high availability architecture, DfR diligence is necessary to assure that the application's high availability architecture runs properly on the virtualized infrastructure, especially that it does not conflict with any automatic recovery or HA mechanism of the virtualized platform itself. Thus, one should explicitly assure that the application's HA mechanism configured on virtualized hardware:
  1. Reliably detects all failures with latency comparable with native configuration.
  2. Maintains acceptable service latency for failover, switchover or other recovery actions on the virtualized configuration.
  3. Properly contains failures to software modules or VM instances.

The risks that were discussed in Chapter 5, "Reliability Analysis of Virtualization," are addressed via qualitative and quantitative analysis (Sections 12.4, "Qualitative Reliability Analysis," and Section 12.5, "Quantitative Reliability Budgeting and Modeling") and validated via robustness testing (see Section 12.6, "Robustness Testing").

12.2.2 Server Consolidation Usage Scenario

Server consolidation increases the reliability and availability risks compared with the hardware independence usage scenario (Section 12.2.1) with the following additional risks:

- *Resource sharing introduces contention risks*—resource sharing between applications—especially when resources are overcommitted to maximize resource utilization—increases the risk of occasionally experiencing significantly longer latencies when acquiring or accessing resources. This increased resource latency can translate to higher service latency for application users. If workloads between consolidated applications are correlated, then service latency could degrade even further in busy periods. While applications and platforms should be configured so that literally no traffic is served with unacceptably long service latency or ever dropped, it is possible that the observed service latency (e.g., 50th or 95th percentile) will be longer in busy server consolidation scenarios compared with simpler hardware independence usage scenarios.
- *Ineffective failure containment* causes error or failure of a coresident application to cascade and impact the other applications.
- *Resource sharing increases recovery latencies*—when multiple applications share a single hardware resource (e.g., computer server hardware), then failure of that shared resource can cause all consolidated (i.e., coresident) applications to simultaneously initiate recovery actions. If those consolidated applications use other common applications, like databases or registries, then failure of the shared hardware may push one or more of those common applications into overload when multiple applications attempt to recover simultaneously. This may cause some or all of the consolidated applications to recover slower than expected as the common database, registry, or other applications take longer to serve the aggregate recovery workload.

These risks are primarily mitigated by proper configuration and operation of the hypervisor. Thorough robustness testing (see Section 12.6) validates that these risks are successfully mitigated.

In server consolidation deployments, the hypervisor may also be able to occasionally boost the resource allocation to a VM instance to give it more resources (e.g., CPU allocation or network bandwidth) than the VM's nominal reservation to address a traffic spike. This brief resource boost may sometimes make it unnecessary to activate application overload controls because sufficient resources are made available to serve the spike without explicitly activating overload controls, like traffic shaping or shifting. To take advantage of this capability of the hypervisor, applications must be prepared for some variation in resource allocation.

12.2.3 Multitenant Usage Scenario

Although multitenant deployment of VMs permits multiple independent instances of a single application to be consolidated onto a single virtualized platform, one must assure that the failure of one tenant's application instance is fully contained so other tenants are not affected. This scenario increases reliability and availability risks beyond the server consolidation case when the consolidated multitenant instances share some resources (e.g., system configuration data) but not other resources (e.g., tenant configuration and user data), so rigid failure containment between tenants can be more challenging. In particular, one must assure that failure of one tenant instance does not cascade to other tenant instances, and that service transition activities properly apply to individual application instances rather than inadvertently impacting multiple tenant application instances. In addition, multitenancy introduces the risk of correlated or synchronized behaviors that can stress the underlying virtualized platform, such as when multiple application instances execute the same recovery action or periodic maintenance actions simultaneously.

12.2.4 Virtual Appliance Usage Scenario

As the virtualized appliance scenario fundamentally changes the entire industrialization model (i.e., software packaging plus supporting materials), all aspects of application industrialization like installation, upgrade, and license management must be re-verified to assure that the virtualized appliance distribution offers the same service reliability, availability, and latency performance as traditional deployment.

12.2.5 Cloud Deployment Usage Scenario

While the earlier application deployments are statically configured, the cloud deployment usage model adds rapid elasticity to the mix, and thus adds the risk of elasticity-related failures. In addition, cloud service provisioning and orchestration, on-demand self-service, security, and other cloud characteristics introduce risks to service reliability, availability, and latency. Cloud deployment risks and mitigations are primarily considered in Chapter 13, "Design for Reliability of Cloud Solutions."

12.3 RELIABILITY REQUIREMENTS

The market generally expects virtualized applications to deliver the same service reliability and availability as that offered by traditional system deployments in which application and platform software is configured directly on physical hardware (rather than above a hypervisor). Thus, if the traditional system is expected to achieve 99.999% product-attributable service availability, then the virtualized application is typically expected to be capable of achieving 99.999% service availability when run on a suitably configured and properly operated virtualized hardware platform. Likewise, if the market expects the traditional system deployment to achieve 99.999% service reliability (i.e., ≤10 defects per million [DPM]), then the virtualized application should be capable of 99.999% service reliability (≤10 DPM) on a suitably configured virtualization platform. This expectation of requirements parity is presumed to apply to all reliability requirements.

This section first considers general service availability requirements, service reliability and latency, and overload requirements, and online capacity growth and degrowth requirements. Requirements for live migration and service transition activities are then considered, and finally georedundancy requirements are discussed.

Regarding requirements notations: the "*[A | B]*" notation is used to indicate that a requirement applies identically to both the "*A*" reading and the "*B*" reading, such as the same requirement applies to both "traditional system" and "virtualized application" configurations. A "*[C]*" notation indicates that "*C*" is optional in the requirement and may be included when applicable, or omitted if not appropriate or not necessary. Angle brackets "<" and ">" indicate that a quantitative value should be included in the requirement; common recommended values are shown as "*<99.999%>*", and unspecified quantitative values are shown as "<X>".

12.3.1 General Availability Requirements

The list below illustrates how the same general reliability requirements that apply to traditional system configurations can also be applied to virtualized application configurations

1. *[Traditional system | virtualized application] high availability configurations shall achieve <99.999%> product-attributable service availability in production operation.*
2. *[Traditional system | virtualized application] shall demonstrate compliance to service reliability requirements (see Section 12.3.2) and complete stability during at least <72> continuous hours of stability/endurance testing of mixed user and operational activities running mostly at engineered capacity.* "Complete stability" means no process failures, stable resource usage, and no degradation in transactional reliability, latency or throughput, and so on.
3. *[Traditional system | virtualized application] high availability configurations shall support deployment with no single point of failure.*
4. *The high availability [traditional system | virtualized application] configuration shall automatically detect and recover from any single hardware or software failure within <X> seconds.* This value <X> will be referred to as the *MaximumAcceptableServiceDisruption* time throughout this chapter.
5. *Testing of the [traditional system | virtualized application] shall demonstrate that all critical component failovers have service impact of less than MaximumAcceptableServiceDisruption.*

6. *Testing of the [traditional system | virtualized application] shall demonstrate that at least <95%> of likely hardware and software failures are detected and recovered in less than MaximumAcceptableServiceDisruption seconds.*
7. *Testing of the [traditional system | virtualized application] shall demonstrate that execution of automatic switchover mechanisms are at least <99%> successful.*

Multitenant applications add the following general requirement:

8. *No failure of one tenant's application instance shall impact service offered by any other tenants' application service.*

12.3.2 Service Reliability and Latency Requirements

While none of the virtualized application usage scenarios introduced in Section 10.1 will fundamentally impact service reliability, server consolidation and other usage models are likely to impact service latency because of resource sharing (discussed in Section 7.1.2, "Slashdot Effect"). The magnitude of the service latency impact will vary based on architecture and configuration of the virtualized platform, as well as operational factors, possibly including the workload of other applications that are consolidated onto the shared hardware platform. Thus, one should quantitatively characterize service latency expectations and carefully measure actual system performance to assure that service latency and service reliability are acceptable when the system operates at engineered load for a particular configuration.

Table 12.1 gives an example of service latency and reliability requirements for a sample application. Two service latency requirement points are specified for each transaction type (maximum 50th and 95th percentile latencies) which define a service latency envelope, as well as a maximum acceptable service latency and an overall service reliability requirement, expressed in defects per million (DPM) transactions when system configuration is running at or below engineered load. While the maximum acceptable service latency and maximum number of DPM operations (service reliability) requirements are likely to be the same for both traditional and virtualized

TABLE 12.1. Sample Service Latency and Reliability Requirements at MP 2

Transaction Type	Maximum 50th Percentile Latency (Milliseconds)	Maximum 95th Percentile Latency (Milliseconds)	Maximum Acceptable Service Latency (Milliseconds)	Maximum Number of Defects per Million Operations (DPM)
Logon	500	1,000	10,000	10
Query	150	300	3,000	5
Update	250	500	5,000	10
Logoff	200	400	4,000	5

deployments, the maximum 50th and 95th percentile service latency requirements may be less aggressive (i.e., longer) for virtualized deployment than for traditional systems to permit the system to be deployed in server consolidation, multitenant, or cloud computing configurations without violating the service reliability requirements.

Formal service reliability and service latency requirements at MP 2 can include:

1. *The maximum acceptable service latency for virtualized application deployment shall be the same as for traditional system deployment (and is shown in Table 12.1).*
2. *The virtualized application shall continuously meet all service reliability and service latency requirements (of Table 12.1) when offered load is less than or equal to the engineered capacity of the system under test.*

Engineered capacity of traditional systems was usually fairly straightforward to specify as it was either:

- *explicitly specified based on well-known hardware configurations*, for example, server configuration "A" with "B" CPU cores, and "C" Gb of RAM can support "D" simultaneous users); or
- *based on an observed characteristic*, for example, maximum engineered capacity is reached when CPU occupancy reaches $X\%$, or when memory usage reaches $Y\%$.

As virtualization weakens the application's linkage to the physical hardware, it is important to explicitly specify the observed behaviors that signify that engineered capacity has been reached for the particular application configuration. Many critical applications will include multiple VM instances—each of which may have different operational characteristics, including indications that engineered capacity has been reached for each instance—so it may be difficult to offer a single one-size-fits-all requirement specifying engineering capacity of virtualized application configurations. Thus, the authors offer the following higher level requirements:

3. *The performance indicator(s) that indicate that a virtualized application instance has reached engineered capacity shall be specified.* These indicators will typically include CPU usage, depth of work queue, and so on. Note that different VM packages (e.g., VMware) provide additional metrics that may be useful to assess resource usage.
4. *Service reliability and latency testing will be executed on a virtualized application instance running near or at engineered capacity.*

12.3.3 Overload Requirements

By decoupling application and platform software from the underlying hardware resources, virtualization complicates hardware dimensioning and capacity planning because engineered capacity is determined by the virtual configuration rather than explicitly via specified physical hardware. This reinforces the need to complete robust overload testing to assure that when offered load exceeds the engineered virtual resource capacity, the application will automatically detect the overload, gracefully control the overload, and resume normal operation promptly after the overload condition clears. Assuming that the overall application deployment includes appropriate perimeter security elements like firewalls, deep packet inspection elements, and so on, DoS/DDoS attacks should be largely controlled by that network infrastructure; however, it may also be appropriate to rate limit the bandwidth of either virtual network adapters or networking infrastructure to further control potential DoS/DDoS attacks on the application.

The following overload requirements are generally appropriate:

1. *The [traditional system | virtualized application] shall have a mechanism for detecting and automatically mitigating overload conditions.* This mechanism shall be based on characteristics such as resource availability (e.g., CPU, memory, and buffers), message priority (e.g., continuing to service emergency calls), disk read and write latency, and transaction "stickiness" (i.e., directing all messages or actions associated with a particular transaction to the same component for processing). Note that the guest OS's view of resource usage may be different from that of the host OS so care should be taken to ensure the correct metrics are used.
2. *The [traditional system | virtualized application] shall endure at least two continuous hours of <4> times the nominal engineering capacity of the configured system without critical failure.* Note that many, and perhaps all, user requests will fail during this period, and service reliability requirements do not apply during overload situations.
3. *The [traditional system | virtualized application] shall revert to normal operation and meet service reliability and service latency requirements within <5> minutes of overload situation clearing.*
4. *The [traditional system | virtualized application] shall raise an alarm when overload controls are activated and clear the alarm when overload controls are deactivated.*

As discussed in the server consolidation usage scenario (Section 12.2.2), hypervisors may occasionally offer VM instances a burst of additional resources when needed, and thus overload controls may not need to activate if the platform delivers a sufficient and timely burst of additional resources to meet the offered load. Likewise, the virtualized platform may be unable to deliver the nominally allocated or requested resource capacity (e.g., because of oversubscription), and thus overload controls may be required to activate at less than nominal workload levels. The risk that the hypervisor may deliver less capacity than requested leads to the following requirement:

5. Overload controls will activate and deactivate to assure minimum user service impact when the hypervisor provides fewer resources (e.g., CPU or network bandwidth) than were requested or reserved and cannot meet the offered load.

12.3.4 Online Capacity Growth and Degrowth

The rapid elasticity that is an essential characteristic of cloud computing is underpinned by the fundamental ability of an application to grow and shrink capacity while the application remains online and serving users. This fundamental online application growth and degrowth feature transforms into rapid elasticity when service orchestration mechanisms coordinate and automate execution of the online application growth and degrowth mechanisms to enable rapid—and perhaps even automated—capacity growth and degrowth to closely track offered load.

As discussed in Section 7.4, "Cloud and Capacity," there are three fundamental application growth strategies: horizontal growth, vertical growth, and out growth. While not all applications will support all three growth strategies, applications should support vertical or horizontal growth (or both) to permit a single application instance to grow, as well as outgrowth in which additional application instances are created (possibly even on multiple infrastructure as a service (IaaS) providers' platforms simultaneously in a cloud bursting scenario).

Applications should consider the following requirements:

1. *It shall be possible to increase application capacity without impacting service to existing application users.*
2. *Mechanisms shall be provided to gracefully release excess resources without causing unacceptable user service impact.* Note that this is often implemented by migrating user sessions or workload away from lightly loaded process or VM instances so that (then unused) processes or VM instances can be terminated and their resources released.

The following requirements enable application providers and IaaS providers to compare and agree upon capacity growth expectations.

3. *The IaaS provider must indicate the speed at which additional capacity could be brought online.*
4. *Application providers must specify the speed at which additional capacity is required.*

12.3.5 (Virtualization) Live Migration Requirements

Proper support of live migration affords the cloud service provider significant operational flexibility for application management. For example, live migration enables sophisticated cloud service providers to migrate running applications (i.e., their memory and execution state and disk storage access), onto a subset of servers during off peak periods and power down the unneeded servers to save electricity and cooling expense. As load increases approaching the next busy period, more servers can be powered on and the workload can be rebalanced over the expanded server pool via live migration. Thus, cost- and/or energy-conscious cloud providers might live migrate many application instances twice a day or more.

Live migration inherently entails a brief window of service disruption when the VM is paused and not executing as the VM instance memory is transmitted from one host to another and has to resynch its state information with the new copy. This period of disruption should be shorter than the maximum acceptable service disruption time so live migration should not cause transactions pending during the live migration event to be delayed so much as to make the responses unacceptably late and thus impact service reliability metrics. Note that since slow live migration might be mistaken by an application's high availability mechanism as a critical application failure, it may be appropriate to implement some interlock between live migration and an application's high availability mechanism to prevent spurious application failovers. If this is necessary, then additional requirements will be applicable to the application's high availability infrastructure.

Virtualized applications should support the following requirements:

1. *In no cases will live migration cause application's high availability mechanism to delay successful heartbeat/keepalive or other responses/protocol exchanges by more than <X> milliseconds.*
2. *In no cases will live migration cause the maximum acceptable service latency (from Table 12.1) or maximum DPM rate (also from Table 12.1) to be exceeded.*
3. *During the live migration event, the 50th and 95th percentile of service latency will be no more than <X> milliseconds more than the normal service latency requirement (from Table 12.1).*
4. *Live migration shall cause no nonvolatile data—including performance counts, provisioning data, or usage/billing information—to be lost.*

Multitenant applications have the following requirement:

5. *No live migration executed on one application instance shall have any user service impact on any other (multitenant) application instances that might be running.*

Compliance with these requirements must be robustly tested and validated as they are key indicators of the availability and reliability of the application.

12.3.6 System Transition Activity Requirements

Enterprises will expect any user service disruption due to service transition activities like software upgrade, update, or retrofit to be no greater than for traditional application deployment. Enterprises may even expect virtualized applications to have less service impact for service transition activities because live migration or geographic distribution (or georedundancy) of application instances can be used to minimize service disruption of service transition activities. Any increase in the expectation should be clearly captured in requirements to assure that architects and developers are aligned, and so testers can methodically verify achievement of higher expectations.

1. *Software upgrade, update, retrofit, and patching of [traditional system | virtualized application] shall have less than <X> seconds of impact on existing users with no loss of state or context, and no more than <Y> seconds of impact on new user sessions.*

A similar requirement should be captured to include service transition activities involving the infrastructure (e.g., hardware, firmware, operating system, and hypervisor):

2. *Upgrade, update, retrofit, and patching of [traditional system | virtualized system, including hardware, firmware, operating system, and hypervisor] shall have less than <X> seconds of impact on existing users with no loss of state or context, and no more than <Y> seconds of impact on new user sessions.*

Multitenant applications have the following requirement:

3. *No service transition activity executed on one tenant's application instance shall have any user service impact on any other tenants' application instances that might be running.*

Inevitably, the IaaS provider will occasionally execute planned or preventive maintenance on the underlying physical hardware, and the IaaS provider should take steps to minimize any application service impact due to the provider's actions. While application support of live VM migration is the first step in minimizing the impact of IaaS service transition downtime, individual IaaS providers may offer additional requirements to minimize the risk of service quality, reliability, or availability impact due to IaaS provider operations.

12.3.7 Georedundancy and Service Continuity Requirements

[Bauer11] offers a complete set of system georedundancy requirements to consider, so this section reviews only the highest level requirements which should be relevant for both traditional as well as virtualized applications.

1. *[Traditional system | virtualized application] shall support georedundant deployment.*
2. *The recovery time objective (RTO) for restoring user service to a redundant system configuration shall be no more than <X> hours/minutes.*
3. *The RTO for restoring administrative, maintenance, and provisioning service to a redundant system configuration shall be no more than <Y> hours/minutes.*
4. *The recovery point objective (RPO) shall be no more than <Z> hours/minutes/seconds.*

12.4 QUALITATIVE RELIABILITY ANALYSIS

Having completed the high availability architecture for the virtualized system, one next completes qualitative reliability analysis to assure that it is feasible and likely that qualitative reliability requirements will be met. If the native high availability mechanism still meets the reliability requirements for the virtualized application, then that mechanism should continue to be used particularly when first virtualizing an application that existed on a traditional architecture. Guidelines for mapping application instances onto servers to support high availability are discussed in Section 11.1.1, "Mapping Software into Virtual Machines."

As the architecture becomes more virtualization aware, it may become beneficial to leverage at least some of the high availability mechanisms offered by the virtualization platform, such as automatically restarting failed VM instances; however, the risk due to the added complexity and possible race conditions between the two high availability mechanisms must be well analyzed and tested before that is considered. In addition, the use of the high availability mechanisms associated with a particular virtualization platform may also lock the application into that particular virtualization platform, thereby reducing deployment flexibility. Several general architectural questions to consider are:

- Does the native high availability mechanism still meet the reliability requirements for the virtualized application?
- Would the virtualization platform's high availability mechanism meet the reliability requirements for the virtualized application?
- Would the use of a particular virtualization high availability mechanism constrain the application to that particular hypervisor or virtualization platform?
- What is the impact of simultaneous use of both native and virtualized high availability mechanism? Is there a risk of race conditions?

In addition, the following reliability analyses should be completed:

- Verify virtualized hardware is not a single point of failure—Section 12.4.1
- Verify that results of FMEA of virtualized system are comparable to native system results—Section 12.4.2
- Verify that live migration should have a negligible service impact—Section 11.3.3
- Verify that elastic growth and degrowth should have negligible service impact—Section 12.4.3
- If applicable, verify that configuration and management of individual tenant application instances has no impact on other tenant instances—Section 12.4.3.1

12.4.1 SPOF Analysis for Virtualized Applications

Unlike traditional systems, virtualized applications are explicitly decoupled from the underlying hardware, so hardware platform components are often not explicitly shown in architectural diagrams. Nevertheless, application suppliers should recommend at least one high availability configuration for critical applications that maps application and platform software components onto virtualized machines configured on different hardware platforms to assure that no single point of failure exists. This means assuring that redundant VM instances are not hosted by a single element (e.g., all active and redundant VMs running on a single server) so that no single hardware or infrastructure failure will impact both primary and redundant VM instances of application components. Clear configuration information should be provided to architects and operations engineers on the engineering rules to assure that the application is deployed in a suitably high availability configuration so that no single (noncatastrophic) hardware, software, or other failure produces an unacceptable service impact. Server anti affinity policy should be set to assure that specific VMs must not be active on the same server.

Figure 12.2 illustrates a fault tolerant architecture that has mapped the various redundant component instances of a sample application (A_1 and A_2; B_1, B_2, B_3, and B_4; C_1 and C_2) across four physical servers hosting VMs (virtual servers 1, 2, 3, and 4). Figure 12.3 shows how a failure that impacts server 1—and hence A_1, B_1, and C_1—leaves redundant VM instances—A_2, B_2, B_3, B_4, and C_2—available to serve users, and thus server 1 (or servers 2, 3 and 4) is not a single point of failure. Readers can imagine how individual failures of virtualized servers 2, 3, or 4 would not be single points of failure either. Note that in the case of load-shared $N + K$ redundancy there does not have to be a one-to-one mapping of VM to server as indicated in Figure 12.2; however,

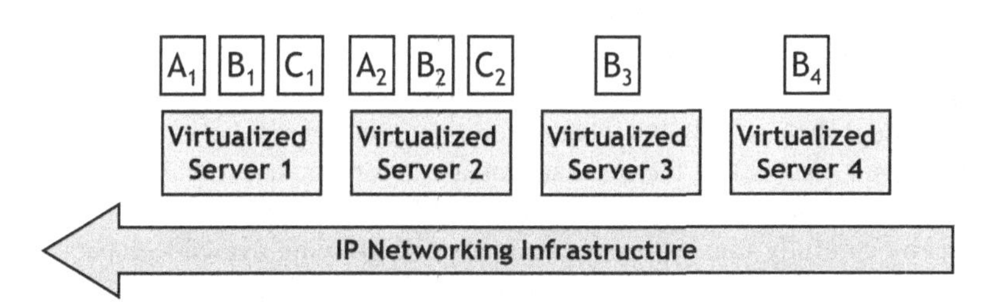

Figure 12.2. Mapping Virtual Machines across Hypervisors.

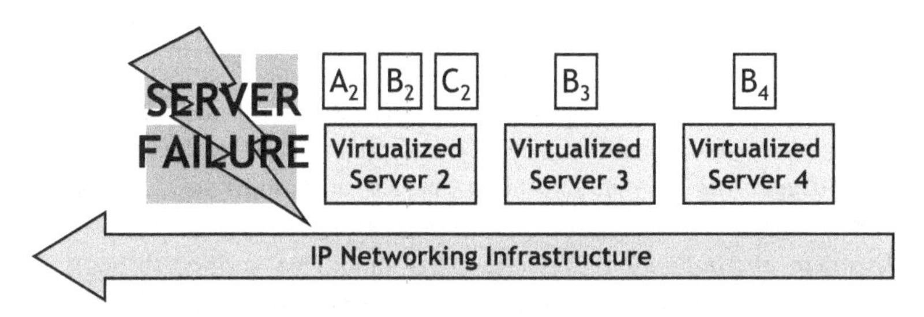

Figure 12.3. A Virtualized Server Failure Scenario.

there does have to be a distribution of VMs that supports the availability of redundant VM instances in the event of a total server failure.

12.4.2 Failure Mode Effects Analysis for Virtualized Applications

Failure mode effects analysis (FMEA; described in Section 5.1.3) for traditional applications generally analyzes down to the level of recoverable software and hardware modules. Virtualization adds two additional levels of complexity to consider:

- Failure and recovery of every VM instance should be considered separately.
- Failure analysis of traditional hardware modules (e.g., compute blades, rack mounted servers, hard disk drives, and network adapters) must be replaced with analysis of virtualized platform components (e.g., virtual CPU, virtual NIC) including the hypervisor.

Depending on the application's virtualization architecture, a single virtualized server failure may impact a greater portion of application functionality than a hardware failure of a traditional application configuration, and thus one must assure that the failure scenario will not produce user impact of longer than the application's maximum acceptable service disruption time.

12.4.3 Capacity Growth and Degrowth Analysis

Application growth and degrowth are facilitated by rapid elasticity mechanisms. In traditional systems, it is often important that application and resource growth are non-service impacting; growth and degrowth are treated as manual service transition activities. The virtualization and cloud mechanisms support traditional nonservice impacting growth and degrowth as well, but what differentiates cloud rapid elasticity mechanisms from traditional ones is that the mechanisms are automatic and usually do not require the physical installation or removal of hardware. They can also mitigate some overload situations by carefully monitoring traffic loads and growing available capacity before the offered load reaches the engineered capacity of the allocated resources. Section 7.4, "Cloud and Capacity," provides background on how the cloud mechanisms impact capacity. Section 11.1.5, "Rapid Elasticity," discusses how to maximize effective usage of the mechanisms to ensure reliable growth and degrowth. Similar to the live migration process above, growth and degrowth scenarios must be analyzed and verified through testing to ensure that they can meet the requirements defined in Section 12.3.4, "Online Capacity Growth and Degrowth." Scenarios to consider include:

- *Growth of Application Capacity.* This should be verified through a manual request for an additional instance of the application, as well as through an automatic trigger (as indicated through service orchestration—see Section 8.3.1, "Role of Rapid Elasticity in Cloud Management").
- *Degrowth of Application Capacity.* This should be verified through a manual request for removal of an instance of the application as well as through an

automatic trigger (as indicated through service orchestration—see Section 8.3.1, "Role of Rapid Elasticity in Cloud Management").

- *Growth of Persistent (Nonvolatile) Storage.* This should be verified through a manual request for more storage, as well as through an automatic trigger (as indicated through service orchestration—see Section 8.3.1, "Role of Rapid Elasticity in Cloud Management").
- *Degrowth of Persistent (Nonvolatile) Storage.* This should be verified through a manual request for removal of storage as well as through an automatic trigger (as indicated through service orchestration—see Section 8.3.1, "Role of Rapid Elasticity in Cloud Management").
- *Overload Conditions.* Applications must be designed and validated for overload control management. This can be tested by applying varying loads on a system to simulate overload conditions followed by low traffic conditions to ensure that rapid elasticity can provide the needed growth and degrowth and mitigate the impact of overload (as defined in the requirements in Section 12.3.3, "Overload Requirements").

The same scenarios should be run in a multitenant configuration to ensure that none of the growth or degrowth activities negatively impacts another tenant's service.

12.4.3.1 Multitenancy Considerations. If the application supports multitenancy, then the service transition activity analysis should also verify that no service transition activity impacts active application instances that are not the explicit target of the activity. In addition to traditional service transition activities, the multitenancy analysis should also verify that there is no service impact on other tenant instances when each and every tenant-specific configuration parameter is changed.

12.5 QUANTITATIVE RELIABILITY BUDGETING AND MODELING

To assure that quantitative service availability requirements are met, the best practice is to create a downtime budget that meets the requirements and construct mathematical availability modeling demonstrating the feasibility of achieving the budget. The following sections will discuss availability analysis. Quantitative service reliability (i.e., DPM) and latency budgeting and modeling are beyond the scope of this book.

12.5.1 Availability (Downtime) Modeling

As discussed in Section 10.3, the best practice is to construct architecture-based availability modeling to analyze the feasibility and likelihood of a system achieving its quantitative service availability expectations. Fine-grained availability modeling will typically consider each component, module, or subsystem included in the FMEA analysis that can impact the primary service being modeled; medium- or coarse-grained modeling will aggregate some of the individual components, modules or subsystems to simplify the mathematics as well as parameter estimation and ultimate model validation and calibration; coarse-grained modeling considers very simple architectural or "black box" view of the system.

While virtualized application suppliers may not supply specific hardware to customers, the application supplier should have one or more reference hardware configurations. It is generally appropriate to use the reference high availability hardware configuration when modeling system availability. This enables the application supplier to offer customers a reference system availability prediction, and one can logically delete the reference hardware from the model by setting all hardware failure rates to 0 (e.g., 0 FITs, which is 0 failures per billion hours). One can then construct a mathematical model of their specific virtualized platform, estimate the service availability of that platform, and then sum that downtime with the hardware-free (i.e., hardware failure rates equal 0 failures per hour) system availability prediction to estimate overall unplanned software downtime for the deployed virtualized application.

12.5.2 Converging Downtime Budgets and Targets

Inevitably, several iterations are necessary to refine downtime modeling input parameters to reach feasible inputs that result in acceptable predicted values. Refinement to the budget—and even system architecture—might be necessary for reasonable modeling input parameters to make achieving the overall service availability requirement feasible and likely. Having reached a set of input parameters that achieve the requirement, the quantitative values should be used as baseline targets for failure rates, failure coverage factors, switchover/failover latencies, and switchover/failover success probabilities. These targets should be considered by architects and developers to guide their designs, and should be verified by unit and system testing.

12.5.3 Managing Maintenance Budget Allocation

Downtime budget must also account for downtime incurred while performing maintenance activities. As has been discussed in Section 11.3, "IT Service Management Considerations," it may be feasible for these activities to be performed reliably with no service downtime. Architects and developers should take maximum advantage of live migration, elasticity, and service orchestration mechanisms and incorporate them in their system architecture and designs to ensure that successful execution of the maintenance activities does not accrue any service downtime. Robust unit and system testing must be performed to validate that user service is not affected by maintenance activities. If successful execution of the maintenance activities accrues no service downtime and the activities are well tested and validated, then budget allocated in this category can be minimal (e.g., 1 minute or less) to account for a rare product-attributable failure in the mechanisms or procedures.

12.6 ROBUSTNESS TESTING

Robustness testing confronts a system with plausible failure scenarios to verify that the failure is properly contained, automatically detected, and rapidly recovered without causing unacceptable service impact to system users. Robustness tests of critical systems with high availability expectations should be completed to assure that inevitable failures will be successfully and automatically mitigated by the deployed system. Robustness testing is considered in the context of each of the 8 product-attributable failure vectors reviewed in Section 4.4, "Risks of Service Models" (and [Bauer10]).

Virtualization technology should reliably emulate application operation on traditional/native hardware configurations, and thus most or all basic robustness testing executed on traditional/native hardware configurations is expected to work properly on virtualized configurations. For applications that are offered on both native and virtualized configurations, some robustness testing can be performed on one of those configurations on the assumption that behavior will be identical on the other configuration; other scenarios are sufficiently different that at least some robustness test cases should be executed on both configurations. Note that in many cases, applications are built on previously tested platform software, so the bulk of robustness testing may have been covered during platform software testing. Nevertheless, some robustness testing is appropriate to verify that application software is properly integrated with high availability mechanisms in platform software and that the application and platform's high availability mechanisms are properly integrated with the hypervisor.

12.6.1 Baseline Robustness Testing

Mapping a traditional application onto a virtualized platform means that integration of automatic failure detection and recovery must be reverified, at least to the VM instance and virtualized server level. In addition to verifying proper failure containment and reliable recovery, all failure detection and latencies and switchover/failover time measurements should be repeated so validated latency inputs can be used as inputs to service availability modeling. Some spot checking of programming error, data error, redundancy, application protocol, and network error robustness testing should be repeated against the virtualized application to assure high-quality integration/adaptation of the application's high availability infrastructure with the virtualized environment.

Server consolidation, multitenant, and cloud computing usage scenarios add the additional risk that failure or abnormal behavior of a coresident application may impact the target application. Since consolidated applications inherently share the same physical processing, networking, and other physical resources, a heavy workload on a consolidated application might affect resources available to the target application. While the virtualization platform should prevent a heavy workload by a consolidated application from impacting actual resource availability (i.e., resource allocation requests should not explicitly fail), a heavy workload may increase resource access latency for other applications. For example, if a consolidated application is enduring a heavy workload or otherwise under stress (e.g., a security attack), then the target application may not enjoy the same favorable CPU scheduling as when consolidated applications are lightly loaded, and less favorable CPU scheduling may translate directly into somewhat increased transactional latency. These risks are sometimes euphemistically referred to as "noisy neighbor" problems because just as a noisy neighbor in an apartment building can impact your ability to sleep, a busy or errant application on a shared IaaS platform can impact the target application. We will lump all server consolidation related robustness scenarios into a new robustness vector called "Neighbor" detailed in Section 12.6.1.1. In addition, virtualization introduces several new failure scenarios, including:

- Failures of virtualized hardware, discussed in Section 12.6.1.2.
- Failures related to virtualized redundancy, discussed in Section 12.6.1.3.
- Failures related to virtualized activities, discussed in Section 12.6.1.4.

12.6.1.1 Neighbor Failures. Consolidated applications—"neighbors" on a shared virtualized platform—share access to essentially the same virtualized resources as the target application. For example, if the workload on neighboring applications increases, then resource-sharing algorithms (e.g., CPU scheduling) implemented by the hypervisor may either make less CPU resources available to the target application or may increase the latency between time windows when the target application is executed on a physical CPU or otherwise alter the resource access that the target application enjoyed when the neighbors were making light resource demands on the shared virtualized platform. Likewise, if the failure or errant behavior of a neighboring application is not fully contained by the hypervisor, then that failure or errant behavior might cascade to other VM instances, including the target application. Neighbor failures to consider in robustness test planning include:

- *CPU Exhaustion by Neighboring Application(s).* What happens to the target application when one or more coresident VM instances go to 100% CPU utilization, due to software defects (e.g., infinite loops and ineffective overload control), spikes in legitimate traffic, security attack, and so on?
- *Delay in Real-Time Notification.* activation of clock or timer handlers is delayed to simulate target application's handler being queued behind one or more other application's interrupt handlers. Thus, the target application's handler might finally execute significantly later than was nominally requested.
- *Network "Receive" Saturation by Neighboring Application(s).* What happens when a coresident VM instance is experiencing a traffic flood or DDoS attack? Is the target application's IP traffic impacted?
- *Network "Send" Saturation by Neighboring Application(s).* what happens when a coresident VM sends massive volumes of network traffic for a window of time (e.g., replicates a massive data set to another application instance)?
- *Disk Read Saturation by Neighboring Application(s).* What happens when a coresident VM(s) needs to read massive data sets from disk?
- *Disk Write Saturation by Neighboring Application(s).* What happens when coresident VM(s) attempt to write massive data sets to disk?
- *Memory Exhaustion by Neighboring Application(s).* What happens when coresident VM instances simultaneously reach their heap memory allocation?
- *Attack by a Neighbor Inducing Any or All of the Above as Well as an Internal Network Saturation.* What happens when one neighbor maliciously attacks another?

Applications that support multiple independent instances executing simultaneously on the same virtualized platform should verify that both unplanned failures and successful and unsuccessful service transition activities do not adversely impact service offered by other applications instances.

12.6.1.2 Failures of Virtualized Hardware. By decoupling application and platform software from the underlying hardware resources, virtualization introduces a risk that hardware failure and status information (e.g., resource load/overload) may not flow properly in virtualized configuration compared with native deployments. As discussed in Chapter 6, "Hardware Reliability, Virtualization and Service Availability," virtualization merely decouples applications from fallible hardware rather than completely eliminating the risk of hardware failure. Thus, robustness testing of applications is necessary to assure that the virtualized platform, application platform, and application software will seamlessly interact to assure that inevitable hardware failures are detected, contained, and service is recovered in less than the MaximumAcceptableServiceDisruption time. Assuring effective containment and rapid automatic detection and recovery from hardware failures is primarily the responsibility of the virtualized platform supplier or IaaS provider.

It is obviously infeasible to verify proper detection and recovery from all possible failures of all potential hardware platform configurations. Thus, one should verify proper rapid and automatic failure detection and recovery from the types of hardware failures that are most likely to occur. The types of hardware failures to consider include:

- *Processor Failure.* Complex and highly integrated devices like microprocessors, digital signal processors, network processors, field programmable gate arrays, and so on are critical to field replaceable unit (FRU) functionality and are often more susceptible to wear out due to environmental-related effects.
- *Disk Failure.* Hard disk drives are built around high performance spinning platters and moving magnetic heads. Over time, moving parts (e.g., lubricated bearings) will wear and eventually fail. Note that IaaS infrastructure, especially redundant array of inexpensive (or independent) disks configurations, may be designed to automatically mitigate the risk of most disk failures.
- *Power Converter Failure.* Board-mounted power modules are used to convert voltages provided on the system's backplane to the voltages required by devices on the board itself.
- *Clock Failure.* Oscillators drive the clocks that are the heartbeat of digital systems.
- *Clock Jitter.* In addition to hard (persistent) clock failures, the clock signal produced by an oscillator can jitter or drift. Clocks can drift as they age for a variety of reasons including mechanical changes to crystal connections or movement of debris onto crystal. This jitter or drift can cause circuitry served by one oscillator to lose synchronization with circuitry served by another oscillator, thus causing timing or communications problems between circuits.

- *Switching/Ethernet Failure.* These devices enable IP traffic to enter and leave the FRU, and thus are critical.
- *Memory Device Failure.* Memory devices are typically built with the smallest supported manufacturing line geometries to achieve the highest storage densities.
- *Parallel or Serial Bus Failure.* High-speed parallel and serial busses are very sensitive to electrical factors like capacitance and vulnerable to crosstalk. Many connector failures can be covered via this error category.
- *Transient Failure or Signal Integrity Issue.* Weak electrical design or circuit layout can lead to stray transient signals, crosstalk, and other impairments of electrical signals.
- *Application-Specific Component Failure.* Application-specific components like optical or radio frequency devices may be more failure prone because of small device geometries, high power densities, and newness of technology or manufacturing process. Components like fans, aluminum electrolytic capacitors, and batteries are also subject to wear out.

Hardware fault insertion (HFI) testing is the best practice for verifying automatic recovery from hardware failures. Virtualization enables one to execute the HFI tests on whatever hardware platform supports HFI, regardless of whether the deployment/production hardware supports HFI testing because virtualization in general and the hypervisor in particular should assure that classes of hardware failures should be presented to applications via VMs in similar ways. Note that different virtualized hardware drivers may cause some hardware failures to be presented to the VM instances hosting applications differently.

12.6.1.3 Virtualization and Redundancy Failures. Virtualization creates the risk that unneeded VM instances will be spuriously spawned or incorrectly remain active or paused, thereby causing application software—especially high availability software—to malfunction. In addition, live migration, online capacity growth, and online capacity degrowth are complex operations that can fail. Specific failure scenarios to consider:

- Spurious/unexpected application VM instance (e.g., a snapshot) is activated.
- Application VM instance spawns very slowly.
- Stale (paused) application VM instance activated.
- Live migration fails to successfully restart a VM instance.
- Live migration slowly restarts a VM instance (e.g., <X> seconds of paused or "lost" time).
- Request to allocate and activate a new VM instance fails.
- Request to allocate more persistent (nonvolatile) storage fails.
- Request to destroy a VM instance fails to complete successfully.

12.6.1.4 Network Errors. Presumably, errors in virtualized network interfaces would be rendered to VM instances either as "normal" NIC errors, or the network packet/data would simply never be presented to the VM instance. Since IP networking expects IP packets to occasionally be lost, protocol and application mechanisms mitigate occasional lost packets. Thus, failures of virtualized network adapters should not be materially different from failures presented by traditional NICs on native hardware configurations, yet it may be appropriate to repeat some testing to verify proper operation of network error mitigation mechanisms.

12.6.1.5 Summary. Figure 12.4 gives an Ishikawa, or "fishbone," diagram for high-level robustness test cases for virtualized applications that augments "Figure 4.8—Software Supplier (and SaaS) Responsibilities for Traditional Error Vectors," with the robustness test cases discussed earlier in this section.

12.6.2 Advanced Topic: Can Virtualization Enable Better Robustness Testing?

Beyond the obvious benefit of potentially shortening execution time of robustness test cases by restarting snapshot VM images rather than requiring testers to wait while slower traditional application startup completes, virtualization technology can offer opportunities for better, cheaper or more effective execution of some robustness test cases. As an analogy, consider that the boundary scan technology (IEEE 1149.1, sometimes referred to as JTAG for the Joint Test Action Group that developed the standard),

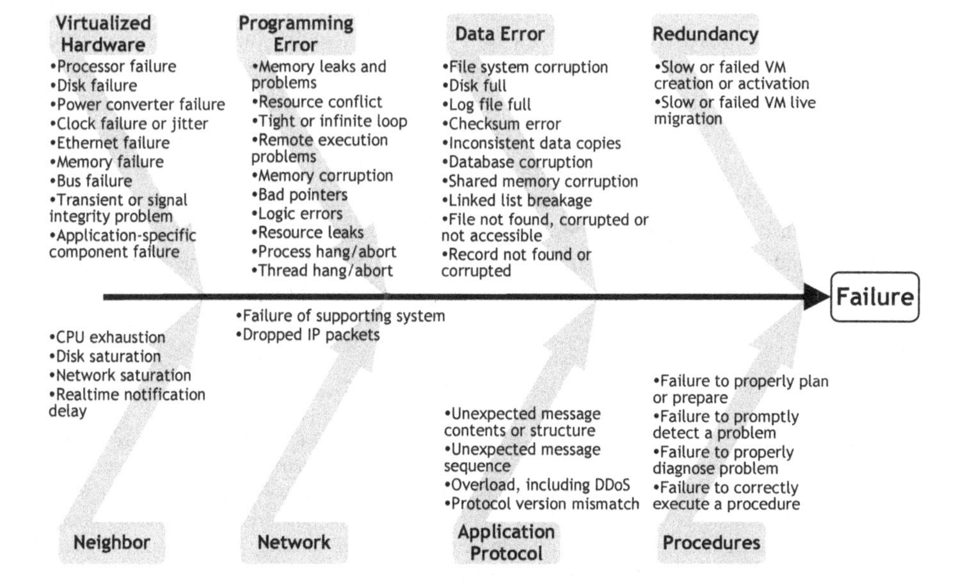

Figure 12.4. Robustness Testing Vectors for Virtualized Applications.

which is primarily used for manufacturing and assembly testing of electronic components and assemblies can also be used to simulate hardware failures.[1] Robustness test cases that are better, cheaper, or faster to execute on virtualized configurations than on native configurations can potentially save application development teams' time and money by executing those robustness test cases on a virtualized configuration instead of the native configuration when appropriate. Below is a list of ways that virtualization might be leveraged to enable better, cheaper, or faster robustness test case execution.

- Suspending application VM instances is an effective way to simulate profound unavailability of a target network element because it makes the entire element completely nonresponsive (e.g., even PINGs to the VM fail).
- Virtual network interface cards could theoretically simulate a variety of network impairments, such as dropped and corrupted IP packets, packets delivered out of sequence, and packet jitter.
- Virtualization can be used to create dummy or simulated systems (often on less powerful hardware, such as older servers or even laptops) which can be used for training and perhaps even to practice operations, administration, maintenance, and provisioning (OAM&P) procedures before executing procedures on live systems. Training via realistic simulation is a best practice for minimizing the risk of human error during "live" execution of operational and maintenance procedures. Appropriate virtualized environments with suitable training materials and scenarios can reduce the risk of human/procedural errors and associated service downtime.

One could even construct a virtualized platform that has fault insertion enabled so that one could easily simulate hardware, networking, and perhaps other failure scenarios to verify robustness of virtualized applications. Presumably, such a fault insertion enabled virtualized platform could be used for robustness testing of literally any application that can be hosted on the platform.

Cycle time for robustness test case execution might also be reduced by restoring system snapshots taken before executing failure scenarios because restoring a system snapshot should be faster and potentially more automated than completing traditional system restoration activities between of each robustness test iteration. Reducing cycle time means that either more robustness test cases can be completed in the same interval (producing a more robust product for the same testing investment), or a fixed number

[1] JTAG enables "tests" to be externally applied to hardware under test via a five-wire JTAG connection, such as driving a boundary scan-enabled component into "HIGHZ" (high electrical impedance) state, which electrically isolates (i.e., disconnects) the component from circuits on a printed wiring board, thus simulating a profound component failure. A variety of other JTAG commands and mechanisms can be used to simulate a variety of hardware failures on unmodified hardware and execute completely independently of normal software and firmware executing on the hardware assembly.

of robustness test cases can be completed in a shorter interval (producing similar robustness with smaller testing investment).

12.7 STABILITY TESTING

Assuming an appropriate stability testing campaign has been successfully completed against the traditional application configuration, stability testing of the virtualized application configuration can focus on the following virtualization-related risk areas:

- *Overload.* Application deployments on virtualized platforms will undoubtedly offer somewhat different service capacities and thus applications may experience a broader range of overload manifestations in virtualized deployment than in traditional deployment. Note that the overload test phase should include both light overload periods (e.g., 105% of engineered capacity) in which the hypervisor may boost the resource allocation to cover the increased workload, as well as true overload (e.g., 150% of engineered load) to assure that overload controls properly activate and later deactivate.
- *Live Migration.* Live migration enables data center operations staff to move active VM instances from host to host to optimize data center operations. If live migration will be used with the virtualized application in field deployment, then stability testing should verify that live migration has no impact on system stability even when occurring during failure situations.
- *Variations in Resource Availability.* In server consolidation, multitenancy, cloud, and virtual appliance deployment scenarios, the virtualized platform resources might be oversubscribed so the resources actually available to an application instance could range from being greater than the nominal resource reservation to somewhat less than the reservation. Stability testing should verify stability even when actual resource allocation differs from the nominal reservation.
- *Growth/Degrowth.* The virtualization platform and the hypervisor in particular manage the allocation and deallocation of resources for new and removed VM instances. Stability testing should verify that growth/degrowth of VM instances has no impact on system stability.

Ideally, a stability testing campaign includes a long-duration endurance test (e.g., 72 hours) which demonstrates complete system stability over an extended period. This endurance test should include a diverse and realistic mix of user operations as well as operational, administrative, maintenance and provisioning tasks. The load on the system under test should vary and include long periods with a system under heavy sustained load. One stability testing strategy is for the test campaign to simulate heavy daily traffic patterns with very heavy user traffic loads during a (possibly extended) busy period and a heavy provisioning load running during a maintenance period and/or during the period of busy user service, depending upon expected traffic patterns of the deployed application.

When possible, stability testing should mirror heavy usage patterns likely to be experienced in production operation. Thus, a stability test run might begin with a heavy (e.g., 80% of engineered capacity) mixed and sustained load of user traffic; a modest OAM&P workload, like adding new users and provisioning existing users, should be running also, as would be expected in production. After a continuous period of at least as long as the longest daily busy period the application is expected to experience in production, a simulated maintenance period can begin with maximum administrative load, such as bulk provisioning of new and existing users and perhaps database backup, with a moderate user workload continuing. The maintenance phase should be somewhat longer than the longest plausible maintenance period in production deployment. After the maintenance phase completes, a series of live migrations with moderate user and OAM&P workloads running can be executed. Periods of light and heavy overload can also be included, as can some simulated failure and recovery scenarios. The stability test generally ends with a final soak phase with moderate to heavy user and administrative workloads to assure that the system is truly stable after all of the activities and loads.

12.8 FIELD PERFORMANCE ANALYSIS

A key of continuous quality improvement is following the Deming cycle of plan/do/check/act to use feedback to close the loop and drive improvements. Figure 12.5 maps the system design for reliability activities against plan, do check, and act phases:

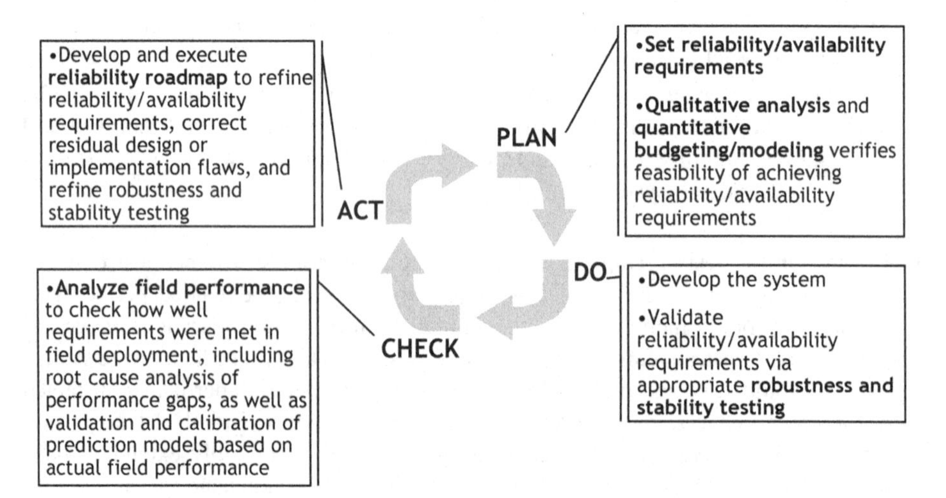

Figure 12.5. System Design for Reliability as a Deming Cycle.

- *Plan* involves setting reliability requirements and completing qualitative analysis and quantitative budgeting and modeling to assure it is feasible for the planned architecture and design to meet the requirements.
- *Do* is developing the system and validating achievement of the reliability and availability requirements via appropriate robustness and stability testing
- *Check* is the purpose of field performance analysis.
- *Act* is developing and executing a reliability roadmap to refine reliability requirements, correct residual design or implementation flaws, and refine robustness and stability testing in subsequent system releases.

Field performance analysis fundamentally involves two actions:

1. Gathering and analyzing service reliability and availability measurements from in-service deployments (discussed in Section 8.2.2, "Service Reliability and Availability Measurements").
2. Root cause analysis of service outages, as well as acute and chronic service latency and service reliability impairments if possible.

The resulting data enable:

- *Assessment of whether or not reliability/availability requirements were met* in the analysis period. Failure to meet requirements in production deployment often prompts aggressive development of a reliability roadmap (discussed in Section 12.9) and investment to execute the roadmap promptly.
- *Identification of residual defects and vulnerabilities* that caused service availability, reliability, and latency impairments during the analysis period. Correcting the residual defects and mitigating vulnerabilities inevitably make up a substantial portion of any reliability roadmap.
- *Validation, calibration, and refinement of predictive models and associated budgeting* to track better with actual performance. These refinements should enable models and budgets of future releases to be more accurate.

12.9 RELIABILITY ROADMAP

As explained in Section 12.2, "Tailoring DfR for Virtualized Applications," existing applications often enhance their support or leverage virtualization and other features across several releases. For example, a preexisting application may initially support virtualization for hardware independence only, and then add support for resource sharing and multitenancy before finally supporting elastic growth and other advanced virtualization and cloud characteristics. Multirelease roadmaps are often a convenient way to manage the often diverse set of reliability and availability feature and test investments that will drive service availability to exceed customer expectations. As with traditional reliability roadmaps, a roadmap to cloud computing may enable one to estimate the service availability at each release on the journey.

12.10 HARDWARE RELIABILITY

Responsibility for hardware-attributed downtime, hardware failures, and related maintenance actions lies with the virtualized platform provider (e.g., cloud service provider) and the hardware system supplier(s), rather than the virtualized application supplier. Thus, hardware reliability diligence should be worked in the context of the virtualized (e.g., IaaS) platform rather than in the context of the virtualized application.

13

DESIGN FOR RELIABILITY OF CLOUD SOLUTIONS

Amazon Web Services' best practices for architecting cloud solutions [Varia] says: "be a pessimist when designing architectures in the cloud; assume things will fail. In other words, always design, implement and deploy for automated recovery from failure." Solution design for reliability (DfR) is a methodical process that addresses the design for failure intent recommended by [Varia], [Hamilton], and elsewhere. This chapter introduces solution DfR, considers each of the solution DfR activities in detail, and discusses several related topics.

13.1 SOLUTION DESIGN FOR RELIABILITY

Solution DfR is visualized in Figure 13.1, and involves the following primary activities:

- *Define Key Service Reliability and Availability Requirements.* Good designs begin with clear and complete requirements. The best reliability and availability requirements include quantitative targets for maximum acceptable service disruption latency, service availability, service reliability, latency, and related behaviors for the target solution. The quantitative targets enable mathematical modeling

Reliability and Availability of Cloud Computing, First Edition. Eric Bauer and Randee Adams.

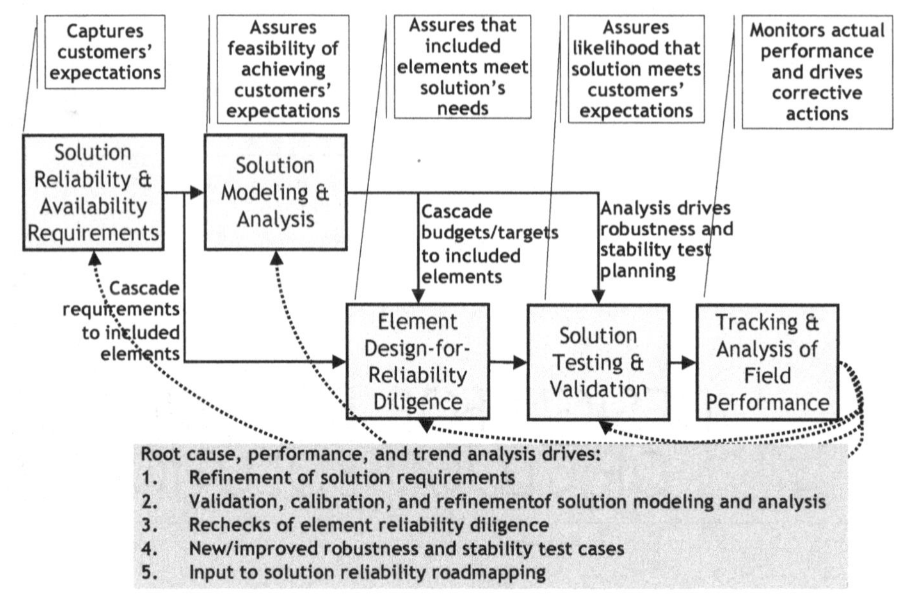

Figure 13.1. Solution Design for Reliability.

and permit richer architectural and design analysis than the qualitative requirements alone. Solution reliability requirements are covered in Section 13.3

- *Perform End-to-End Modeling and Analysis of Key Availability and Reliability Performance Metrics to Assure Feasibility of Meeting Targets.* Designing a solution is inherently complex because a large and diverse set of functional and nonfunctional requirements must be met along with strict cost and schedule constraints. These constraints drive architects to select most or all of the hardware and software (i.e., systems), environments, networking, power, policies, and humans to be preexisting or "off the shelf" (also known as commercial off-the-shelf, or COTS). The modeling and analysis step checks whether it is feasible that a potential "paper" design will meet the reliability and availability requirements over the long term. As various arrangements of components are considered to maximize the design goals and requirements while simultaneously not exceeding cost and schedule targets, modeling and analysis enables these options to be quickly assessed from a reliability and availability perspective so the project team can select the overall optimal solution architecture and design. Solution modeling and analysis is covered in Section 13.4.
- *Execute DfR Diligence on Included Elements.* Ideally, the DfR diligence of systems considered for inclusion will be assessed and considered before deciding whether or not to include a particular system in the solution. Just as the cost of correcting a defect found later in the development and deployment cycle is higher than if it is found earlier, it is generally cheaper to simply avoid using systems

with unacceptable field performance or inadequate DfR diligence than it is to attempt to address issues after the element is selected. Element reliability diligence is discussed in Section 13.5.

- *Validate Solution to Assure Service Reliability and Availability Targets Are Likely to Be Met.* Testing is necessary to validate that the solution meets both its functional requirements as well as nonfunctional requirements, like security, quality, reliability, and availability. Solution testing and validation are covered in Section 13.6.
- *Track Field Performance Against Key Service Reliability and Availability Targets, and Drive Appropriate Corrective Actions if Those Targets Are not Consistently Met or Exceeded.* It is well known that what isn't measured can't be managed well. Thus, well-run enterprises define key quality indicators (KQIs) covering service quality, reliability, and availability, establish quantitative targets for those metrics, measure those values, and compare to targets. Metrics are reported to enterprise leaders on a weekly, monthly, quarterly, and/or annual basis, and corrective actions are expected for any KQIs that fail to meet targets. Best practice is to tie enterprise compensation to achieving these KQIs so the financial interests of enterprise personnel are better aligned with the interests of end users. Beyond addressing root causes of specific service impairment incidents (e.g., fixing the particular defect that triggered a particular outage), it is often appropriate to refine the planned DfR diligence for the next solution release, such as:
  - *Adding or refining requirements* related to automatic detection and recovery for defects that escaped to the field and produced service impairments
  - *Calibrating, validating, and refining mathematical modeling* so predictions of future releases are more accurate.
  - *Rechecking DfR diligence of included elements* associated with service impairments
  - *Adding or refining robustness and/or stability tests* to reduce the risk of similar defects escaping from solution testing into production deployment.

Tracking and analysis of solution field performance is covered in Section 13.7.

13.2 SOLUTION SCOPE AND EXPECTATIONS

Detailed solution architecture and design begins by bounding the scope of the end to end service delivery path that is in-scope for the solution architecture and design, and what equipment, facilities and other components are outside of scope (i.e., not open for redesign) and thus must be accepted "as is." For example, if service users will access the application via their choice of browsers on their own personal device via their own wireless carrier's network, then the likely operational characteristics of that equipment and those facilities should be accepted as a given. Having defined the scope of the solution, one can frame the high-level KQI expectations for service quality, reliability, and availability as seen by end users across the end to end solution. Often, one also

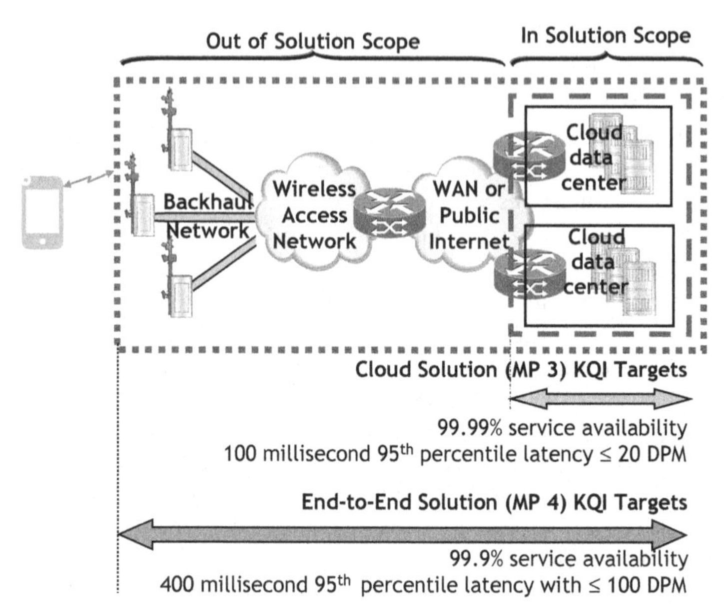

Figure 13.2. Sample Solution Scope and KQI Expectations.

sets target KQIs for end users that include facilities and equipment beyond what the cloud supplier is strictly accountable for. While this true end user KQI target is generally beyond the control of the enterprise, considering the broader perspective, as well as the narrower accountability perspective enables better solution architecture and is ultimately likely to produce a better—and perhaps more cost effective—solution because the performance expectations and assumptions of other elements have been explicitly considered.

Consider the sample end to end solution example of Figure 13.2. A client application runs on a mobile device which communicates across a wireless network and the public internet, to one of the infrastructure as a service (IaaS) service provider's data centers hosting the target application. Assume end users expect 99.9% service availability, less than 400 millisecond transactional latency for at least 95% of their operations, and less than 100 defects per million (DPM) service reliability for their transactions. A solution architect can then estimate the likely service availability, reliability and latency across the out-of-scope solution elements and facilities, and select a budget for in-scope solution components that makes it feasible to meet end to end performance targets over the long term. In this example, we assume the MP 3 targets for the solution to present 99.99% service availability to the public internet with less than 100 milliseconds of latency 95% of the time and no more than 20 DPM. While the cloud consumer has little or no control over the service quality, reliability, or availability of the public Internet, the end users' wireless access network, or the users' wireless devices, having considered the overall solution the architect can set balanced expectations for both the in-scope cloud consumer and cloud provider solution elements, as well as the out-of-scope elements and facilities.

13.3 RELIABILITY REQUIREMENTS

This section considers the following categories of solution reliability and availability requirements:

- service availability requirements (Section 13.3.1);
- service reliability requirements (Section 13.3.2);
- disaster recovery requirements (Section 13.3.3); and
- elasticity requirements (Section 13.3.4).

Note that these requirements make extensive use of the MP 3 and MP 2 measurement points introduced in Section 10.6.1.

13.3.1 Solution Availability Requirements

Solutions typically have several types of users who interact with the solution for different reasons, and often via different protocols, applications, and systems. In addition to end users, there are often maintenance engineers who operate, back up, and maintain the solution and included components. There may also be provisioning or data entry staff that add, modify, and delete application data and/or user account information. There may be software programs that autonomously interact with the solution, and there are often business support systems that extract usage data that is used to measure service and often charge cloud consumers based on the resources they actually used. There may even be regulatory or compliance officers or systems who monitor the overall solution or individual components. Each of these user types may access different types of service from the solution and may have different key quality expectations for service reliability and service availability. Solution availability requirements begin by specifying the highest level expectations of primary solution users across multiple cloud data centers (KQI 3), such as:

1. End user service across multiple [cloud] data centers is at least 99.995% available (MP 3 measurement).
2. Cloud consumer's maintenance staff shall experience operations, administration, maintenance, and provisioning (OAM&P) service across multiple [cloud] data centers at least 99.995% available (MP 3 measurement).

Requirements can also be set for single data center service availability (MP 2), such as:

3. End user service across a single [cloud] data center is at least 99.97% available (MP 2 measurement).
4. Cloud consumer's maintenance staff shall experience OAM&P service across a single [cloud] data center that is at least 99.97% available (MP 2 measurement).

Alternately, the MP 2 target(s) can be set during the architecture and analysis phase of solution design from modeling results that make MP 3 requirements feasible and likely.

13.3.2 Solution Reliability Requirements

Service reliability requirements are often specified as defective transactions per million service operations (DPM) in conjunction with a maximum acceptable service latency, so it is clear exactly when a slow service response is considered a failed transaction. Service latency targets (e.g., median and 95th percentile service latencies) can also be specified at the solution level, as they were in Section 12.3.2, "Service Reliability and Latency Requirements," for the application level. Thus one can construct service reliability and latency summaries like Table 13.1:

Formal service reliability and service latency requirements can include:

1. End users shall experience an average service reliability across multiple [cloud] data centers of less than 100 defective transactions per million attempts (DPM), which is at least 99.99% reliable
2. OAM&P transactions shall experience an average service reliability across multiple [cloud] data centers of less than 100 defective transactions per million attempts (DPM), which is at least 99.99% reliable
3. The maximum acceptable solution service latency is shown in Table 13.1.
4. The solution shall continuously meet all service reliability and service latency requirements (of Table 13.1) when offered load is less than or equal to the engineered capacity of the solution under test.

Note that the MP 1 service reliability requirements for application elements in the service delivery path accumulate to create the solution MP 2 value. While MP 2 service

TABLE 13.1. Sample Solution Latency and Reliability Requirements

Transaction Type	Maximum 50th Percentile Latency (Milliseconds)	Maximum 95th Percentile Latency (Milliseconds)	Maximum Acceptable Service Latency (Milliseconds)	Maximum Number of Defects per Million Operations (DPM)
Logon	3,000	6,000	15,000	20
Query	500	1,000	5,000	10
Update	2,000	4,000	10,000	20
Logoff	500	1,000	6,000	10

latency and reliability might be better than a straight sum of MP 1 values, one must always remember that solution component performance must generally be significantly better than the solution requirement. Thus, a five 9's single data center solution (MP 2) is not built by integrating a series of five 9's components (MP 1) because downtime and defective operations typically accumulate across the components of the solution.

While service availability requirements or targets consider both MP 2 and MP 3, service reliability requirements apply only to MP 2 on the assumption that a user will be served by a single cloud data center throughout a single session unless service from that cloud data center becomes unavailable (hence impacting availability MP performance) or the user is explicitly migrated to another cloud data center (considered in Section 13.3.4, "Elasticity Requirements").

13.3.3 Disaster Recovery Requirements

Beyond requirements for service reliability and availability, solution requirements should also cover disaster recovery time objectives (RTO) and recovery point objectives (RPO), along with any special disaster recovery considerations. For example:

1. The disaster RTO to restore user service to a georedundant [cloud] data center shall not exceed 2 hours.
2. The disaster RPO for user data following georedundant recovery shall not exceed 10 minutes.
3. The disaster RPO for operations and provisioning data following georedundant recovery shall not exceed 5 minutes.
4. *Cross-border disaster recovery shall be supported.* This requirement gives the cloud consumer and cloud service provider(s) more flexibility, such as being able to recover service for users served by an impacted cloud data center in the United Kingdom to an alternate cloud data center in Eastern Europe or perhaps North America. Note that the flexibility of cross-border disaster recovery may raise a variety of data privacy, regulatory/compliance, and other issues that must be worked in addition to the regular technical and operational challenges of disaster recovery.

13.3.4 Elasticity Requirements

Rapid elasticity enables the resources available to an application instance to be increased or decreased while the application is online, rather than the traditional model of requiring the application to be shutdown, reconfigured, and restarted. As online elastic growth and elastic degrowth often require different architectures, designs, and procedures than offline growth and degrowth do, requirements should explicitly specify the behavior of both supported online growth and degrowth. Moreover the growth and degrowth will have limits and require finite time to complete. Thus, elastic growth requirements might be structured as:

1. A solution instance deployed to a particular [cloud] data center shall support online capacity growth in increments of <X> [users | capacity units] that complete in no more than <Y> seconds/minutes per increment to a maximum of <Z> [users | capacity units].
2. Additional solution capacity can be brought online by instantiating the service in another [cloud] data center in no more than <X> hours/minutes, with additional capacity to serve at least <Y> [users | capacity units].
3. It shall be possible to distribute new users across [elastically grown] new data center instances of the solution without impacting existing users.

Elastic degrowth requirements might be structured as:

4. It shall be possible to gracefully reduce resource usage by an application instance as traffic decreases.
5. It shall be possible to gracefully (i.e., with minimal or no service disruption) migrate traffic away from a solution data center instance so that a solution data center instance (e.g., resulting from a cloudburst event) can be taken offline to reduce online capacity as traffic load decreases without impacting existing users.

The primary online elasticity expectation is that elasticity events should not produce unacceptable service impact for users, thus requirements similar to those for live migration can be applied:

1. In no cases will elastic growth or degrowth cause the maximum acceptable service latency (from Table 12.1) or maximum DPM rate (also from Table 12.1) to be exceeded for existing users.
2. During the elastic growth or degrowth event, the 95th percentile of service latency will be no greater than twice the applicable normal 95th percentile service latency requirement (from Table 12.1).
3. Elastic growth or degrowth shall cause no nonvolatile data—including performance counts, provisioning data, or usage/billing information—to be lost.

Note that elastic de-growth is generally more complicated than elastic growth since service for existing users should not be impacted.

13.3.5 Specifying Configuration Parameters

Achieving solution failure detection and recovery requirements often requires various protocol timers, maximum retry counts and other configurable parameters to be set appropriately. These configuration parameters may be necessary to assure the feasibility of the solution meeting its service reliability and availability requirements, and often impact the configuration of one or more individual elements of the solution. Some teams will capture these configurable parameter settings in solution requirements, architecture, installation, or configuration guides; some will push them to the individual elements' requirements or configuration guides. Regardless of a solution team's specific process, it is often important to explicitly capture the configuration details necessary for a solution's high availability mechanisms and make sure the configuration details are communicated to impacted elements in order to perform properly and so that solution reliability and availability expectations can be met when the service is deployed.

13.4 SOLUTION MODELING AND ANALYSIS

Having established KQI expectations and requirements, one constructs a reliability block diagram of the solution to identify elements in the critical service delivery path (Section 13.4.1) and to verify no single points of failure. One completes both failure mode effects analysis (FMEA) (Section 13.4.2) and service transition activity analysis (Section 13.4.3) to assure that solution requirements can be met including inevitable failures and planned activities. One also completes mathematical modeling to assure the feasibility of meeting primary data center (MP 2) service availability requirements and aggregate data center (MP 3) service availability. One completes a paper georedundancy analysis to assure that timing of data backup, replication, synchronization, and so on, assures that the RPO requirements can be met, and that element configurations (especially heartbeat timers, retry counts, and failure recovery strategies) and architectures make it feasible and likely that RTO requirements can be met (Section 13.4.6).

This section will use the sample application from Figure 10.1, which is repeated as Figure 13.3.

13.4.1 Reliability Block Diagram of Cloud Data Center Deployment

The first step in solution reliability analysis is to complete a reliability block diagram of all service impacting solution components in a single cloud data center. This reliability block diagram should highlight whatever relevant redundancy is deployed within a single cloud data center. If some cloud data centers have materially different configurations—such as different redundancy arrangements—then create RBDs for

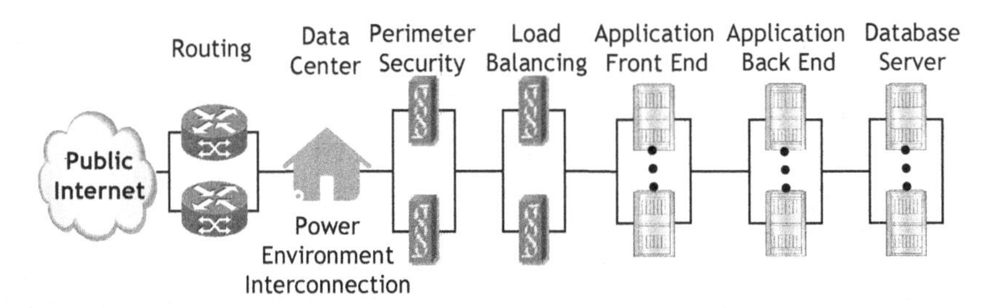

Figure 13.3. Sample Cloud Data Center RBD.

each different cloud data center configuration. Having constructed RBDs of cloud deployment architectures, one verifies that there are no single points of failure.

13.4.2 Solution Failure Mode Effects Analysis

As described in Section 5.1.3, a solution FMEA considers the service impact of the failure of any component in the data center reliability block diagram. The FMEA table should include one row for each solution component, and columns for the expected impact to primary service for primary solution users (e.g., end users, maintenance engineers, provisioning staff, regulatory/compliance systems, or users).

Failure of the primary data center itself is mitigated via georedundancy and is considered in the context of MP 3 modeling and analysis.

13.4.3 Solution Service Transition Activity Effects Analysis

One must consider the service impact for all maintenance activities, including elastic online growth, elastic online degrowth, and software patching and upgrade for each component in the solution. For each growth, degrowth, or software change (e.g., patch application, OS upgrade) event the maintenance activity analysis should list:

- strategy for minimizing service disruption, such as whether traffic must be migrated to operational components before service transition action, or whether any changes need to be made to service orchestration (e.g., new or updated policies);
- service impact if operation is successful; and
- likely service impact and recovery technique if operation is unsuccessful.

One can organize a service transition analysis like a failure mode analysis except specific activities appear as rows (e.g., horizontal growth of front-end server instances and degrowth of front-end server instances) with columns for each class of primary users, and individual cells capture the service impact on that particular class of user when the particular service transition action is executed.

13.4.4 Cloud Data Center Service Availability (MP 2) Analysis

Service availability from a single cloud data center (i.e., MP 2 availability) can generally be modeled algebraically by summing the expected annual downtime for each element (i.e., MP 1) in the cloud data center RBD. Figure 13.4 shows predicted annualized service downtime for each component of Figure 13.3, as well as the sum of predicted downtime across all components and facilities. Inevitably individual components will perform better or worse in any particular month, quarter or annual measurement period. For example, the Uptime Institute [UptimeTiers] reports that Tier IV data centers are likely to experience one 4-hour failure in a 5-year period, rather than accrue 48 minutes of service downtime every single year. Nevertheless, annualized downtime

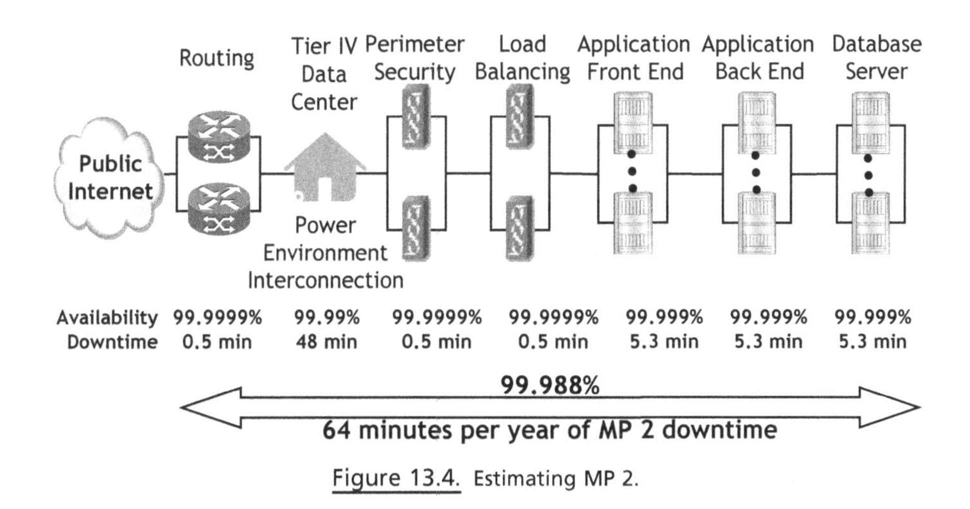

Figure 13.4. Estimating MP 2.

predictions are standard and they are a useful tool for analysis, comparison, and planning.

13.4.5 Aggregate Service Availability (MP 3) Modeling

MP 3 considers the overall service availability offered across a pool of two or more geographically redundant cloud data centers hosting the target solution. In the most general case, users served by an impacted data center will be redistributed across several operational data centers that are geographically close enough to the impacted users to deliver service with acceptable service quality and latency. Distributing the recovery load across multiple operational sites may produce shorter recovery times because the workload to reregister/reauthenticate, reestablish sessions, and rebuild context of impacted users would be naturally spread across multiple data centers. While the high availability mechanisms generally protect an application instance on a single site by rapidly recovering volatile data for active users and sessions to minimize user visible service impact of failure and recovery actions, georedundant recoveries do not generally automatically include volatile user data, so visible impact of failure and recovery is often greater for those services that use volatile data. Thus, georedundant recovery may be inappropriate for partial capacity or partial functionality outages where some or all users of the affected site have at least partial service because a georedundant recovery may have a more negative impact on the otherwise affected users than continuing attempts to recover on the affected site itself.

As described in Section 9.5, "Georedundancy Recovery Models," and [Bauer11], there are three fundamental georedundant recovery strategies: manually controlled, server driven, and client initiated; estimating the availability benefit of each is considered separately.

13.4.5.1 Estimating MP 3 for Manually Controlled Georedundant Recovery. The latency for manually controlled georedundant recovery is a function of:

- How rapidly maintenance engineers detect the failure and decide to initiate manual georedundant recovery,
- How long it takes for user service to be recovered when manual georedundant recovery procedure is executed,

These factors can be summed to estimate the per event service downtime when manually controlled recovery is executed. After the failure of the primary data center is corrected, the primary data center will typically be returned to serving user traffic. If there is any service impact when the recovered application instance in the primary data center is returned to serving user traffic, then that service downtime should also be included.

One then considers the predicted MP 2 service unavailability events to identify which of those events should be mitigated via manually controlled georedundant recovery, and replaces the MP 2 downtime estimate for those events with the sum of estimated manually initiated recovery and switchback latencies. Since manually controlled georedundant recovery is likely to impact all users served by the target data center, georedundant recovery is generally undesirable for mitigating partial outages, which have not impacted 100% of the users of the target data center. Likewise, for some failures, it will be faster with less overall service impact to simply repair (e.g., restart some software module) and recover service in the original data center rather than switching service to one or more georedundant sites, and later recovering it back to the recovered site.

13.4.5.2 Estimating MP 3 for Server-Driven Georedundant Recovery. The service availability benefit of automatic server-driven georedundancy can generally be estimated using appropriate traditional redundancy models (e.g., active–standby and $N + K$ load sharing) with appropriate input parameters and appropriate corrections for failures not recovered via georedundancy; see [Bauer11] for more details. Do not be surprised if server-driven recovery offers only a modest predicted benefit compared with manually controlled georedundant recovery.

13.4.5.3 Estimating MP 3 for Client-Initiated Georedundant Recovery. Solution clients (e.g., smartphone, laptops, and tablets) can implement client-initiated georedundancy mechanisms to efficiently detect service unavailability and autonomously initiate service recovery to an alternate application instance in a georedundant data center. Figure 13.5 illustrates the canonical configuration: client "A" accesses service offered by a pool of cloud data centers "B1" through "Bn." Assume the client establishes a service session with cloud data center "B1," a failure occurs that renders service unavailable from data center "B1," so the client "A" must detect unavailability of "B1," identify an alternate data center offering the service (e.g., via DNS), establish a session with an alternate cloud data center, and restore service state/context with redundant application instance before the client can resume using the service. Unlike

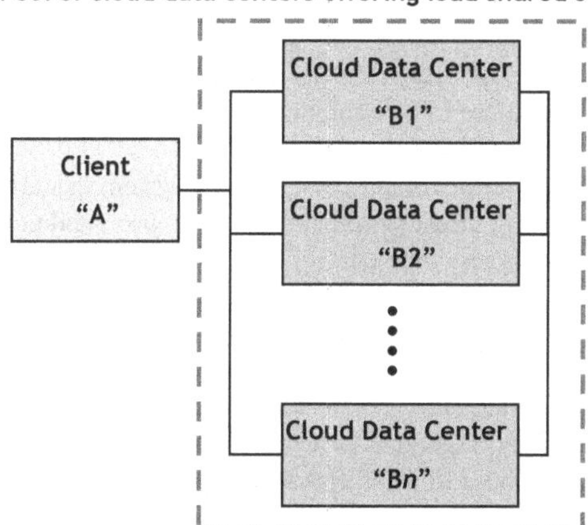

Figure 13.5. Modeling Cloud-Based Solution with Client-Initiated Recovery Model.

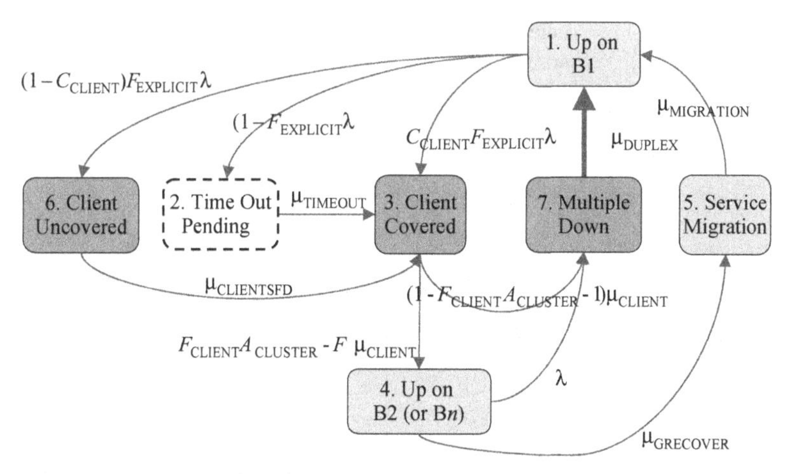

Figure 13.6. Client-Initiated Recovery Model from [Bauer11].

both manually controlled and server-initiated georedundant recoveries, where there is essentially one party controlling the recovery, control in client-initiated recovery is inherently distributed across the entire pool of client users because each client is responsible for its own recovery. In addition, since the client is driving the recovery action, the client can proactively store and rebuild session context to minimize user-visible impact of the failure event and recovery action.

The availability of service protected via client-initiated recovery can be predicted via the Markov client- initiated recovery model from [Bauer11] shown in Figure 13.6;

TABLE 13.2. Modeling Input Parameters from [Bauer11]

Symbol	Description
λ	Critical failure rate of solution hosted by target data center instance experienced by a client
$F_{EXPLICIT}$	Portion of critical failures of primary system that are explicitly reported to client systems (rather than profound nonresponse, which triggers time outs)
C_{CLIENT}	Portion of error responses from critical failures that trigger client-initiated recovery
μ_{CLIENT}	Automatic client failover (or takeover) rate
$A_{CLUSTER-1}$	Service availability of alternate data center(s) after solution service offered by target data center is unavailable
$\mu_{CLIENTSFD}$	Rate for client to determine that uncovered-to-client failures (e.g., failures signaled with wrong return code) are detected
F_{CLIENT}	Portion of automatic client recoveries that succeed
$\mu_{TIMEOUT}$	Overall time for client to time out from nonresponsive server
μ_{DUPLEX}	Duplex system recovery rate
$\mu_{GRECOVER}$	Rate (mathematical reciprocal of duration) of orderly service migration from alternate georedundant site back to primary data center
$\mu_{MIGRATION}$	Rate (mathematical reciprocal of duration) of service disruption on orderly service migration to alternate data center

Table 13.2 describes the modeling input parameters. [Bauer11] describes how to estimate input parameters and integrate the result with standard modeling results to create an overall estimate. The two particular considerations when applying this model to cloud data centers are:

- λ (critical failure rate experienced by clients) inherently integrates all causes of failure for both service components in the primary data center, as well as supporting infrastructure like power and network interconnections.
- $A_{CLUSTER-1}$ (service availability offered across pool of data centers after the client's selected or primary data center is unavailable) is generally the MP 2 value of the georedundant data center hosting an alternate instance of the target application or solution. This value will be somewhat higher than MP 2 when multiple alternate data centers are available to serve clients following the failure of one data center, but operational considerations, like how fresh and accurate the information provided to clients regarding available data centers is and the extent of how excess online capacity maintained ready to serve recovering clients, impacts the effective $A_{CLUSTER-1}$.

13.4.6 Recovery Point Objective Analysis

A recovery point analysis identifies:

- all nonvolatile data maintained by an application;
- the offsite backup, replication, or mirroring strategy; and
- the scheduled frequency of those backup, replication, or mirroring events.

One then verifies that the scheduled frequency assures that the RPO requirement is not violated.

13.5 ELEMENT RELIABILITY DILIGENCE

Solution architecture and budgeting should establish feasible and reasonable service reliability and availability requirements for included components, and modeling should verify that if these budgets are met by the components then the solution-level requirements should be met. The purpose of the element reliability diligence is to assure that it is both feasible and likely that each component in the service delivery path of the solution will meet the reliability and availability requirements of the solution.

Design for reliability of traditional information and computer-based systems is detailed in [Bauer10] and DfR of virtualized applications is covered in Chapter 12, "Design for Reliability of Virtualized Applications." Ideally, DfR diligence will be completed on all elements before they are included in the solution, and hopefully that diligence will assure that it is feasible and likely for the element to meet the quantitative KQI expectations cascaded from the solution KQI budget discussed in Section 10.6, "Solution Service Measurement." If the status of an element's DfR diligence is unknown, or field data suggest elevated service quality, reliability, and/or availability risks, then a high-level DfR assessment of the element is recommended. Chapter 15 entitled "Appendix: Assessing Design for Reliability Diligence" in [Bauer10] gives details on completing such an assessment.

13.6 SOLUTION TESTING AND VALIDATION

All components included in a solution should have been thoroughly tested at the component level, so solution-level testing can focus on verifying both the functional requirements for solution features, as well as nonfunctional requirements, like service reliability. This section considers validation of solution reliability and availability requirements, especially robustness testing, reliability testing, georedundancy testing, elasticity and orchestration testing, and stability testing. Adequate predeployment testing should enable occasional in-service testing (e.g., periodic disaster drills to verify georedundant recoveries) to be executed to assure the deployed solution is likely to achieve its reliability and availability requirements.

13.6.1 Robustness Testing

Robustness testing at the element level verifies that likely component related failures do not cause unacceptable service impact to MP 1. Robustness testing at the solution level verifies that inevitable element failures do not cause unacceptable service impact to MP 2 and MP 3. Component level robustness testing should address the ordinary hardware, programming, data, power, and other system specific failure scenarios of "Figure 4.6—Traditional Error Vectors" and "Figure 12.4—Robustness Testing Vectors for Virtualized Applications." Solution robustness testing verifies that the solution level service impact of these and other failures does not cause unacceptable service impact. Thus solution level robustness testing should consider at least the following scenarios:

- catastrophic failure (i.e., total unavailability) of every individual solution component;
- IP connectivity degradation (i.e., packet loss), disruption, and failure between solution components;
- network latency and jitter between solution components;
- overload of individual solution components; and
- inconsistent real-time clock settings on the physical servers hosting solution components.

Data center unavailability and WAN failures are considered in georedundancy testing (Section 13.6.3).

13.6.2 Service Reliability Testing

Solution-level testing should verify that service reliability (i.e., DPM) and service latency requirements (Section 13.3.2) are met for key service operations to primary solution users. Many million attempts of high volume transactions should be measured to accurately characterize the service reliability and latency. As it may be infeasible to attempt millions of iterations of high latency or nominally low volume transactions, a reasonable number of repetitions should be completed to assure that those transactions are also acceptable reliable and rapid.

13.6.3 Georedundancy Testing

Georedundancy testing verifies prompt service recovery to an alternate data center with acceptable service impact. Depending on the solution architecture and design, geo-redundant recovery can be activated in at least the following scenarios:

1. Catastrophic unavailability of the primary data center
2. WAN unavailability prevents at least some users' client devices from communicating with the primary data center

3. Catastrophic (e.g., duplex) failure of component instances at the primary data center prevent timely service recovery by primary data center
4. Orderly service migration to drain solution traffic from primary data center, prior to execution of a profound reconfiguration or maintenance action
5. Orderly service migration back to the primary data center from a georedundant data center

Georedundancy testing should verify that service disruption for each of these scenarios is within specification, that is, RTO and RPO requirements are met for disaster recovery scenarios, and the service disruption for nondisaster scenarios does not exceed the maximum acceptable service disruption for that scenario.

13.6.4 Elasticity and Orchestration Testing

Elasticity and orchestration testing should both verify that service reliability requirements are met during successful elastic growth and degrowth testing, and that service orchestration and elasticity failures are automatically detected and recovered without producing unacceptable service disruption. "Figure 7.14—Elasticity Failure Model" offers a handful of general failure scenarios to explicitly consider. At least the following adversarial elasticity scenarios should be considered for formal verification via solution level testing:

- *Slashdot scenario*—(see Section 7.1.2) traffic spikes faster than the elasticity slew rate can grow online capacity.
- Orchestration infrastructure and/or IaaS provider is nonresponsive.
- *IaaS resource stock-out*—a resource allocation request fails outright.
- *IaaS resource shortage*—IaaS provider offers less resource than was requested.
- *IaaS provider is slow* to respond to allocation requests.
- *IaaS allocation requests nominally succeed*, but allocated resource is unavailable or otherwise unusable
- Wide and rapid fluctuations in offered load.

13.6.5 Stability Testing

Stability testing is to verify that the cloud-based solution is completely stable and meets service reliability requirements while enduring a sustained period of heavy and mixed usage (often at least 72 hours). Ideally, stability testing will include periods of heavy load lasting longer than the longest typical daily busy period. For example, if an application normally serves 12 continuous hours of heavy load per day, then the stability test should include periods of heavy sustained loads for significantly longer (e.g., 16 or 18 hours). Provisioning and operational activities should also be included in the stability test. Ideally, the stability test will include online elastic growth and degrowth, as well as live migration. Service reliability and latency should be measured throughout the stability test to assure that requirements are met throughout the test.

13.6.6 In Service Testing

Practicing disaster drills once or twice a year to verify that business continuity plans, georedundancy configurations, manual procedures, and so on all function properly is a recognized best practice. In addition to verifying the configurations and procedures, disaster drills give staff valuable experience executing disaster plans so they will be more familiar with procedures and thus be more likely to execute them correctly in an emergency.

Just as disaster drills verify proper behavior of disaster recovery plans, limited failure scenarios can be induced in some solutions' production environments (subject to market's expectations for transactional and service reliability and availability) to verify efficacy of automatic failure detection and recovery mechanisms. Techniques that inject random failures into the system, such as those discussed in Section 11.6.5, "Operational Considerations," should be used to help verify those mechanisms. Among other things, solution validation and testing should assure that the recovery scripts are reliable and robust enough that they can be periodically tested on production systems to verify the robustness mechanisms and the policies, documentation, and training of the human staff that operate the solution.

13.7 TRACK AND ANALYZE FIELD PERFORMANCE

While service availability of individual elements can be averaged across the total population of elements to create a broad average, individual solutions are sometimes unique enough that it is inappropriate to attempt to create a useful and actionable analysis by simply aggregating field performance of all other cloud consumers' solutions. For example, simply because two different commercial airlines happen to operate the same type of passenger jets does not mean that they will both achieve similar on-time departure and arrival performance. At the solution level, operational policies, solution architectures, and other factors become increasingly important.

As explained in Chapter 3, "Service Reliability and Service Availability," the following metrics have traditionally been widely used:

- *Service reliability* (see Section 3.4), especially defective transactions per million attempts. Sophisticated enterprises will track user service reliability for specific transactions (e.g., data query and data update) or specific service scenarios (e.g., service accessibility and service retainability).
- *Service latency* (see Section 3.5), especially median or average service latency. Sophisticated customers will also consider a tail latency like the 95th percentile service latency or the percentage of traffic exceeding a fixed latency target (e.g., greater than 500 milliseconds of latency).
- *Service availability* (see Section 3.3).

As discussed in Section 10.6.1, MPs 1, 2, 3, and 4 are generally applicable service measurement points, which can be used as follows:

- *MP 1: Component Instance Availability.* Can be measured via a service probe installed in the data center hosting the target application instance. MP 1 can also be computed from trouble tickets, assuming those tickets capture sufficient detail (e.g., outage and service recovery times, number of users or portion of service capacity impacted, and portion of functionality lost).
- *MP 2: Primary Data Center Availability.* Can be measured via a service probe installed outside of the target data center. Note that the MP 2 service probe must be configured to access only the target data center so client-initiated and other georedundant recovery must be inhibited for the MP 2 probe.
- *MP 3: Aggregate Service Availability.* must be measured via a service probe that uses the same client-initiated recovery logic and configuration (e.g., time outs and maximum retry counts) as the client application(s) used by end users.
- *MP 4: End-to-End Service Availability.* Should be measured from actual users devices or client software, ideally by characterizing actual end user experiences via software running on client devices. Raw MP 4 data may be read from individual client applications if they record appropriate performance metrics and make that data remotely accessible. Alternately, a service probe application can be installed on some or all clients to explicitly probe and characterize service quality, reliability, and availability, and return the data to cloud consumers or service providers for offline analysis.

13.7.1 Cloud Service Measurements

The rapid elasticity essential characteristic of cloud computing complicates service reliability and availability measurements because application instances will be dynamically added and deleted, and user traffic will be dynamically balanced across a varying pool of online application instances. Thus, if a particular user executes a particular application transaction repeatedly (e.g., viewing a particular web page or making a particular service query every hour or every day), then it is possible that at least some of those requests will be served by different application VM instances, and may even be served by different data centers. This dynamic and elastic nature of cloud-based services adds uncertainty to service measurements at the network element level, because it may be difficult both to trace unsuccessful client requests (i.e., slow or defective transactions) to specific application instances, as well as to deduce the end user impact of specific application instance failures.

Service reliability and service latency are inherently user-oriented metrics that naturally scale with elastic growth, geographic distribution, and reconfiguration of cloud-based applications. After all, defective transactions per million operations metrics naturally normalize when data for one hundred, one thousand, or one million users is considered from one or more application instances. Likewise, these metrics can be scaled down to individual application instances. For example, one can meaningfully analyze the median service latency or reliability of an individual application VM instance, as well as the median service latency and DPM for a particular data center.

In contrast, traditional service availability is measured and normalized on a per system or per network element basis, and partial capacity loss outages are normalized against the engineered capacity of the affected system instance. Service availability of elastic applications might not be measured the same as service availability of native, deployments is measured. If the native application uses three server process instances and failure of one process instance is deemed a 33% capacity loss outage, then the same prorating can be applied to a non-elastic (i.e., non-cloud) virtualized deployment. Service availability measurements of cloud deployments are more challenging because the "rapid elasticity" characteristic of cloud computing translates to an elastic normalization factor to prorate outages against. While catastrophic failures that render all VM instances of a cloud-based application utterly unavailable are obviously deemed total capacity loss outages, partial capacity loss outages (e.g., impacting a single VM instance, see Section 7.6.2, "Partial Capacity Failure") are more common; thus, the key question is how to prorate these more likely partial capacity loss events. Theoretically, a partial capacity outage event of a cloud-based application could be prorated via one of the following strategies:

- *Normalize by Maximum Contracted Service Capacity.* This is likely to understate the availability impact because some services will only rarely operate at maximum contracted capacity. If the event occurs at an off-peak period when engaged capacity is only a small fraction of contracted service capacity, then the normalized event will appear very small if considered at all.
- *Normalize by Total (Engaged Plus Spare) Online Capacity the Moment Before the Critical Failure Occurred.* This has the adverse impact of making availability impacts look smaller for conservative enterprises, which maintain more spare online service capacity. In addition, the total capacity might be in flux if the failure is caused by or correlated with fluctuations in traffic volume that lead the cloud to add or release capacity.
- *Normalize by the Engaged Capacity the Moment before the Critical Failure Event Occurred.* Engaged capacity prior to the failure might not be known accurately, and the engaged capacity may well be impacted by the primary failure event itself, such as during elasticity related failures.
- *Normalize by the "Average" Engaged Capacity for the Duration of the Outage Event.* For example, one can determine average service utilization at the same time on the same day of the week for the past few (e.g., 4) weeks. This works well for established services with regular and stable traffic patterns, but might not work well for services with rapidly growing or unpredictable traffic volumes, or new services with insufficient historic data to reasonably characterize "average" engaged capacity.

These imperfect options for normalizing partial capacity loss outages can be avoided by adopting probed accessibility style metrics that consider the probability that a "typical" user can successfully establish a new session or complete a new request at any particular instant, averaged across the entire measurement period, such as an entire month. These metrics are often measured from a service probe that launches requests at the target application on a regular basis, such as every few minutes. If the probe client is successfully served, then the application is deemed to be up; if the probe client is not served successfully for several sequential attempts, then the application is deemed to be down. Service availability is computed by normalizing successful service responses to the probe client against the total number of service attempts executed by the probe client in the measurement period.

13.7.1.1 On-Demand Self-Service Measurements. Conceptually, on-demand self-service is analogous to end user service in that it can be neatly characterized with service reliability and service latency metrics. For example, the service reliability (e.g., DPM) of successfully allocating and engaging additional "elastic" resource capacity within a maximum acceptable time and the service latency of successful allocation and engagement are obvious and useful metrics to characterize one aspect of rapid elasticity. Thus service reliability and latency metrics can be useful for frequently executed self service actions. Enterprises should define a service measurement architecture that enables solution KQIs to be accurately measured and for operations policies to be deployed so the performance data are examined with suitable regularity and appropriate corrective actions are taken if performance falls below target.

Note that for on-demand self service actions that are rarely executed it is generally more effective to focus on troubleshooting and correcting the individual executions which failed or experienced unacceptable service latency rather than struggling with statistical analysis of tiny data sets. After all, it is impractical to consider the service reliability of an operation that is performed only a couple of times per year.

13.7.2 Solution Reliability Roadmapping

If a deployed solution is not consistently meeting its reliability and availability expectations, or if the expectations are rising, then one can construct and execute a roadmap of reliability- and availability-improving features, testing enhancements, and other changes. While individual failure events should be subjected to root cause analysis and corrective actions, occurrence of more than a very small number of reliability and availability impacting failures suggests that a deeper analysis of the solution architecture and design should be performed, resulting in recommendations for improvement. Solution reliability roadmaps often include one or more of the following work items:

- Changes or enhancements to the failure detection and recovery mechanisms (e.g., tuning timers, adding more explicit error messages) implemented for the interfaces between elements.
- Addition or enhancement of products, tools, or documentation to facilitate detection and troubleshooting of system failures.
- Enhancements to maintenance activity tools and procedures to reduce system downtime or mitigate risks associated with performing those activities.

13.8 OTHER SOLUTION RELIABILITY DILIGENCE TOPICS

Three additional topics worth considering are:

- service-level agreements (SLA) (Section 13.8.1);
- cloud service provider selection (Section 13.8.2); and
- written reliability plan (Section 13.8.3).

13.8.1 Service-Level Agreements

Service-level agreements or SLAs are for business remedies, not expectations of actual performance. For instance, a retailer that offers an "unconditional guarantee" doesn't imply that customers will be 100% satisfied with products purchased from them, merely that they will replace a product or refund the purchase price if the customer is dissatisfied. Likewise, a "100%" uptime SLA simply means that the service provider is prepared to offer some remedy for any covered incident. The "fine print" of generally surrounds the definition of covered events and the offered remedy. While customers might like significant remedies that make them "whole" after an incident (think homeowner or automobile insurance), the service provider may offer only nominal remedies (e.g., a modest service credit) as a standard part of their offering. Various papers like [InfoWeek] offer practical information on constructing cloud computing SLAs. The key "real" option that cloud consumers should expect is the option to terminate a (long-term) contract/agreement without penalties because of major and ongoing SLA breach.

Thus, one should carefully consider the following when evaluating suppliers' SLAs:

- *Does the Measurement Metric Actually Model How the Service Will Be Needed by Users of the Target Solution?* For example, measuring service availability by probing a data center or application every 5 minutes doesn't mean service was available every second or every minute, merely that it wasn't down for more than 5 minutes (or longer if the SLA requires sequential failures of two or more 5 minute probes to trigger remedies).
- *Does the Offered Remedy Provide Meaningful Relief if SLA Is Missed?* Not charging customers for the time that a service is unavailable is polite, but does little to mitigate unavailability of critical services. Unlike insurance companies, service providers are unlikely to provide what make customers "whole" after a failure, but meaningful remedies to consider are:
  - Root cause analysis and corrective actions for any SLA violation.
  - Add additional customer support staff and/or replace existing support team.
  - Be ineligible to bid for new contracts for other enterprise projects unless all SLAs are met.
  - Right to cancel contract without penalty for any SLA violation.

As important as the SLA itself are:

- *Is It Technically Feasible and Likely for the Service to Meet the SLA Over the Long Term?* Solution architects should design around the feasible and likely estimated long-term performance levels rather than potentially misleading SLA claims.
- *What Performance Level Has Been Demonstrated in the Past?* As with financial products, past performance is no guarantee of future results, but it is a baseline. More importantly, if the supplier is unable or unwilling to provide extensive data on historic performance, then you should learn why. Is performance data unavailable because the service is new (which raises certain risks), or because the service provider does not actually measure performance (which raises other risks), or because the service provider does not share performance data with prospective customers (which raises still other risks).

More important that the availability SLA offered by a cloud service provider is the best estimate of the likely long-term average service availability. While past performance is no guarantee of future results, historic performance is far more credible than a weak SLA metric with nominal remedies.

13.8.2 Cloud Service Provider Selection

The IaaS, PaaS, or SaaS service provider that a cloud consumer selects has a profound impact on the service quality, reliability, and availability that will be experienced by end users because the cloud service providers have direct control of virtually all the ingredients that comprise a cloud-based service. In addition to controlling the hardware, power, operational environment, IP networking data center maintenance staff, and policies governing operation of the data center, the service provider brings at least some platform software into the solution and at least some application protocol support (e.g., DNS, HTTP/HTTPS, and SNMP). Thus, cloud consumers should carefully determine the service quality, reliability, and availability targets of considered service providers, and verify the feasibility and likelihood of those targets being met over the long term. ODCA SLA levels (i.e., bronze, silver, and gold, platinum) may provide a useful framework to use when discussing service quality, reliability, and availability expectations with XaaS service providers.

13.8.3 Written Reliability Plan

A best practice is to create a written reliability plan in the planning phase of a solution development to lay out the program of reliability diligence in advance. Depending on the organization's development methodology, the reliability plan may either reference other documents (e.g., requirements and test plans) and artifacts (e.g., modeling spreadsheets and reliability reports), or the plan may actually include or embed those items.

The reliability plan may be a document, or a presentation, or a spreadsheet, or a wiki, or some other scheme for organizing plans and artifacts. The exact representation is not particularly important; the thoroughness of the plan, care taken in executing the steps, and promptly notifying members of the project team when risks exceed acceptable levels is most important.

The reliability plan, possibly in conjunction with the quality plan and the overall project plan, should cover the following topics:

- *Solution Scope.* What is the end-to-end scope of the solution, and exactly what components and facilities are in-scope.
- *Solution KQI Targets for Service Reliability and Availability.* What are the key quality indicators for this solution and what are the targets for those KQIs? Note that these targets can be framed relative to previous releases, other deployments or competitive offers, or technologies. For example:
  - *Cloud-based solution will offer same service accessibility, retainability, reliability, and availability as traditional deployment.*
  - *Solution will offer equivalent service availability and service reliability as market leading offering* <X>.
- *Verifiable Solution Reliability and Availability Requirements.* Specific requirements that will drive robustness and stability testing of solution and key components of the solution.
- *Plans for Reliability Analyses (e.g., FMEA).* Give plan for what reliability analyses and reviews will be done by who and when, and what document or artifact will contain the final analysis results.
- *Solution Modeling and Budgeting of Primary Quantitative KQI's to Assure Feasibility of Meeting Targets/Requirements.* Constructing mathematical modeling is a foundation for analyzing the feasibility and likelihood of achieving quantitative KQIs. Given some mathematical model, one can create allocations or budgets of key impairments or results across solution components and facilities, which solve the model and meet the requirements, and then manage individual components and facilities to those targets to assure the feasibility of meeting those requirements over the long term.
- Data on field reliability, availability, and quality of both previous solution release(s) and all included components.
- *Plans or Results of DfR Assessments of Components Included in the Solution.*
- *Enumeration of Features and Testing Expected to Improve Service Reliability, Availability, and Latency, along with Brief Rationale for Any Expectations of Improvement.* Optionally, the reliability or availability improvement for the feature or tested can potentially be estimated via changes to input parameters or structure to mathematical modeling.
- *Plans to Report on Feasibility and Likelihood of Meeting Reliability Requirements with Plan of Record and Committed Resources to Project Leaders and Decision Makers.* As there is inherent uncertainty in assessing risk, and more time often offers more information to assess risk, there is frequently a temptation

to postpone raising a risk to the project team or decision makers in the hope that more time and more data will reveal that the risk is acceptable. To minimize the tendency to postpone reporting of bad news, it is often best to explicitly schedule regular updates on the reliability risk, such as at every project decision review or every month. Planning this in advance enables the reliability prime to plan to obtain appropriate updated information prior to each report on reliability risk, thereby assuring that decision makers and project team see fresh and realistic assessment of the reliability risk.

- *Plan for Measuring Service Reliability, Availability, Latency, and Quality KQIs of the Deployed Solution.*

Best practice is to name an individual as the reliability prime for the solution and make that individual responsible for assessing and reporting the reliability risk to both decision makers and the project team. As this individual knows they are expected to present and defend the reliability risk assessment—potentially charged and project-impacting—information regularly and held accountable for field performance after release, then they will be highly motivated to assure that the reliability plan is both complete and methodically executed. Thus, the reliability prime is the obvious primary author for the reliability plan.

14

SUMMARY

Cloud computing is a business model that enables computing to be offered as a utility service, thereby shifting computing from a capital intensive activity to an expense item. Just as electric utilities and railroad companies freed consumers of power and land transportation from the capital expense of building private infrastructure, cloud computing enables consumers to focus on solving their specific business problems rather than on building and maintaining computing infrastructure. The U.S. National Institute of Standards and Technology (NIST) offers five essential characteristics of cloud computing:

1. on-demand self-service;
2. broad network access;
3. resource pooling;
4. rapid elasticity; and
5. measured service.

A handful of common characteristics are shared by many computing clouds, including virtualization and geographic distribution. Beyond shifting computing from a capital expense topic to a pay-as-you-go operating expense item, rapid elasticity and other

Reliability and Availability of Cloud Computing, First Edition. Eric Bauer and Randee Adams.

characteristics of cloud computing enable greater flexibility and faster service deployment than traditional computing models.

Virtualization is one of the common characteristics of cloud computing. Virtualization decouples application and operating system software from the underlying software by inserting a hypervisor or virtual machine manager above the physical hardware, which presents a "virtual" machine to the guest operating system and application software running on that guest operating system. Virtualization technology can boost resource utilization of modern server hardware by permitting several application instances executing in virtual machines to be consolidated onto a smaller number of physical machines, thereby dramatically reducing the number of physical systems required. Applications generally leverage virtualization in one or more of the following usage scenarios:

- *Hardware Independence.* Virtualization is used to enable applications to be deployed on different (e.g., newer) hardware platforms.
- *Server Consolidation.* Virtualization is used to enable several different applications to share the same physical hardware platform, thereby boosting utilization of the underlying physical hardware.
- *Multitenancy.* Virtualization is used to facilitate offering independent instances of the same application or service to different customers from shared hardware resources, such as offering distinct instances of an e-mail application to different companies.
- *Virtual Appliance.* Ultimately, software applications can be offered as downloadable "appliances" that are simply loaded onto virtualized platform infrastructure. While a commercially important application may not yet be as simple to buy and install as a household appliance, like a toaster, virtualization can streamline and simplify the process for customers.
- *Cloud Deployment.* Virtualization is used to enable applications to be hosted on cloud providers' servers, and take advantage of cloud capabilities, such as rapid elasticity growth and degrowth of service capacity.

14.1 SERVICE RELIABILITY AND SERVICE AVAILABILITY

Failures are inevitable. The service impact of failure is measured on two dimensions:

- *Extent.* How many users or operations are impacted.
- *Duration.* How many seconds or minutes of service impact accrues.

While extent of failure linearly affects service impact (e.g., impacting 100 user sessions is nominally twice as bad as impacting only 50 sessions), the duration of impact is not linear because of the way modern networked applications are implemented. Failure impacts that are very brief (e.g., less than 10 or perhaps a few hundred milliseconds) are often effectively concealed from end users via mechanisms like automatic protocol message retries for transactions and lost packet compensation algorithms for streaming media; these brief events are often referred to as *transient*. Failures that are somewhat longer (e.g., less than a few seconds) will often cause some transactions or sessions to present signs of failure to users, such as web pages failing to load successfully, call attempts failing to complete and returning ringback promptly, or noticeable impairments to streaming media sessions. Service users will often accept occasional service failures and retry failed operations, such as canceling a stuck webpage and explicitly reloading the page, or redialing after a failed call attempt. If the first (or perhaps second) retry succeeds, then the failed operation will typically count against service reliability metrics as a defective operation; since service was only impacted briefly, the service will not have been considered "down" so the failure duration will not count as outage downtime. However, if the failure duration stretches to many seconds, then reasonable users will abandon service retries and deem the service to be down, so availability metrics will be impacted. This is illustrated in Figure 14.1.

Since failures are inevitable, the goal of high availability systems is to automatically detect and recover from failures in less than the maximum acceptable service disruption time so that outage downtime does not accrue for (most) failure events, and ideally service reliability metrics are not impacted either, as shown in Figure 14.1. The maximum acceptable service disruption target will vary from service to service based on user expectation, technical factors (e.g., protocol recovery mechanisms), market factors (e.g., how reliable alternative technologies are), and other considerations. Thus, the core challenge of service availability of cloud computing is to assure that inevitable failures are automatically detected and recovered fast enough that users don't experience unacceptable service disruptions.

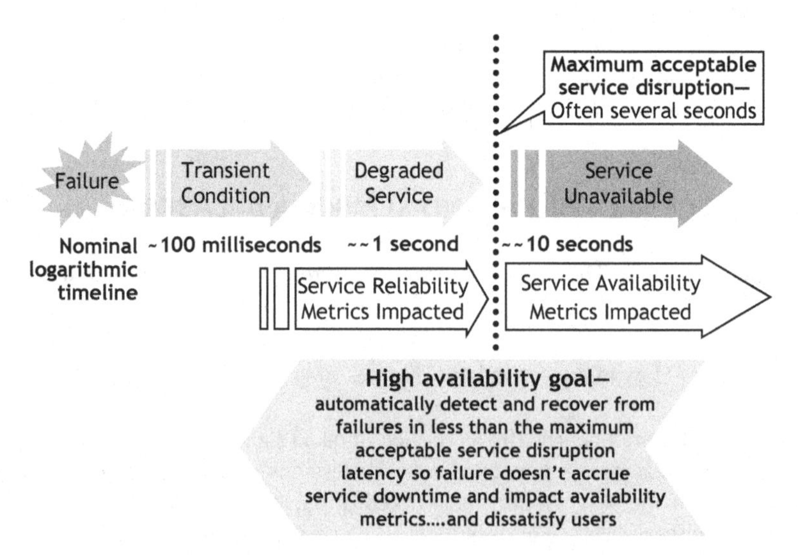

Figure 14.1. Failure Impact Duration and High Availability Goals.

14.2 FAILURE ACCOUNTABILITY AND CLOUD COMPUTING

Information systems require a handful of fundamental ingredients to function. Computing *hardware* executing application *software* interworks with client devices via application protocol *payloads* across IP *networks*. The computing hardware is installed in a suitable data center *environment* and must be provided with suitable electrical *power*. Application software and the underlying hardware inevitably require application, user and configuration *data* to provide useful service. *Human* staff is required to provision and maintain the data, software, hardware, and supporting infrastructure; enterprise *policies* define the interactions between ingredients and guide the actions of human staff. In addition to ordinary single failures of, say, hardware components, physical systems are vulnerable to force majeure or disaster events, like earthquakes, fires, and floods, which can simultaneously impact multiple ingredients or components. This is illustrated in Figure 14.2, repeated from Figure 3.4. All of these ingredients are subject to risks, which can compromise ingredient availability, thereby impacting end user service.

Traditionally, enterprises broadly factored accountability for failures and outages into three broad buckets: product attributable, customer (or enterprise or user) attributable, and externally attributable. Figure 14.3 (repeated from Figure 4.1) visualizes the traditional factorization of accountability by ingredients. Accountability for each ingredient maps across the three traditional categories as follows:

- *Product suppliers* are primarily accountable for the hardware and software they supply, and the ability of that hardware and software to interwork with other systems via defined application protocol payloads.
- *Customers* are accountable primarily for the physical security, temperature, humidity, and other environmental characteristics of the facility where the hardware is installed, as well as for providing necessary electrical power and IP network connectivity. Customers are also responsible for their application, user, and configuration data, as well as for the operation policies and human staff.

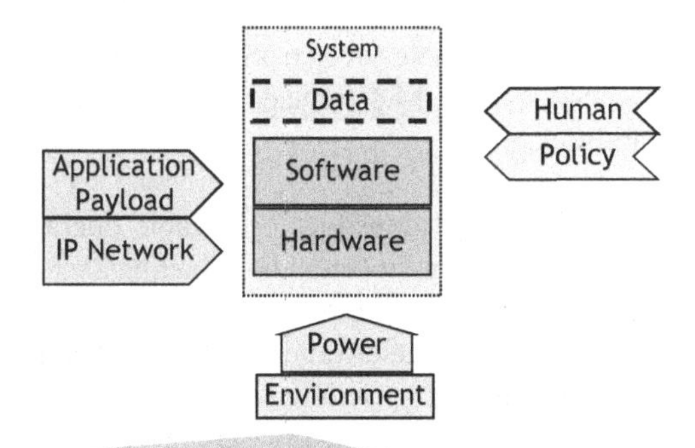

Figure 14.2. Eight-Ingredient Plus Data Plus Disaster (8i + 2d) Model.

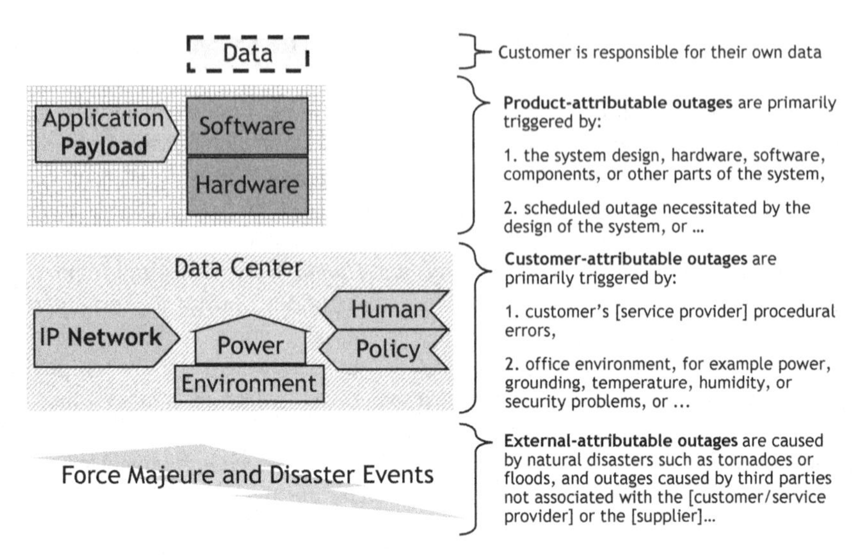

Figure 14.3. Traditional Outage Attributability.

- *External*: Data centers are inherently vulnerable to force majeure and disaster events like earthquakes, tornadoes, fires, and so on. As these risks are not appropriately attributed to either the customer or the supplier, they are placed in this "external" attributability category.

Cloud computing fundamentally changes the accountability model because there is no longer a monolithic customer who buys and operates equipment. Instead, there is a cloud service provider who owns and operates cloud computing facilities (analogous to a landlord), and a cloud consumer who leases those computing facilities (analogous to a rental tenant). Thus, the accountabilities that were solely the responsibility of the "customer" in the traditional model must now be split between the cloud consumer and the cloud service provider. The exact division is determined by the particular cloud service model (i.e., infrastructure as a service [IaaS], platform as a service [PaaS], and software as a service [SaaS]); Figure 14.4 (repeated from Figure 10.7) gives a typical breakdown of accountabilities.

- *Cloud service providers* are responsible for reliable operation of the compute, storage, and networking equipment (including load balancers and security appliances) in their data centers which host their cloud service offering. Along with responsibility for the hardware itself and the base software, the cloud service provider has responsibility for providing electrical power, highly reliable IP networking, and maintaining a physically secure data center with acceptable temperature, humidity, and other environmental conditions. The cloud service provider is also responsible for the human maintenance staff, contractors, and suppliers who support that data center, as well as the operational policies

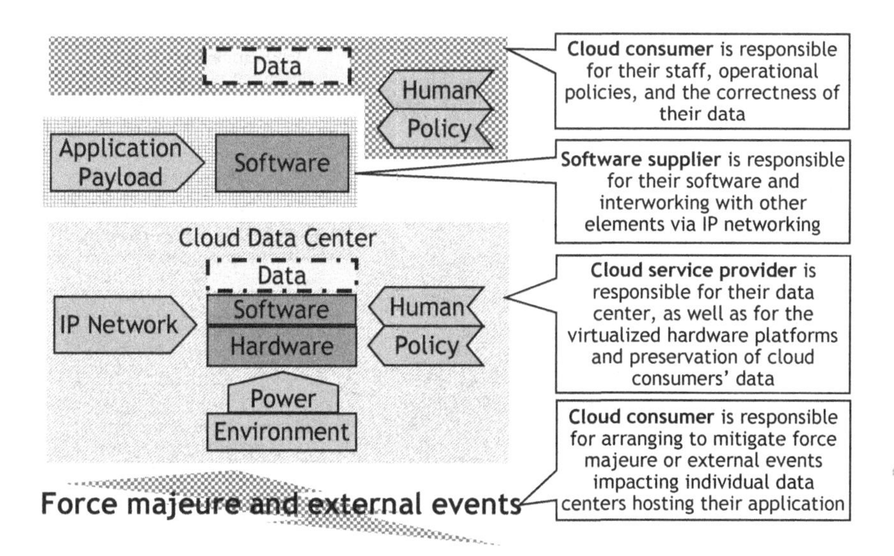

Figure 14.4. Sample Outage Accountability Model for Cloud Computing.

followed by those people when designing, operating, and maintaining the data center.

- *Software suppliers* are responsible for delivering and supporting the software running on the virtualized platform. Note that there are often many software suppliers contributing different platform and application software components, and some of the software may even be supplied by the cloud service provider, the cloud consumer, or both.
- *The cloud consumer* is the enterprise (or individual) who pays for the cloud services. The cloud consumer is responsible for their own enterprise or application data (e.g., user records and inventory data) and service configuration, such as firewall settings and load balancer policies. The cloud consumer is also responsible for the staff that provisions their data (e.g., adding users and manipulating enterprise data) and operates the application. In addition, the cloud consumer is responsible for assuring that force majeure risks are adequately mitigated via service continuity and disaster recovery planning, as well as georedundancy. While cloud service providers offer the services necessary to construct robust georedundancy configurations and disaster recovery plans, the cloud consumer has primary responsibility for business continuity planning of their service and data.

14.3 FACTORING SERVICE DOWNTIME

The outage accountability model of cloud computing can also be factored based on process areas as shown in Figure 14.5 (repeated from Figure 4.5). The risks to software,

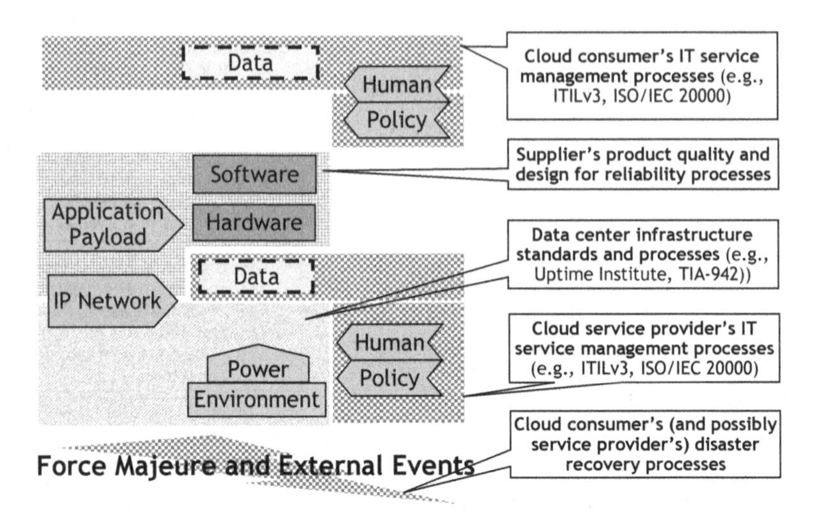

Figure 14.5. Outage Responsibilities of Cloud by Process.

hardware, and payload ingredients are primarily managed and mitigated via the design for reliability and quality processes of the equipment and application suppliers. The risks to power, environment, and IP networking infrastructure are primarily considered in the context of data center best practices and standards like [Uptime] and [TIA942]. The data, human, and policy ingredients are addressed via IT service management best practices and standards like ITIL, ISO/IEC 20000, and COBIT. And the risk of force majeure events is mitigated via business continuity and disaster recovery plans. While IT service management generally considers service continuity, it is useful to explicitly consider disaster recovery because it presents different risks and challenges to high availability mechanisms. Note that it is essential that cloud consumers and cloud service providers have aligned and interlocked their roles and responsibilities for IT service management and disaster recovery to assure that service is rapidly restored after any failure or disruption event.

A side benefit of the risk by process area analysis of Figure 14.5 is that is offers a simple and actionable factorization of cloud downtime into four categories:

- *Product-attributable downtime* for events primarily attributable to application software and virtualized compute and storage hardware. Traditional service availability claims (e.g., "five 9's") referred exclusively to product attributable service downtime.
- *Data center-attributable downtime* for events primarily attributable to power, environment, and IP networking facilities and infrastructure. Note that one should explicitly decide if IP equipment, like routers, end-of-row and top-of-rack switches, security appliances, and load balancers, are lumped into data center-attributable or product-attributable downtime categories. Each of these components can be reasonably covered in either product-attributable or data

center-attributable category; the key is to make sure that these crucial elements are not overlooked.

- *IT service management-attributable downtime* for downtime primarily attributable to elastic growth and degrowth, software release management, and for downtime prolonged by ineffective problem/event/incident management procedures.
- *Disaster attributable downtime*: Force majeure events that destroy a data center or render it inaccessible or unavailable are fortunately so rare that it is inappropriate to estimate an average annualized service downtime attributable to disasters. Instead, rare disaster events are expected to be recovered within the target RTO and lose no more data than the target RPO.

The best practice to manage application service availability is to create a feasible downtime budget and manage architecture, design, testing, deployment, and operations to maximize the likelihood that the budget is not exceeded. Table 14.1 (repeated from Table 10.4) shows how a sample "five 9's" budget for a traditional application morphs for virtualized deployment and for cloud deployment. The key insights are of this sample budget are:

- Software downtime is likely to remain essentially the same across all three deployment options.
- Hardware related downtime changes in virtualized and cloud deployments; it does not vanish.
- Product attributable planned and procedural or IT service management downtime remains a significant contributor to service downtime. Consumer-attributable IT service management downtime (e.g., due to human provisioning mistakes) is in addition to this downtime budget allocation.

14.4 SERVICE AVAILABILITY MEASUREMENT POINTS

To analyze and understand the reliability and availability of cloud computing, the authors recommend considering four measurement points (MPs) illustrated in Figure 14.6 (repeated from Figure 10.10):

- *Component Instance Level (MP 1).* Solutions are built from instances of various components like firewalls, load balancers, application servers, and databases that together deliver valuable services to end users. A fundamental MP is of the service delivered by each component instance. If an individual component instance is protected by a redundant component instance, then it is best to consider the overall availability of component service across the pool of instances.
- *Primary Data Center Level (MP 2).* An ensemble of component instances will generally be arranged in one or more data centers and integrated to offer services

TABLE 14.1. Evolution of Sample Downtime Budgets

Sample "Five 9's" Product or Application Attributable Downtime Budgets			Annual Down Minutes	%
Traditional Deployment	Virtualized Deployment	Cloud Deployment		
Hardware related—target: 30 seconds = 0 minute 30 seconds				
Hardware failure downtime—service downtime triggered by hardware failures.	*Virtualized hardware platform downtime*—service downtime attributed to virtualized hardware resources (e.g., virtual CPU, memory, disk, and networking).	*Application downtime recovering from ordinary XaaS failures*—service downtime for application to detect and recover from ordinary XaaS platform failures.	0.50	10
Software attributable—target: 225 seconds = 3 minute 45 seconds				
Application software failures—service downtime due to software failures of platform and/or application software.			3.75	71
Procedural and maintenance attributable—target: 60 seconds = 1 minute 0 second				
Successful scheduled activities—service downtime "by design" for successful upgrade, update, retrofit, hardware growth and other scheduled or planned maintenance activities. *Unsuccessful procedural activities*—service downtime attributed to unsuccessful or botched maintenance activities, such as upgrade, update, retrofit, hardware growth, and provisioning.	*Application software-related planned and procedural downtime*—product-attributable service downtime attributed to successful and unsuccessful planned and procedural activities associated with application software. *Virtualized platform-related planned and procedural downtime*—product-attributable service downtime attributed to successful and unsuccessful planned and procedural activities associated with the virtualized hardware platform.	*Product-attributable cloud maintenance activities*—chargeable service downtime for: • elastic capacity growth and degrowth; • software upgrade, update, retrofit, and patching; • live migration; and • other IT service management activities.	1.00	19
		Total	5.25	
		Availability	99.999%	

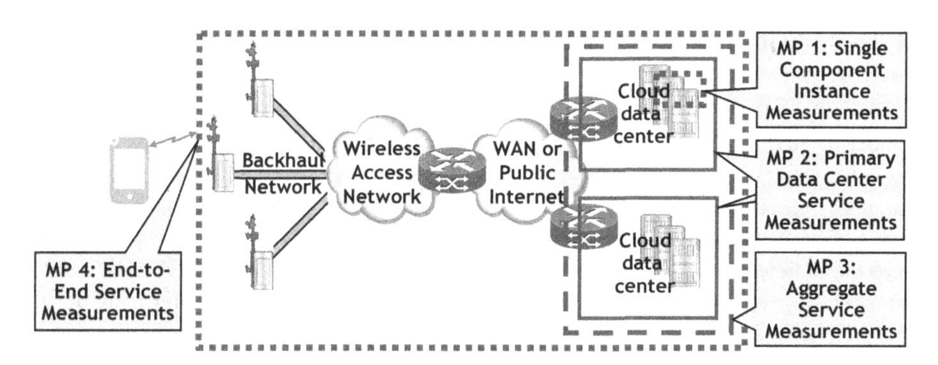

Figure 14.6. Measurement Points (MPs) 1, 2, 3, and 4.

to end users. MP 2 captures the overall performance of the solution services offered by the ensemble of component instances in a particular (nominally primary) data center, as well as the data center itself and the impact of IT service management.

- *Aggregate Service Level (MP 3).* Often solutions are georedundantly deployed across several data centers so that one data center can recover service for users of another data center when a disaster or catastrophic failure renders service from the users' primary data center unavailable. MP 3 captures the overall aggregate service performance across a pool of data centers, excluding user service impairments from WAN and access network equipment and facilities.
- *End-to-End Service Level (MP 4).* End users of solution services are rarely located in the data center hosting the ensemble of component instances implementing the solution. Instead, end users access solution services via some client device like a smartphone, laptop, tablet, set top box, etc., across a wireless or wireline access network and wide area network to communicate with the data center hosting the component instances implementing the service. These access, backhaul and wide area networking equipment and facilities are subject to failures and impairments that can compromise the users' quality of experience. MP 4 integrates the impact of these access, backhaul and WAN considerations along with the aggregate service of MP 3.

Application suppliers, cloud consumers, and cloud service providers share accountabilities for component instance availability (MP 1), primary data center availability (MP 2), and aggregate service level availability (MP 3). End-to-end service availability (MP 4) introduces accountability for numerous communications service providers and others who are often outside of the control of the cloud consumer and the cloud service provider. Thus, one must carefully consider the service MPs when setting service level expectations and accountabilities.

14.5 CLOUD CAPACITY AND ELASTICITY CONSIDERATIONS

Historically, enterprises have had to predict the expected traffic levels for an application months in advance so the organization could acquire and install sufficient hardware and software resources, arrange for sufficient networking bandwidth and intermediate systems (e.g., security appliances and load balancers), as well as install and configure application software and data to assure sufficient service capacity was available to serve the anticipated load. Deploying all of these resources often entailed great capital and operating expenses, and carried a huge financial risk for the enterprise. If the enterprise was too pessimistic in their predictions, then users would saturate the system and traffic would be turned away, thereby impacting customer goodwill and perhaps revenue; and if the enterprise was too optimistic, then all of the excess capital expense and operating expense for the unneeded capacity would be carried forward by the enterprise, possibly for years, because it was hard to release and reuse resources that were no longer needed.

The essential cloud computing characteristic of rapid elasticity can eliminate virtually all capacity planning risks and capacity planning work itself because cloud computing enables cloud consumers to request or release resources on-the-fly, and then pay for the resources actually used. As a result, cloud expenses can track with the actual workload, rather than being a function of installed capacity (which was driven by capacity plans completed months earlier). Note that rapid elasticity makes resources available promptly (e.g., in hours), but not instantly (e.g., in seconds), so rapid elasticity is not an alternative to redundancy for high availability, and it does not eliminate the need for overload control mechanisms. Careful application monitoring and management should minimize the frequency of overload events (and associated service reliability impact) by elastically growing online service capacity ahead of offered load. Naturally, the complexity of rapid elasticity introduces a variety of reliability risks, and these were considered in Chapter 7. Growing—or degrowing—the online service capacity of an application or solution requires careful coordination of added (or deleted) compute resources, storage resources, network bandwidth, load balancer configurations, and application software and configuration. Automating these tasks so growth or degrowth operations complete rapidly and reliably is the purpose of service orchestration. Service orchestration and the reliability risks of orchestration were considered in Chapter 8.

14.6 MAXIMIZING SERVICE AVAILABILITY

While cloud computing slightly increases the risk of critical failures due to the added complexity of virtualization, rapid elasticity and increased resource sharing, as well as the associated IT service management risks, cloud computing does present opportunities to mitigate both the preexisting and new reliability risks to potentially offer higher service availability than traditional deployment scenarios. Consider the opportunities to reduce service downtime in each of the four general categories of service downtime as described in the next sections.

14.6.1 Reducing Product Attributable Downtime

Virtualization should have a minimal impact on failure rates: reliability of application software executed in virtual machines should be comparable with reliability of execution in native environments, and the failure rate of the underlying physical hardware should be nominally the same when hosting a virtualization manager and application as when executing an application natively (i.e., without virtualization). To assure comparable service availability of applications in virtualized configurations, it is important that failure detection and recovery times of virtualized deployments be comparable with native deployment. While virtualized implementations of traditional redundancy strategies (e.g., active/standby, active/active) should offer comparable performance to native (nonvirtualized) deployments, virtualization offers some new redundancy options, such as activating paused or snapshot images, as well as enabling different redundancy models. For example, traditionally one might assume a 4-hour mean time to repair (MTTR) for a hardware failure of the server natively hosting a critical application, but virtualization can support (offline) migration of an impacted application from failed hardware to an operational hardware platform that has sufficient spare capacity in far less time, thus improving application service availability by shortening the effective (i.e., "virtualized") hardware MTTR. A detailed analysis of the software reliability risks and high availability options of virtualization are considered in Chapter 5, the hardware reliability risks of virtualization and their impact on service availability are discussed in Chapter 6, and virtualization's impact on service transition activities is addressed in Section 4.5.

14.6.2 Reducing Data Center Attributable Downtime

Cloud computing makes it easy for cloud consumers to shop around to pick an IaaS or PaaS service provider who offers the desired data center performance at the lowest total cost for the cloud consumer.

14.6.3 Reducing IT Service Management Downtime

Reduction of downtime associated with service management activities, such as software upgrade or patch, are discussed in Section 11.3, "IT Service Management Considerations." In summary, service transition tools and procedures should have the following requirements:

- Automation (e.g., service orchestration) to replace manual procedures and make use of mechanisms such as live migration when appropriate.
- Ability to fulfill the reliability requirements associated with the service transition activity (e.g., x seconds of service downtime).
- Use of the Open Virtualization Format (OVF) or similar to provide configuration information so that it can be clearly defined and validated by the tools.
- Ability to create and configure an updated instance of the application while running the old version, and to seamlessly stop the old version and activate the

new version once is ready. This may be performed on the same server or on a different server dependent upon the type of maintenance activity and the availability of resources.

- Clear, accurate documentation and training is provided for those managing the service transition activities.
- Thorough testing of the maintenance procedures must be performed to ensure the procedures meet the reliability requirements for service transition activities.

14.6.4 Reducing Disaster Recovery Downtime

Geographic distribution is a common characteristic of cloud computing, but geographic distribution does not automatically mean geographic redundancy and support for disaster recovery. Leveraging geographic distribution to create geographically distributed redundancy (a.k.a., georedundancy) for disaster recovery requires careful planning, configuration, and testing to assure that user service can be recovered fast enough (i.e., the recovery time objective or RTO) with acceptably fresh application data following a disaster (i.e., the recovery point objective or RPO) and to meet the needs of the business. Disaster recovery time objectives are typically measured in hours or days, and thus disaster recovery mechanisms alone generally offer limited mitigation for catastrophic failure events. Appropriately engineered applications and solutions can leverage spare online capacity in other data centers to mitigate catastrophic failure events by rapidly detecting failures and redirecting workloads to the spare online capacity much more quickly than the traditional (manual) disaster recovery plans. Geographic distribution, georedundancy and disaster recovery are considered in Chapter 9.

14.6.5 Optimal Cloud Service Availability

It is tempting to assume that given the vast pool of cloud computing resources available, service downtime should vanish because somewhere there is an instance of the application that is available and capable of providing service for each user. While there may be one or more instances of the target application available for service somewhere in the cloud, it is not practical to achieve 100% service availability for reasons including:

- *Noninstantaneous Failure Detection.* After clicking a button or icon, users generally wait for an operation to complete. For certain types of failures, one must simply wait for the request to time out to determine that the application is not available for service.
- *Noninstantaneous Service Recovery.* Recovering authenticated services often requires reauthenticating the user, with the redundant server or application instance entailing security credentials to be exchanged and validated. Recovering session-oriented and stateful services requires rebuilding or recreating context to minimize user visible service impact. Both of these activities take time, during which service is unavailable to the impacted user.

- *Noninstantaneous Access to the User's Data.* User service often requires running software, application data, and user data. For example, while a running e-mail server instance is necessary for e-mail service, users also expect access to their personal inbox data to be accessible via that e-mail server. Thus, service recovery involves restoring access to the user's individual data, and that data may not be "instantly" available to the alternate server instance.

Cloud computing makes it feasible to consider redundant compute arrangements, such as where a client application maintains authenticated sessions with two or more application instances, sends each individual request to each of those applications instances simultaneously, and uses the first correct response, thereby mitigating at least some application downtime. Unfortunately, determining the first "correct" response may not be trivial, and assuring consistency and correctness of data across a pool of servers operating in parallel can be challenging. While it is fine to have any DNS server instance return an IP address for a particular domain name independent of all other DNS server instances, one does not want multiple instances of your bank's online application to permit independent application instances to make simultaneous and overlapping withdrawals from your account to unknowingly overdraft your bank account. It is certainly feasible to leverage new redundancy options offered by cloud computing to boost service availability, but maximizing these potential service availability benefits will probably require enhancements to service architectures, application protocols, and application and client software.

14.7 RELIABILITY DILIGENCE

Highly reliable and highly available services can be implemented and deployed through appropriate reliability diligence. The authors presented a cloud solution design for reliability process in Chapter 13, which is visualized in Figure 14.7 (which is the same as Figure 13.1). Many readers will recognize close similarities with the service strategy, service design, service transition, and continual service improvement activities of IT service management processes, like ITIL (service operation is purely an IT service operations activity and thus is not covered by design for reliability diligence). Regardless of whether reliability diligence is worked in the context of an R&D activity, in an IT service management activity or some other workflow process, the key activities of Figure 14.7 and Chapter 13 should be addressed:

- Capture customers' service reliability and availability expectations in requirements.
- Perform analysis and modeling to assure it is feasible and likely that the requirements can be met with the target architecture, feature set, and proposed project plan.
- Assure that appropriate design for reliability diligence is completed on solution elements to assure that those elements will meet the solution's requirements.

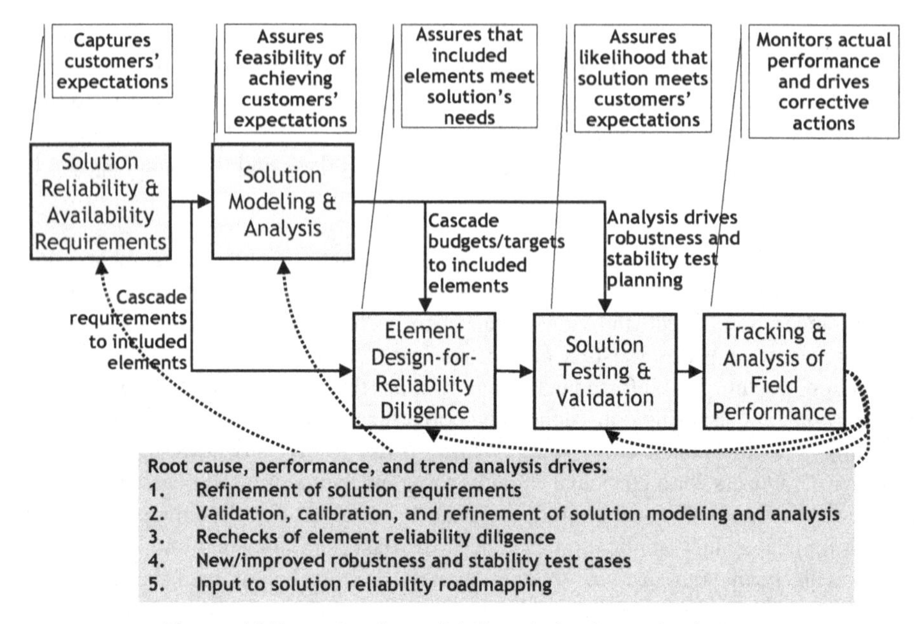

Figure 14.7. Design for Reliability of Cloud-Based Solutions.

Chapter 12 reviewed the recommended design for reliability diligence for virtualized applications.

- Test the solution to assure that service reliability, latency, quality, and stability are acceptable, and that robustness and recovery mechanisms function properly and meet all corresponding reliability requirements.
- Track and analyze field performance to drive continual service improvement, and to validate and calibrate predictive models to be used in future solution releases.

14.8 CONCLUDING REMARKS

Cloud computing is a compelling business model for delivering information services; many new applications will be explicitly developed for cloud deployment, and many preexisting applications will evolve to cloud deployment. The dynamic and flexible characteristics of cloud computing provide the basis for highly reliable, always available services. The careful analysis of the reliability and availability risks and architectural opportunities presented in this book offers guidance on how to develop cloud-based solutions that meet or exceed service reliability and availability requirements of traditional deployments.

ABBREVIATIONS

3G	Third-generation wireless network (e.g., UMTS)
4G	Fourth-generation wireless network (i.e., LTE)
ACID	Atomicity, consistency, isolation, and durability
API	Application programming interface
APM	Application performance management
ARP	Address resolution protocol
BASE	Basically available, soft state, eventual consistency
BRAS	Broadband remote access server
CapEx	Capital expense
COTS	Commercial off the shelf
CPU	Central processing unit
CSA	Cloud Security Alliance
CSP	Cloud service provider
DAS	Direct attached storage
DDoS	Distributed denial of service (attack)
DfR	Design for reliability
DHCP	Dynamic host configuration protocol
DMTF	Distributed Management Task Force
DNS	Domain name system
DoS	Denial of service (attack)
DR	Disaster recovery
DSL	Digital subscriber loop, a copper access technology
DSLAM	Digital subscriber loop access module
EOR	End-of-row Ethernet switch
FAA	U.S. Federal Aviation Administration
FIT	Failures in time (10^9 hours)
FMEA	Failure mode effects analysis

Reliability and Availability of Cloud Computing, First Edition. Eric Bauer and Randee Adams.

FRU	Field replaceable unit (hardware)
GPON	Gigabit passive optical networking, an optical access technology
GR	Geographic redundancy
HFI	Hardware fault insertion
IaaS	Infrastructure as a service
IC	Integrated circuit
ICT	Information and communication technology
IETF	Internet Engineering Task Force
IP	Internet protocol
IS	Information systems
iSCSI	Internet Small Computer System Interface
ISP	Internet service provider
IT	Information technology
ITIL	Information Technology Infrastructure Library
ITSCM	IT service continuity management
ITSM	Information technology service management
ITU	International Telecommunications Union
KPI	Key performance indicator
KQI	Key quality indicator
LAN	Local area network
LTE	Long-term evolution, a fourth-generation wireless networking standard
MOP	Methods of procedure
MOS	Mean opinion score
MP	Measurement point
MTBCF	Mean time between critical failures
MTBF	Mean time between failures
MTTR	Mean time to repair
MTTRS	Mean time to restore service
NAS	Network attached storage
NIC	Network interface card
NIST	U.S. National Institute of Standards and Technology
OAM	Operations, administration, and maintenance
OAMP	Operations, administration, maintenance, and provisioning
ODCA	Open Data Center Alliance
OpEx	Operating expense
OS	Operating system
OVF	Open Virtualization Format

PaaS	Platform as a service
PC	Personal computer
RAID	Redundant array of inexpensive (or independent) disks
RBD	Reliability block diagram
RPO	Recovery point objective
RTO	Recovery time objective
RTP	Real time protocol
SaaS	Software as a service
SAN	Storage area network
SIP	Session initiation protocol
SLA	Service-level agreement
SLR	Service-level requirement
SPOF	Single point of failure
TIA	Telecommunications Industry Association
TOR	Top-of-rack Ethernet switch
VBF	Vital business function
VLAN	Virtual local area network (LAN)
VM	Virtual machine instance
VMM	Virtual machine manager
VNIC	Virtual network interface card
VPN	Virtual private network
WAN	Wide area network
XaaS	Refers to one or more of: software as a service, platform as a service, and infrastructure as a service.

REFERENCES

[Arbor] Worldwide Infrastructure Security Report, Volume VI, 2010, Arbor Networks, http://www.arbornetworks.com.

[AWS08] Amazon Web Services Launches "Elastic IPs"—Static IPs for Dynamic Cloud Computing, March 27, 2008, http://www.businesswire.com/portal/site/google/?ndmViewId=news_view&newsId=20080327005155&newsLang=en.

[AWSFT] Amazon Web Services Building Fault-Tolerant Applications on AWS, May 2010, http://d36cz9buwru1tt.cloudfront.net/AWS_Building_Fault_Tolerant_Applications.pdf.

[Bauer10] *Design for Reliability: Information and Computer-Based Systems*, Eric Bauer, 978-0470604656, Wiley-IEEE Press, 2010.

[Bauer11] Eric Bauer, Randee Adams, and Dan Eustace, *Beyond Redundancy: How Geographic Redundancy can Improve Service Availability and Reliability For Computer-Based Systems*, Wiley-IEEE Press, 2011.

[Bigtable] Fay Chang et al., Bigtable: A Distributed Storage System for Structured Data, http://static.googleusercontent.com/external_content/untrusted_dlcp/labs.google.com/en/us/papers/bigtable-osdi06.pdf.

[BT.500] Methodology for the Subjective Assessment of the Quality of Television Pictures, International Telecommunications Union Recommendation ITU-R BT.500-12, 09/2009.

[CASS] Avinash Lakshman and Prashant Malik, *Cassandra—A Decentralized Structured Storage System*, Cornell University, http://www.cs.cornell.edu/projects/ladis2009/papers/lakshman-ladis2009.pdf, November 13, 2009.

[CSAa] Top Threats List, https://cloudsecurityalliance.org/topthreats/csathreats.v1.0.pdf.

[CSAb] Guidelines, https://cloudsecurityalliance.org/guidance/csaguide.v2.1.pdf.

[CSAc] Cloud Controls, https://cloudsecurityalliance.org/guidance/CSA%20Cloud%2 Controls%20Matrix%20(CCM)_R1.1_FINAL.xlsx.

[CSAd] Cloud Provider Assessment Questions, https://cloudsecurityalliance.org/guidance/CSA-CAI-Question-Set.1.0.xlsx.

[DSP0102] Architecture for Managing Clouds, http://dmtf.org/sites/default/files/standards/documents/DSP-IS0102_1.0.0.pdf.

[DSP1050] Ethernet Port Resource Virtualization Profile, Distributed Management Task Force, DSP1050, Version 1.0.0, October 21, 2010.

Reliability and Availability of Cloud Computing, First Edition. Eric Bauer and Randee Adams.

[DSP1057] DMTF DSP1057 Virtual System Profile, http://www.dmtf.org/sites/default/files/standards/documents/DSP1057_1.0.0_0.pdf, Version 1.0.0, April 22, 2010.

[DSP2017] Open Virtualization Format White Paper, Distributed Management Task Force, DSP2017, Version 1.0.0, February 9, 2009, http://dmtf.org/standards/ovf.

[Edberg] Jeremy Edberg, Customer Best Practices for Surviving a Cloud Outage, Virtual Cloud Connect conference, September 29, 2011.

[ENISAa] Cloud Risk Assessment, http://www.enisa.europa.eu/act/rm/files/deliverables/cloud-computing-risk-assessment/at_download/fullReport.

[ENISAb] Cloud Assurance Framework, http://www.enisa.europa.eu/act/rm/files/deliverables/cloud-computing-information-assurance-framework/at_download/fullReport.

[FAA-HDBK-006A] Federal Aviation Administration Handbook: Reliability, Maintainability, and Availability (RMA) Handbook, FAA-HDBK-006A, January 7, 2008.

[GR2841] Generic Requirements for Operations Systems Platform Reliability, Telcordia Technologies System Documentation, GR-2841-CORE, Issue 1, June 1994.

[Hamilton] James Hamilton, On Designing and Deploying Internet-Scale Services, http://www.usenix.org/event/lisa07/tech/full_papers/hamilton/hamilton_html/.

[InfoWeek] Jonathan Shaw, 4 Steps to Cloud Quality, Information Week Reports, November 8, 2011, http://reports.informationweek.com/abstract/5/8539/Cloud-Computing/4-steps-to-cloud-quality.html.

[ISO24762] ISO/IEC 24762:2008, Information Technology—Security Techniques—Guidelines for Information and Communications Technology Disaster Recovery Services.

[ITILv3CSI] *ITIL® Continual Service Improvement 2011 Edition*, Cabinet Office, TSO, 2011, ISBN-13: 978-0-11-331308-2.

[ITILv3SD] *ITIL® Service Design 2011 Edition*, Cabinet Office, TSO, 2011, ISBN-13: 978-0-11-331305-1.

[ITILv3SO] *ITIL® Service Operation 2011 Edition*, Cabinet Office, TSO, 2011, ISBN-13: 978-0-11-331307-5.

[ITILv3SS] *ITIL® Service Strategy 2011 Edition*, Cabinet Office, TSO, 2011, ISBN-13: 978-0-11-331304-4.

[ITILv3ST] *ITIL® Service Transition 2011 Edition*, Cabinet Office, TSO, 2011, ISBN-13: 978-0-11-331306-8.

[ITU-T G.114] Series G: Transmission Systems and Media, May 2003. Digital Systems and Networks.

[Kundra] Vivek Kundra, U.S. Federal Chief Information Officer, et al., State of Public Sector Cloud Computing, May 20, 2011, http://www.cio.gov/pages.cfm/page/State-of-Public-Sector-Cloud-Computing.

[Linden] Greg Linden, Make Data Useful, http://www.scribd.com/doc/4970486/Make-Data-Useful-by-Greg-Linden-Amazoncom.

[Microsoft] Reliability Overview, October 10, 2008, http://technet.microsoft.com/en-us/library/cc506068.aspx.

[MOF] Microsoft Operations Framework, http://technet.microsoft.com/en-us/solutionaccelerators/dd320379.aspx.

[Netflix10] 5 Lessons We've Learned Using AWS, Netflix Tech Blog, December 16, 2010, http://techblog.netflix.com/2010/12/5-lessons-weve-learned-using-aws.html.

[Netflix11] The Netflix Simian Army, http://techblog.netflix.com/2011/07/netflix-simian-army.html, July 11, 2011.

[NIST-800-145] The NIST Definition of Cloud Computing (Draft), Special Publication 800-145 (Draft), January 2011.

[NIST-B] Peter Mell and Tim Grance, Effectively and Securely Using the Cloud Computing Paradigm, NIST, Information Technology Laboratory, October 7, 2009, http://csrc.nist.gov/groups/SNS/cloud-computing/cloud-computing-v26.ppt.

[NIST-C] Cloud Computing Taxonomy; Preliminary Draft, NIST CCRATWG 003 v2, http://collaborate.nist.gov/twiki-cloud-computing/pub/CloudComputing/Meeting1TCloudTaxonomy111211/NIST_CCRATWG_007_CloudTaxonomy_011011.pdf.

[NIST-D] 800-144—Guidelines for Security in Public Cloud, http://csrc.nist.gov/publications/drafts/800-144/Draft-SP-800-144_cloud-computing.pdf.

[NORS] Network Outage Reporting System User Manual, Version 6, April 9, 2009, U.S. Federal Communications Commission, http://transition.fcc.gov/pshs/outage/nors_manual.pdf.

[ODCA] http://www.opendatacenteralliance.org/.

[ODCA-SUoM] Open Data Center Alliance Usage: Standard Units of Measure for IaaS, 2011, http://www.opendatacenteralliance.org/document-sections/category/71-docs?download=458%3Astandard_units_of_measure.

[Oppenheimer] David Oppenheimer, Archana Ganapathi, and David A. Patterson, Why Do Internet Services Fail, and What Can Be Done About It? 4th Usenix Symposium on Internet Technologies and Systems (USITS '03), 2003, http://roc.cs.berkeley.edu/papersusits03.pdf.

[OVF] Open Virtualization Format Specification, Distributed Management Task Force, DSP0243, version 1.1.0, January 12, 2010.

[P.800] Methods for Subjective Determination of Transmission Quality, International Telecommunications Union ITU-T Recommendation P.800, 8/96.

[Rauscher06] Karl F. Rauscher, Richard E. Krock, and James P. Runyon, Eight Ingredients of Communications Infrastructure: A Systematic and Comprehensive Framework for Enhancing Network Reliability and Security, *Bell Labs Technical Journal*, 2006, 10.1002, John Wiley & Sons, Ltd.

[RFC3060] Policy Core Information Model, Version 1, February 2001.

[RFC3198] Terminology for Policy-Based Management, November 2001.

[RFC3261] SIP: Session Initiation Protocol, Internet Engineering Task Force Request for Comment 3261, June 2002.

[ROC] Recovery-Oriented Computing, http://roc.cs.berkeley.edu/.

[SHARE] http://www.share.org/.

[Slashdot] http://en.wikipedia.org/wiki/Slashdot_effect.

[TIA942] Telecommunications Infrastructure Standard for Data Centers, ANSI/TIA-942-2005, Approved: April 12, 2005.

[TIPS0340] Seven Tiers of Disaster Recovery, http://www.redbooks.ibm.com/abstracts/tips0340.html.

[TL9000] TL 9000 Quality Management System Measurements Handbook 4.5, Quality Excellence for Suppliers of Telecommunications Forum (QuEST Forum), 2010, http://tl9000.org.

[Uptime] http://www.uptimeinstitute.org/.

[UptimeTiers] Data Center Site Infrastructure Tier Standard: Topology, prepared by Uptime Institute Professional Services, LLC, 2010.

[Varia] Jinesh Varia, Architecting for the Cloud: Best Practices, January 2011, http://jinesh-varia.s3.amazonaws.com/public/cloudbestpractices-jvaria.pdf.

[Vishwanath] Kashi Venkatesh Vishwanath and Nachiappan Nagappan, Characterizing Cloud Computing Hardware Reliability, Microsoft Research, http://research.microsoft.com/pubs/120439/socc088-vishwanath.pdf.

[Webster] Isochronal, http://www.merriam-webster.com/dictionary/isochronal.

[Wikipedia] Disaster Recovery, http://en.wikipedia.org/wiki/Disaster_recovery.

[X805] X.805: Security Architecture for Systems Providing End-to-End Communications, International Telecommunications Union, ITU Recommendation X.805, October 2003.

ABOUT THE AUTHORS

ERIC BAUER is reliability engineering manager in the Software, Solutions, and Services Group of Alcatel-Lucent. He currently focuses on reliability and availability of Alcatel-Lucent's cloud-related offerings, IP Multimedia Subsystem (IMS), and other solutions. Before focusing on reliability engineering topics, Mr. Bauer spent two decades designing and developing embedded firmware, networked operating systems, IP telephony Internet platforms, and optical transmission systems. He has been awarded more than a dozen U.S. patents, authored *Design for Reliability: Information and Computer-Based Systems*, coauthored *Beyond Redundancy: How Geographic Redundancy Can Improve Service Availability and Reliability of Computer-Based Systems* and *Practical System Reliability*, and published several papers in the *Bell Labs Technical Journal*. Mr. Bauer holds a BS in Electrical Engineering from Cornell University and an MS in Electrical Engineering from Purdue University. He lives in Freehold, New Jersey.

RANDEE ADAMS is a consulting member of the technical staff in the Software, Solutions, and Services Group of Alcatel-Lucent. She originally joined Bell Labs in 1979 as a programmer on the new digital 5ESS switch. Ms. Adams has worked on many projects throughout the company (e.g., software development, trouble ticket management, load administration research, software delivery, systems engineering, software architecture, software design, tools development, and joint venture setup) across many functional areas (e.g., database management, recent change/verify, common channel signaling, operations, administration, and management, reliability, and security). Currently, she is focusing on reliability for ALU products. She has given talks at various internal forums on reliability. Ms. Adams coauthored *Beyond Redundancy: How Geographic Redundancy Can Improve Service Availability and Reliability of Computer-Based Systems*. Ms. Adams holds a BA from the University of Arizona and an MS in Computer Science from the Illiniois Institute of Technology. She lives in Naperville, Illinois.

Reliability and Availability of Cloud Computing, First Edition. Eric Bauer and Randee Adams.

INDEX

Reliability and Availability of Cloud Computing, First Edition. Eric Bauer and Randee Adams.

Printed in the United States
By Bookmasters